Aircraft Conceptual Design Synthesis

Aircraft Conceptual Design Synthesis

by

Denis Howe
PhD (Cranfield), SM (MIT), FRAeS, FIMechE

Professor Emeritus and formerly Head of the College of Aeronautics, and Dean of Engineering, Cranfield University

Professional Engineering Publishing Limited
London and Bury St Edmunds, UK

First published 2000

ISBN 1 86058 301 6

A CIP catalogue record for this book is available from the British Library.

I applied myself to the understanding of wisdom.....
and learned that this is chasing after the wind.

Ecclesiastes, Chapter 1, verse 17

Dedication

This volume is dedicated firstly to the staff and students of Cranfield University, who, by their suggestions and probing questioning, have made an inestimable contribution to the final form of the work, and secondly to my wife for her continued help and encouragement.

Related Titles of Interest

IMechE Engineers' Data Book – Second Edition	Clifford Matthews	1 86058 248 6
Developments in Turbomachinery Design	Edited by John Denton	1 86058 237 0
Optimizing the Super-turbocharged Aeroengine	Julian R Panting	1 86058 080 7
Military Aerospace Technologies – FITEC '98	Conference volume	1 86058 167 6
Application of Multi-Variable System Techniques (AMST '98)	Edited by R Whalley and M Ebrahimi	1 86058 128 5

Contents

Chapter 7 - Performance estimation

Chapter 9 - Analysis of concept design

Addendum 1 - Landing gear considerations

Notation

The listed notation covers the main text. The notation used in the spreadsheets is defined locally as is that used in the Addenda. All units are kilograms, metres, seconds with forces in Newtons unless otherwise stated.

a	Speed of sound
a	Number of aisles in a passenger cabin (Para. 4.5.3.3)
a_1	Aerofoil lift curve slope
A	Aspect ratio of lifting surface ($A = b^2/S$)
A	Propeller disc area [Eq (3.9h)]
A^I	Factor in definition of lifting surface mass coefficient C_1 [Eq (6.22b) and Table 6.7]
$\bar{A}$	Thrust factor used in variable Mach number climb prediction [Eq (7.34c)]
A_f	Factor defining aerofoil compressible flow characteristics ($A_f = A_F - 0.1t/c$) [Eq (6.13a)]
A_F	Factor defining aerofoil compressible flow characteristics (Para. 5.2.2.4)
A_p	Engine intake face area [Eq (6.17c)]
ASL	Accelerate-stop length at take-off [Para. 7.2.3.2 and Eq (7.3a)]
b	Lifting surface span (perpendicular to flight direction)
B	Maximum width of body (fuselage)
B^I	Factor in definition of lifting surface mass coefficient, C_1 [Eq (6.22b) and Table 6.7]
B^*	Average body width [Eq (6.17c)]
$\bar{B}$	Zero lift drag factor used in variable Mach number climb prediction [Eq (7.34d)]
c	Powerplant specific fuel consumption
c_l	Fraction of wing chord over which there is laminar flow [Eq (6.13a)]
$\bar{c}$	Mean chord of a lifting surface (area, S, divided by span, b)
$\bar{\bar{c}}$	Aerodynamic mean chord [Eq (8.7d)]
c_0	Wing chord on centreline
c^I	Factor in expression for specific fuel consumption [Eq (3.12a)]
$(c)_{OD}$	Off design specific fuel consumption [Eq (3.12b)]
$(c)_p$	Specific fuel consumption of piston engine [Eq (3.15a)]

$(c_T)_S$	Propeller static thrust coefficient [Eq (3.9h)]
C_1	Coefficient of lifting surface mass [Eq (6.22) and Table 6.7]
$\bar{C}_1$	Function of lifting surface mass coefficient [Eq (6.24)]
C_2	Coefficient of fuselage mass [Eq (6.20) and Table 6.6]
C_3	Coefficient of powerplant mass (Table 6.8)
C_4	Coefficient of systems mass (Table 6.9)
C_5	Ratio of total lifting surface mass to wing mass (Table 6.10)
$\bar{C}$	Induced drag factor used in variable Mach number climb prediction [Eq (7.34e)]
C_D	Drag coefficient (Para. 5.1)
C_{DI}	Drag coefficient due to lift (induced) [Eq (6.12a)]
C_{DW}	Wave drag coefficient at zero lift [Eq (6.17a)]
C_{DZ}	Zero lift drag coefficient [Eq (6.17b)]
$\bar{C}_{DW}$	Function of wave drag coefficient [Eq (6.17b)]
$(C_D)_{CO}$	Effective zero lift drag coefficient in climb out condition [Eq (6.15)]
$(C_D)_{GO}$	Effective zero lift drag coefficient at baulked landing [Eq (6.16b)]
C_L	Lift coefficient (Para. 5.1)
C_{La}	Approach lift coefficient (Para. 6.2.4)
C_{LC}	Cruise lift coefficient (Para. 6.2.4)
C_{LH}	Maximum lift coefficient of low aspect ratio wing at high angle of attack (Para. 6.2.5.2 and Table 6.2)
C_{LL}	Maximum lift coefficient of low aspect ratio wing at low speed (Para. 6.2.5.3 and Table 6.2)
C_{LMAN}	Maximum lift coefficient available for manoeuvre (Para. 6.2.4)
C_{LMAX}	Maximum lift coefficient (Para. 6.2.4)
C_{LMD}	Lift coefficient for minimum total drag [Eq (7.14b)]
C_{LUS}	Lift coefficient at take-off unstick condition (Para. 6.2.4)
C_M	Pitching moment coefficient (Para. 5.1)
d	Effective body diameter [Eq (6.17c)]
D	Drag
D_p	Propeller diameter
$\bar{D}$	Ratio of zero lift drag to weight [Eq (7.13e)]
$\bar{D}_a$ $\bar{D}_b$	Functions of drag used in transonic acceleration prediction, $0.9 < M_N < 1.0$ and $1.0 < M_N < 1.2$ respectively [Eqs (7.44a) and (7.44b)]
$\bar{E}$	Factor in evaluation of transonic acceleration, function of thrust and drag [Eq (7.45c)]
F_F	Flap drag factor [(Eq (6.15b)]
F_r	Ratio of sea level static thrusts with and without afterburning [Eq (3.8)]
F_{LE}	Leading edge high lift device criterion [Eq (5.2)]

F_{OP}	Operating items mass factor (Para. 6.4.2.3.)
$\bar{F}$	Factor in evaluation of transonic acceleration, function of drag [Eq (7.45c)]
g	Gravitational acceleration
g	Number of galleys along length of passenger cabin (Para.4.5.3.4)
h	Geopotential height, used in definition of energy height [Eq (7.7)]
h_e	Energy height [Eq (7.7)]
H	Maximum depth of fuselage (body)
$\bar{H}$	Energy content of fuel [Eq (7.47)]
H_1	Actual heights at beginning and end of a given climb phase
H_2	(Para. 7.6) (usually km)
J	Propeller advance ratio [Eq (3.9a)]
k_e	Engine factor in take-off performance [Eq (7.1a)]
k_m	Ratio of mass in a specific flight case to take-off value [Eq 7.13f)]
K_F	Fuselage wave drag factor[(Eq (6.17b)]
K_{i_T}	Factors in specification of powerplant thrust relative to sea level static value (i = 1 to 4) [Eq (3.7) and Table 3.2]
K_0	Overall aircraft shape wave drag factor [Eq (6.17b)]
K_V	Induced drag factor [Eq (6.12b)]
K_W	Lift (wing) wave drag factor [Eq (6.18)]
$\bar{K}$	Factor in prediction of varying Mach number climb, depends upon induced drag [Eq (7.34e)]
KE	Climb kinetic energy correction factor to allow for variation of speed of sound with altitude [Eq (7.11c)]
l	Effective body length used for wave drag evaluation [Eq (6.17b)]
l_A	Distance between midpoints of ailerons ($l_A = 2y_A$) (Para. 8.10.5)
l_{APEX}	Distance of leading edge of centreline wing chord aft of body nose
l_{CG}	Distance of aircraft centre of gravity aft of body nose
l_F	Distance of centre of gravity of fuel aft of body nose (see below)
l_{FUEL}	Distance of centre of gravity of fuel aft of body nose (may be stated relative to wing reference as Δl_{FUEL})
l_H	Distance from centre of pressure of horizontal tail to aircraft centre of gravity (Para. 8.10.3)
l_{MG}	Distance of centre of gravity of main landing gear aft of body nose
l_{NG}	Distance of nose landing gear aft of body nose
l_{OPIT}	Location of mean position of operational items aft of body nose
l_{PAY}	Location of payload centre of gravity aft of body nose

l_{PP}	Distance of powerplant centre of gravity aft of body nose (may to be stated relative to wing reference as Δl_{PP})
l_{SYS}	Mean location of systems masses aft of body nose, excludes landing gear
l_{TAIL}	Mean location of tail mass aft of body nose
l_V	Distance from centre of gravity of vertical tail to aircraft centre of gravity (Para.8.10.3)
l_W	Distance of wing structure centre of gravity aft of body nose (may be stated relative to wing reference as Δl_W)
L	Lift
L	Overall fuselage (body) length
L_L	Factor in prediction of landing length [Eq (7.6a)]
LL	Factored landing length (Para.7.4.2)
m	Powerplant mass flow [Eq (3.2b)]
m_a	Longitudinal acceleration factors in transonic acceleration, $0.9 < M_N < 1.0$
m_b	and $1.0 < M_N < 1.2$ respectively [Eqs (7.44a) and (7.44b)]
m_F	Fuel mass flow [Eq (3.2b)]
$\bar{m}$	Mass of propulsive medium[(Eq (3.1)]
M	Aerodynamic pitching moment
M_L	Landing mass
M_0	Take-off mass
$(Mg)_0$	Take-off weight
$(Mg/S)_0$	Take-off (maximum design) wing loading
M_{ENG}	Mass of engine [Eq (6.25)]
M_{FIXED}	Total mass of absolute (predetermined) items [Eq (6.19)]
M_{FUEL}	Mass of fuel
M_{FUS}	Mass of fuselage (body) [Eq (6.20)]
$M_{LIFTSUR}$	Mass of lifting surfaces [Eq (6.22)]
M_N	Mach number
M_{NCRIT}	Critical Mach number [Eq (5.1a)]
M_{N1}/M_{N2}	Mach numbers at beginning and end of flight phase, such as climb
M_{OP}	Mass of operational items (Para. 6.4.2.3)
M_{PAY}	Payload mass
$M_{POWERPT}$	Mass of installed powerplant, also used as M_{PP} [Eq (6.25) and Ch 8]
M_{SYS}	Mass of systems, including equipment and landing gear [Eq (6.27)]
M_{TAIL}	Mass of horizontal and vertical tails [Eq (6.22) and Table 6.10]
$M_{VARIABLE}$	Total value of variable mass items [Eq (6.19)]
M_W	Wing structure mass [Eq (6.22) and Table 6.10]
$\bar{M}_N$	Number climb [Eqs (7.34f) and 7.(34h)]
M_{N2}/M_{N1}	Ratio of masses at end and beginning of flight phase, such as cruise
n	Propeller rotational speed, rev/s [Eq (3.9a)]
n	Normal manoeuvre (acceleration) factor (Para 1.2.3.2)

n_l	Limit value of n (Paras 1.2.3.2 and AD4.2.1.2)
$(n)_{INST}$	Instantaneous manoeuvre factor [Eq (7.43a)]
N	Ultimate design (manoeuvre or gust) factor [Eq (5.8a)]
N_e	Number of engines located over the top of the wing
N_E	Number of engines
$\bar{N}$	Corrected value of N to allow for overall trim effect [Eqs (6.22a) and (AD4.1e)]
p	Cabin differential pressure design value, bar [Eq (6.20a)]
p_j	Static pressure in jet efflux [Eq (3.2b)]
p_0	Freestream static pressure [Eq (3.2b)]
p_S	Specific excess power (SEP), sec [Eq (7.13)]
P	Total number of passengers
P	Power [Eq (3.5)]
P_0	Maximum static shaft engine power, kW
(P_0/A)	Propeller static disc loading [Eq (3.9h)]
$(P_0)_{EQ}$	Equivalent shaft turbine static power [Eq (3.10f)]
P_r	Piston engine power in given flight condition [Eq (3.15a)]
P_1	Shaft turbine power at 3 km altitude and Mach number M_{N1} [Eq (3.10e)]
$(P)_{3k}$	Nominal shaft turbine power in static condition, referred to 3 km altitude [Eq (3.10e)]
PAY	Payload
q	Dynamic pressure ($\rho V^2/2$)
Q_M	Factor in constant Mach number climb prediction [Eq (7.22b)]
Q_V	Factor in constant equivalent airspeed climb prediction [Eq (7.15)]
Q_{VM}	Factor in varying Mach number climb prediction [Eq 7.34b)]
R	Powerplant bypass ratio (nominally static)
R_W	Ratio of total aircraft wetted area to reference wing area, S [Eq (6.13a) and Table 6.3]
s	Range
s	Engine altitude thrust lapse power [(Eq (3.7a) and Table 3.2]
s	Passenger cabin seat pitch (Para. 4.5.3.4)
s_A	Ground distance covered in flight from lift-off to height datum (air distance) (Para. 7.2.3.2)
s_F	Ground distance covered during landing flare (Para. 7.2.4.2)
s_G	Ground distance from start of take-off to lift-off (Pa. 7.2.3.2)
s_G	Ground distance covered in climbs and descent [Eqs (7.7), (7.21),(7.29), (7.33), (7.38) and (7.56)]
S	Reference wing area (nominally total area obtained by extending root leading and trailing edges to aircraft centreline)
S_H	Area of horizontal tail (Para. 8.10.3)

S_V	Area of vertical tail (Para. 8.10.3)
SP	Wing structural parameters [Eq (5.8a)]
t	Time, seconds
t	Number of toilets along the length of a passenger cabin (Para. 4.5.3.4)
t/c	Aerofoil thickness to chord ratio
T	Thrust
$(T/Mg)_0$	Static thrust to weight ratio at aircraft total weight
T_f	Aircraft type factor in evaluation of zero lift drag [Eq (6.13a) and Table 6.4]
T_a	Thrust terms used in prediction of transonic acceleration
T_b	[Eqs (7.44a) and (7.44b)]
T_{CO}	Thrust in climb out condition [Eq (7.4a)]
T_D/T_W	Ratio of sea level static thrusts in the dry and wet (afterburning) conditions respectively [Eq (3.8)]
T_{OD}	Thrust in engine off-design condition [Eq (3.12b)]
T_S	Thrust in sea level static condition, propeller engines [Eqs (3.9g) and (3.11g)]
T_{SP}	Specific thrust [Eq (3.3)]
T_{SS}	Propeller engine thrust in second segment climb (equivalent to T_{CO}) [Eqs (3.10b)]
ToL	Take-off length, factored (Para. 7.2.3.3)
V	Velocity
V_a	Velocity at approach to land (Para. 7.2.4)
V_j	Jet (exhaust) velocity [Eq (3.2a)]
V_0	Free steam velocity [Eq (3.2b)]
V_A	Manoeuvre velocity, the lowest speed at which the maximum limit normal acceleration can be achieved (Para. 7.2.2)
V_C/M_C	Design cruise velocity and corresponding Mach number (Para.7.2.2)
V_{Cl}	Velocity in climb, piston engine powered [Eq (3.11c)]
V_{Cr}	Velocity in cruise, piston engine powered [Eq (3.11g)]
V_{CRIT}/M_{CRIT}	Critical velocity/mach number (Para. 7.2.2)
V_D/M_D	Maximum structural design velocity and corresponding Mach number (Para. 7.2.2)
V_{EAS}	Equivalent airspeed (Para. 7.2.2)
V_{EF}	Most critical velocity for engine failure (Para. 7.2.3)
V_{LOF}	Speed at which aircraft leaves the ground in take-off, effectively the same as V_{US} (Para. 7.2.3)
V_{MCA}	Minimum velocity at which aircraft can be controlled while in the air (Para. 7.2.3)
V_{MCG}	Minimum speed at which aircraft can be controlled while on the ground (Para. 7.2.3)
V_{MD}	Speed at which the total drag is a minimum (Para. 7.2.2)

V_{Mo}/M_{Mo} Maximum operating speed and corresponding Mach number, often equivalent to V_C/M_C (Para. 7.2.2)
V_{MU} Minimum speed at which aircraft can safely lift off, a safety margin less than V_{LOF} (Para. 7.2.3)
V_{NE}/M_{NE} Velocity and corresponding Mach number which must not be exceeded in flight (Para. 7.2.2)
V_R Velocity at which aircraft rotates to take-off (Para 7.2.3)
V_S Velocity at which aircraft stalls in a given condition (Para. 7.2.2)
V_{TAS} True airspeed (Para. 7.2.2)
V_{TO} Velocity at touch down on landing (Para. 7.2.4)
V_{US} Velocity at which aircraft leaves the ground in take-off, see also V_{LOF} (Para. 7.2.3)
$(V_{US})_{HH}$ As V_{US} but in design high altitude/high temperature case [Eq (3.10b)]
V_V Vertical velocity in climb (Para. 7.6)
$(V_V)_C$ Required residual vertical velocity at ceiling (Para. 7.6)
V_1 Decision velocity after engine failure at take-off (Para. 7.2.3)
V_2 Velocity at which decision is made to rotate aircraft at take-off (Para. 7.2.3)
$\bar{V}$ Horizontal tail volume coefficient [Eq (8.8a)]
$\bar{V}_A$ Aileron volume coefficient [Eq (8.10)]
$\bar{V}_V$ Vertical tail volume coefficient [Eq (8.9a)]

w Number of cross aisles in a passenger cabin (Para. 4.5.3.4)
W Weight
W_F Fuel weight

$x_{1/4}$ Location of the quarter mean aerodynamic chord on the centreline chord [Eq (8.7a)]
X_1 Correction for propeller engines in constant equivalent airspeed climb [Eq (7.15a)]
X_2 Correction factor for approach to ceiling (11-20 km altitude) [Eq (7.26)]
X_3 Correction factor for approach to ceiling (above 20 km altitude) [Eq (7.30)]

y_A Distance from centreline of aircraft to midpoint of aileron [Eq (8.10)]
Y Correction factor for type of engine in prediction of climb performance [Para. 7.6 and Eqs (7.16) and (7.18)]

z Number of propeller blades
Z Ratio of climb velocity to minimum drag velocity [Eq (7.15)]

α Wing angle of attack (Ch 5)
α Climb out angle after take-off [Eq (7.4a)]
α Factor used in defining fuel used in climb[(Eq (7.17)]

β	$(1 - M_N^2)^{1/2}$
β	Factor used in defining fuel in climb [Eq (7.17)]
$\bar{\beta}$	Product of zero lift drag coefficient, C_{DZ}, and induced drag factor, K_V [Eq (7.13b)]
γ	Climb out angle [Eq (7.4a)]
γ	Descent angle [Eq (7.6a)]
γ	Factor used indefining fuel used in climb [(Eq (7.17)]
Δ	Zero lift drag factor in varying Mach number climb prediction [Eq 7.34j)]
Δ_S	Increment in range [Eq (7.47)]
ΔC_{DT}	Increment in drag due to flaps in take-off condition [Eq (6.15)]
ΔC_L	Increment in lift coefficient [Eq (5.3a)]
Δ_H	Increment in lift coefficient due to low aspect ratio configuration at high angles of attack [Eq (6.7) and Table 6.2]
Δ_L	Increment in lift coefficient due to low aspect ratio configuration at low angle of attack [Eq (6.8) and Table 6.2]
Δ_{LEL}	Increment in lift coefficient due to leading edge high lift devices in landing position [Eq (6.2)]
Δ_{LET}	Increment in lift coefficient due to leading edge high lift devices in take off condition [Eq (6.4)]
ΔM	Increment in mass [Eq (7.47)]
Δ_{TEL}	Increment in lift coefficient due to trailing edge high lift devices in landing position [Eq (6.2) and Table 6.1]
Δ_{TET}	Increment in lift coefficient due to trailing edge high lift devices in take-off position [Eq (6.4) and Table 6.1]
η	Propeller efficiency [Eq (3.4)]
η_{Cr}	Propeller efficiency in cruise
η_o	Overall propulsive efficiency [Eq (7.47)]
θ	Angle of thrust deflection [Eq (7.5)]
λ	Taper ratio of lifting surfaces, tip chord divided by centreline chord
$f(\lambda)$	Taper ratio factor in definition of induced drag [Eq (6.14b)]
Λ_E	Effective sweep [Eq (6.22a)]
Λ_{LE}	Sweep of lifting surface leading edge
Λ_{STRUCT}	Sweep of lifting surface structure [(Eq (6.22a)]
$\Lambda_{1/4}$	Sweep of lifting surface quarter chord line
μ_G	Ground friction (drag) coefficient [Eq (7.6)]

ρ	Air density
σ	Relative air density
τ	Thrust factor, value at given Mach number and altitude relative to sea level static (dry) value [Eq (3.7)]
τ_M	τ value in specific Mach number condition [Eq (7.13f)]
$\bar{\tau}$	Correction factor for τ in varying Mach number climb analysis [Eq (7.34b)]
$\bar{\tau}$	Correction for aerofoil thickness to chord ratio on zero lift drag coefficient [Eq (6.13b)]
ψ	Turn rate [Eq (7.39a)]

Suffixes

Cr	Cruise conditions
Cl	Climb conditions
0	Sea level static conditions, that is take-off conditions nominally
1,2	Conditions at beginning and end of flight phase, for example climb
2D/3D	Two-dimensional and three-dimensional lifting surface values

Preface

The process of aircraft design is a complex combination of numerous disciplines which have to be blended together to yield the optimum configuration to meet a given requirement. This is inevitably an iterative procedure which consists of alternative phases of synthesis and analysis. Many compromises are inevitable. Aerodynamic performance has to be matched across the operating speed range, blended with powerplant characteristics, yield satisfactory control and stability and at the same time not unduly penalise structural considerations and the consequent mass penalties. The integration of advanced flying control systems with the aerodynamic and structural design is a unique and vital procedure. At the same time as the technical aspects of the design are under consideration it is essential to allow for manufacturing considerations. The whole design process must make use of the most advanced tools available in computer aided design and manufacturing.

There are a number of excellent texts which cover many of the various facets of initial aircraft design. Some concentrate on one or two of the implied disciplines to the detriment of the others. Although some include consideration of the initial, synthesis, phase of the design the usual emphasis is on analytical methods. Certain of these texts are also restricted, implicitly or explicitly, to particular classes of aircraft. Where the conceptual design process is covered by these texts it is usually based on a direct application of statistical data derived from past designs. This approach can be restrictive in that it cannot safely be applied to novel concepts.

Long experience in the teaching of aircraft design at graduate and postgraduate levels has convinced the author that the great majority of students have a real problem in progressing from a written requirement to a first visual layout of the aircraft. A reasonable attempt at this phase is essential to avoid the all too common difficulty of embarking upon a detailed analysis with no guide as to the integration of the whole. The present book seeks to overcome this problem and is devoted to the first stage of aircraft design, which is defined as the interpretation of a requirement into a preliminary layout of an aircraft. The text outlines a design process derived by simplifying the real, complex, parallel development into an essentially sequential one so that a logical step by step approach may be employed. This utilises modules for the representation of propulsion, lift, drag, mass and performance which are original. As far as possible these modules are theoretically based, but they are simplified by the introduction of empirical data and the inclusion only of carefully selected dominant parameters. Further, an attempt has been made to state the data in a form which enables it to be applied to all classes of fixed wing aircraft.

The design procedure makes use of a spreadsheet approach. The output is optimised to yield a minimum mass solution although the fundamental importance of cost implications is also discussed. The method used is applicable to novel concepts. The necessary feedbacks during the design process are minimised but where they are essential their introduction is facilitated by detailed cross-referencing of the paragraphs in the text. Some of the existing texts require the production of a sketch of the concept design in order to initiate the process and this is frequently a difficulty for inexperienced students. For this reason the current method has been conceived to avoid this necessity although it is required to make an estimate of the overall dimensions of the fuselage, possibly by means of a layout in certain cases. A final addendum illustrates the procedure by covering worked examples of several types of aircraft ranging from a small, light, piston engined type to a supersonic interceptor, the more conventional subsonic transport having been treated as an example in the derivation of the process in the main text.

The spreadsheet approach has been chosen, rather than a program as such, in the belief that the openness enables the procedures and the effect of parameter changes to be more readily understood. Further, it does enable the use of optimisation techniques which are simple to apply and, of especial importance, easy to check. This is not to dismiss the value of formal programs but they are considered to be less appropriate in the present context. Likewise, there is no emphasis on the use of computer aided design. While it is accepted that CAD/CAM procedures are an essential adjunct to the aircraft synthesis process in real applications, it is not considered to be necessary in the present context. There are several reasons for this. One reason is that the subject is adequately covered by other texts and experience has shown that most students can readily learn and apply computer aided design as required. Another reason is the belief of the author that there is still a place, and probably always will be, for the rapidly produced hand sketch or drawing so useful in the initial investigation of alternative layout possibilities and that students should be encouraged in this direction. It was never true that an aircraft can be designed on the back of a menu card during an after-dinner discussion, but some thoughts can be crystallised in this way.

A working knowledge of basic aeronautics is assumed to avoid the need for coverage of elementary material. The information presented in the main text is limited to that which is essential to the application of the procedure outlined. Some subjects of fundamental importance are only covered superficially on the basis that other texts cover them adequately. The procedure for the subsequent, analysis, phase of the design process is outlined and a comprehensive bibliography is included. Certain information, particularly relevant to the analysis phase, is included in the addenda.

No doubt some experienced specialists will be critical of what may be regarded as the gross over-simplifications and the questionably justified assumptions. However, the author believes that the end justifies the means. Hopefully the "systems" engineer will appreciate the logic of the approach. Clearly the text is primarily intended for the use of aeronautical students, especially those in the final undergraduate year and at postgraduate level. However, it is likely that the book will also appeal to the aeronautical enthusiast including those interested in home-built aircraft. Maybe it is not too much to hope that even the experienced aircraft project designer may find the odd item of useful information, but it should be noted that the software is not validated.

Chapter 1

The design process

1.1 Introduction

Engineering design is a non-unique iterative process, the aim of which is to reach the best compromise of a number of conflicting requirements. Whether the need is for a totally new item or for a development of an existing one the design procedure commences with an interpretation of the requirements into a first concept. This is essentially a synthesis process which involves decision making. Once the first concept has been derived it can be analysed in the context of the requirements. The concept is refined by an iterative synthesis/analysis/decision-making sequence until an acceptable solution is achieved. Within one set of assumptions it is possible to derive an optimum solution but in reality the need for continual refinement of the assumptions implies the curtailing of the process when an "acceptably good" result is reached. The rapidity with which the iterations converge is a measure partly of the complexity of the design and partly of the skill and experience of the design team in using the aids available. When the project is a complex one the time needed to build up sufficient experience is extended.

Aircraft design is no exception and is a complex engineering task. In practice the most successful designs are those developed by teams that have considerable experience with similar classes of aircraft. In many of these successful examples the design is a direct development of an earlier type. The reason why past experience is so important is clear. To be commercially successful a new aircraft must represent a technical improvement upon its predecessors and this implies that technology has to be extrapolated. The risks associated with this are less when it is based on a large data bank of previous knowledge.

Past experience plays a major role in the initial synthesis of the concept. It is also a vital ingredient in the somewhat more straightforward analytical stages where the choice of analytical techniques and the interpretation of the results is greatly influenced by the practical knowledge of the design team. This importance of past knowledge is widely recognised and attempts are being made to encapsulate it in the form of knowledge-based computational techniques. Ultimately it is theoretically possible that artificial intelligence programs will become available which will be capable of completing the whole process of synthesis and analysis. However, there are those who doubt whether it will ever be possible to completely replace the initiative and flair of the human brain. Nevertheless the availability of computational techniques of varying degrees of complexity is an essential ingredient of modern aircraft design.

The reliability of the results of the initial synthesis process is critical to the ultimate outcome of the project whatever design procedure is adopted.

1.2 The aircraft design process

1.2.1 General

1.2.1.1 Introduction

An understanding of the whole aircraft design process is helpful in appreciating the importance of the initial synthesis phase. Figure 1.1 is a somewhat simplified representation of the whole aircraft development process and shows the sequence of the main stages which lead to the aircraft becoming operational. Design synthesis occurs in the feasibility, project and, at a different level, in the detail design stages. Unjustified assumptions or mistakes made at the feasibility stage may well result in difficulties in meeting the requirements during the project definition phase. Poor design concepts at this latter stage are likely to result in the need for fundamental changes during detail design or, worse, during prototype testing.

Decisions made at the project definition phase inevitably have a major impact upon manufacturing procedures. It has been suggested that something like 85% of the life cycle costs of a production aircraft are determined at the project stage with a further 10% decided during detail design. Chapter 9 is primarily concerned with the analysis of a derived concept but it does outline the main topics which have to be covered in the design process.

1.2.1.2 The integrated design process

Recent developments in information technology have had a major impact upon the aeronautical design process. The emphasis has become one where the whole design procedure from concept to the derivation of manufacturing data is covered by an integrated information technology system. Computer aided design techniques are used to produce digital data for, for example, application to numerically controlled machining and assembly jig definition. Virtual reality concepts are employed to visualise three-dimensional installation and operational aspects, at least in part replacing full-scale mock-ups. While these procedures are particularly relevant to the detail phases of the design process their application commences as soon as a requirement is defined and numerical data are derived from it. It is theoretically possible to visualise a "paperless" process where all the data are conveyed electronically although in practice there is always likely to be a need for hard copies for record purposes. A knowledge of the existence and practice of these techniques is vital to an understanding of the working of the total design process. However it is not considered that it is essential to become involved with them at the very limited, initial, conceptual synthesis phase.

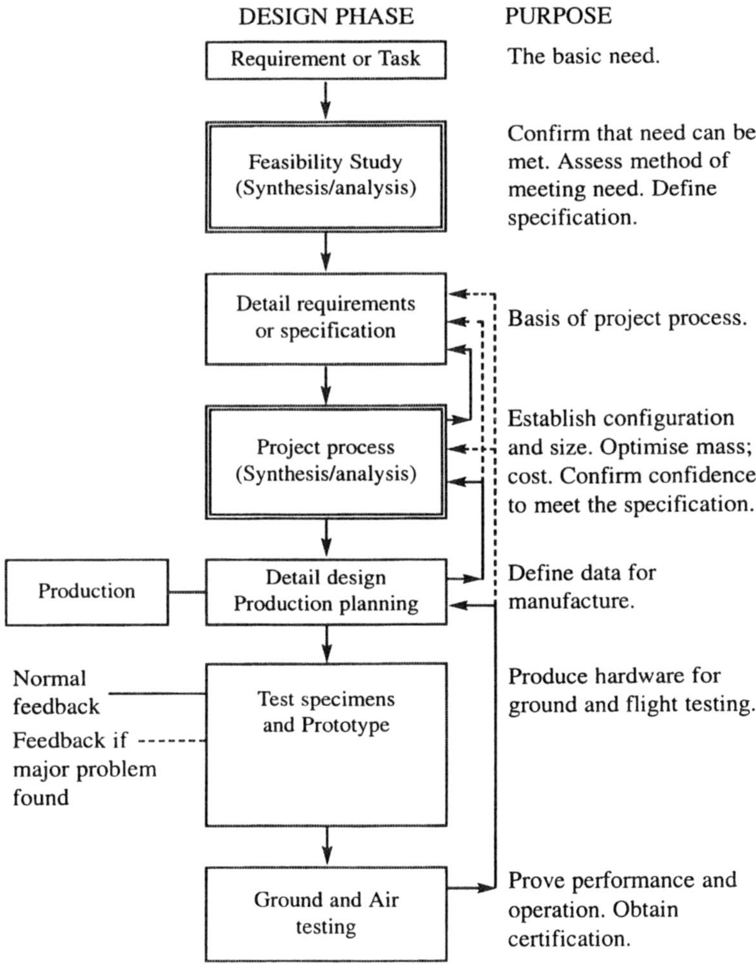

Figure 1.1 The design process

1.2.2 The basic requirement

A new design is launched when it is perceived that there is a requirement to fulfil a need beyond the capability of existing aircraft. In many aeronautical applications the need arises because an existing aircraft is coming towards the end of its useful life or its design has been overtaken by developments in technology.

There are some occasions when a completely new requirement arises as a result of operational experience or when a potentially exploitable, unfulfilled, need is identified. The statement of the need may be defined as a basic requirement or target.

The identification of the need may originate from within a manufacturing organisation or from a potential operator. While the former is more usual for civil aircraft the latter is by no means unknown. Potential manufacturers of a civil type will consult with operators to enable the requirement to be refined to give maximum market potential. Military types most frequently result from a target established by defence organisations. In many cases the initial statement of the basic requirement may be brief, essentially identifying the class of aircraft needed together with its dominant performance characteristics. It is usual for this basic requirement to be considered widely by interested parties. The originators may discuss their concepts with relevant branches of their own organisations as well as with potential manufacturers.

1.2.3 General requirements

1.2.3.1 Design codes

As a result of many years of experience, several codes of practice exist which are applicable to various classes of aircraft. These sets of general requirements are both a guide to designers and a basis for the eventual clearance of the aircraft for its intended operations.

The most important of these documents for civil and general aviation are the FAR/JAR 25 and 23 respectively. FAR are the United States requirements and JAR the equivalent European ones. While they are written in identical format and the aim is eventually to make them identical in content there are, at present, some differences. Similar requirements exist for other classes of civil aircraft.

The situation is less consistent for military types. In the United Kingdom these are covered by Def-Stan 00-970 and until recently by MIL-SPECS in the United States. International collaborative projects often use a special set based on the above two. However, the MIL-SPECS are now being replaced by a new system where individual manufacturers become responsible for defining the requirements.

1.2.3.2 Design codes and conceptual design

The various design codes cover a wide range of topics but it is possible to identify those which have a particular influence at the conceptual design phase. They fall into three categories.

a) Performance requirements. While the specification for the aircraft defines the required overall performance, see paragraph 1.2.5, the design codes include stipulations concerning such matters as the definition of take-off and landing field lengths, residual climb capability subsequent to an engine failure and performance when a landing approach is abandoned. Clearly these have a major impact on such design parameters as powerplant thrust and wing configuration. Chapter 7 considers the relevant performance stipulations in more detail.

b) Flight requirements. The design codes specify criteria for a range of flight characteristics. These include static and dynamic stability, control characteristics and effectiveness and manoeuvre capability at critical flight phases. They effect the size and geometry of the secondary lifting surfaces and flight controls amongst other parameters. Chapter 8 makes further reference to some of these matters and a more complete consideration is given in Addenda 2 and 3.

c) Structural design. Broadly structural design requirements may be classified under two headings, namely stiffness and strength.

The stiffness stipulations are intended to ensure that an airframe will not distort, either statically or dynamically, in such a way which would adversly compromise the performance or safety of the aircraft. These so-called aeroelastic considerations imply a need for adequate stiffness of the airframe in certain modes of distortion, for example wing torsion. The critical modes are influenced by such parameters as lifting surface sweep, span and thickness.

The strength requirements are very detailed but may be summarised by reference to a flight envelope or "n-V" diagram. This envelope represents the extremities of the combinations of normal, that is perpendicular to the flight path, manoeuvre factor, n, and speed, V, to be used to calculate structural loads. A typical flight envelope for an aerobatic light aircraft is shown in Figure 1.2. The critical speeds are defined in relation to the operating requirements as given in the aircraft specification. For example V_C represents the maximum normal operating speed while V_d, the design speed, allows a margin for safety above this. The normal acceleration factors are defined according to the class of aircraft being related to the way in which it is intended to be operated. The value of the maximum normal acceleration factor, n_l, ranges from 2.5 for a large airliner up to 9 for a highly manoeuvrable combat type and possibly higher for an uninhabited aircraft. The aircraft also has to be designed for the loads resulting from flight through a turbulent atmosphere. It is possible to relate these loads due to gust cases to the manoeuvre loads as is illustrated in Figure 1.2. As defined in the design codes the loads at the extremities of the envelopes are the maximum values anticipated to be experienced in actual flight to a given probability and are known as the "limit" loads. It is usual to factor the limit loads by 1.5 to give an "ultimate" load, the 1.5 value effectively being a safety factor. The aircraft must be capable of sustaining the ultimate load for a specified time without collapse. The repetition of loads which give rise to the possibility of structural fatigue is also of considerable importance. Further reference is made to structural requirements in Chapter 5 paragraph 5.4, Chapter 6 paragraph 6.4.3 and Addendum 4 especially in relation to mass prediction.

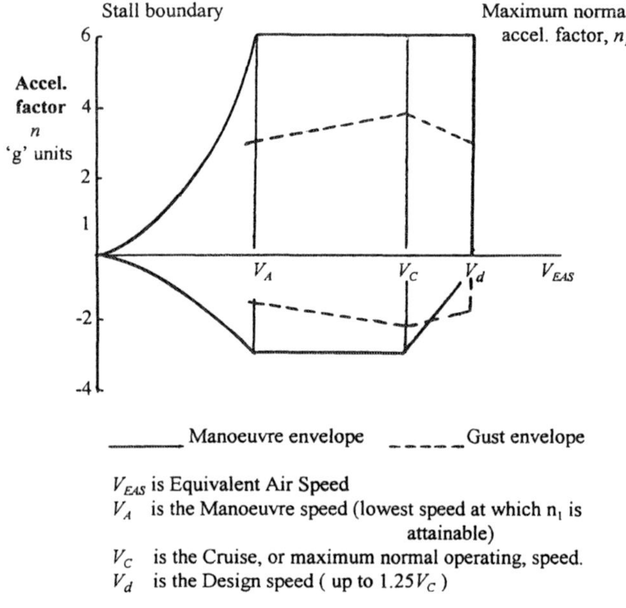

Figure 1.2 Typical flight envelope

1.2.4 The feasibility study

1.2.4.1 General

The complexity of aeronautical projects is such that it is almost inevitable that the statement of the basic requirement will be followed by some form of feasibility study. As the name suggests a primary reason for the feasibility study is to assess whether the need can be met with the technology available. However, in practice the feasibility study is used to much greater purpose than this basic aim.

1.2.4.2 Meeting the basic requirement

There are usually various ways of meeting a basic requirement each of which must be analysed at the feasibility stage. These may be identified as:

a) An adaption or special version of an existing aircraft. This is most likely to involve a change in equipment and the cost of airframe alterations is often relatively small.

b) A major modification or direct development of an existing type. This may well involve expensive major changes to the airframe such as an extended fuselage, new wing or alternative powerplants, as well as equipment update.

c) A completely new design. This is clearly the most expensive option and the one which carries the greatest risk with it. Although completely new designs are much less frequent than major modifications it is essential to embark upon them periodically to maintain a long-term manufacturing or operational capability. Development costs of advanced new projects are usually measured in several billions of dollars.

1.2.4.3 Review of the basic requirement

As a consequence of the feasibility study it is possible to review the statement of the basic requirement. It may be considered to be totally unrealistic, but more likely there may be some revision of the performance characteristics as initially specified.

1.2.4.4 Detail requirements - the specification

One of the outputs of the feasibility study is the definition of a detailed set of requirements, or a specification, for the aircraft. The probable content of this is discussed in paragraph 1.2.5.

1.2.4.5 Cost prediction

The feasibility study may also be used to provide data for the first, reasonably reliable, prediction of the costs of the project. This is clearly of considerable significance as an unfavourable result in this context could lead to the abandonment of the project. See also paragraph 1.3.

1.2.5 The specification

The specification for an aircraft covers many detail aspects of the design, but not all of them are of significance in the synthesis phases of the project. It is convenient to place the main contents of the specification in a number of categories. Some or all of the following items will be included.

a) Performance

i) Range, or sortie pattern, with basic payload mass; probably also alternative range/payload combinations and fuel reserves

ii) Maximum, or maximum normal, operating speed

iii) Take-off and landing field length limitations

iv) Climb performance, such as time to a given height, and service ceiling or operating altitude

v) Point performance covering manoeuvre/acceleration requirements

b) Operational considerations

i) Size limitations, such as for naval aircraft

ii) Mass limitations including runway loading

iii) Crew complement

iv) Occupant environment considerations

v) Navigational/communication fit needed

vi) Payload variations and associated equipment

vii) Maintenance/availability targets

viii) Geographical environment requirements

ix) Low observability (stealth) aspects for combat aircraft

x) Extended engine failed allowances (ETOPS - see page 24) for civil transports

c) General

i) Growth potential

ii) Cost targets

iii) Airframe life

iv) Airworthiness requirements, such as FAR 25/JAR 25 for civil transports

1.2.6 Project definition process

In some respects the project definition process can be considered as an extension of the feasibility study and in practice there is considerable similarity between the two. The major difference is that the aim of the project definition phase is to produce a design having characteristics which are assured sufficiently to enable it to be offered to potential customers with guaranteed performance. This requires a much more thorough and detailed study than is usual in feasibility work, although the basic procedures are similar. These are dealt with in more detail in paragraph 1.5.

1.2.7 Detail design

The detail design is the most extensive phase in the whole design process. Its purpose is both to verify the assumptions of the project definition stage and to produce the data necessary for the manufacture of hardware. In the case of a sophisticated aircraft many tens of thousands of drawings, or computer generated equivalents, are needed to define the aircraft adequately. It is highly desirable for production and operation specialists to be involved to ensure that each item designed represents the best solution in terms of performance, manufacturing costs and operations, see paragraph 1.2.1.2.

1.2.8 Testing and certification

Ground and flight test hardware is manufactured from the data produced by the detail design process. Ground testing includes the use of wind tunnels, structural specimens, and systems rigs, and is intended to investigate critical technical areas and verify as much of the design analysis as possible.

Flight testing is primarily aimed at verifying the performance and flight characteristics of the actual aircraft, although it is also a tool for resolving any particular difficulties which may arise. As the flight testing proceeds advantage is taken of the opportunity to investigate some of the longer term operational matters such as maintenance. A certificate, or operational flight clearance, is issued when the calculations, ground and flight testing of the design have demonstrated to the satisfaction of the appropriate airworthiness authority that it meets all the relevant airworthiness requirements. A customer will also require a demonstration of the performance capability.

1.2.9 Project life cycle

The design phase leading to the certification of an aircraft occupies several years. In some cases where the feasibility study phase is extended it may be as long as a decade. However once the project definition phase begins it is desirable to reach certification as quickly as possible to minimise development costs. The flight testing phase especially incurs high costs and must be completed quickly.

The manufacturer has a continuing responsibility to support the aircraft throughout its operational life. Recent experience suggests that when the type is particularly successful the operational life can be a very long time, probably of the order of half a century. The significance of this continuing responsibility must be recognised during the earlier synthesis phases of the design as decisions made at that time may have long-term consequences.

1.3 Cost considerations

1.3.1 Cost prediction

Consideration of costs is of fundamental importance at all phases of the aircraft design process. The ultimate success or failure of a project is significantly dependent upon the costs associated with its initial acquisition and operation. The selling price of an aircraft is largely determined by market forces and to be profitable for the manufacturer it must be possible to produce it for less than the market price. Likewise in the operation of transport types the economics must be such that expenses are less than revenue. A consideration of the various aspects of aircraft costs is thus essential at the very outset of a project. One difficulty is the paucity of availability of reliable, relevant, cost data as this information is closely guarded by both manufacture and operator. However, some useful references do exist, for example in texts by Raymer and Roskam, see Chapter 9, Appendix items I5 and I6, part VIII.

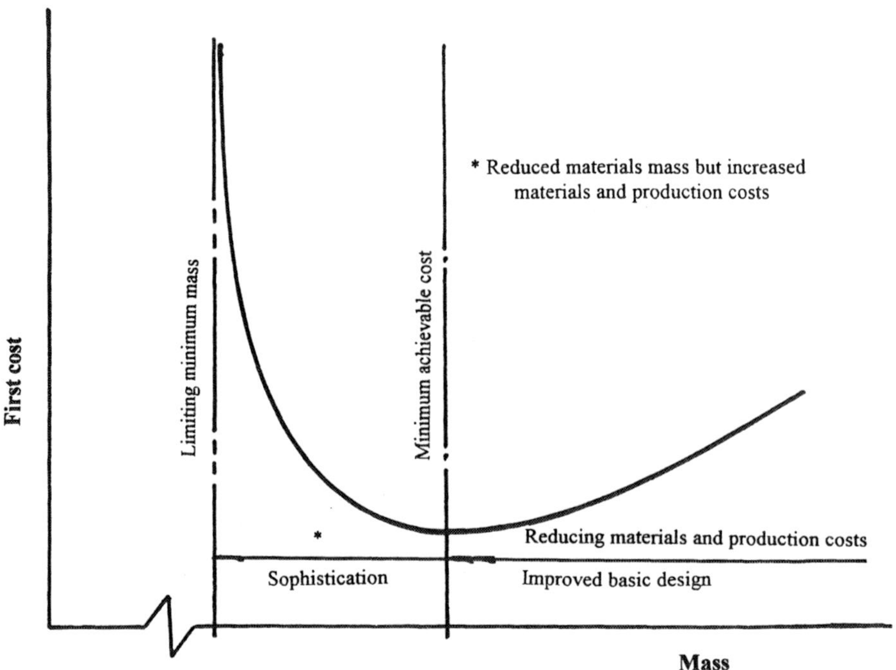

Figure 1.3 Trade off between cost and mass

1.3.2 First cost

It is necessary to distinguish between the cost of producing an aircraft and its selling price. The latter is primarily a function of market forces. The former is the sum of the actual cost of making the aircraft and the share of the development costs allocated to it. At some time during a long production run of a given type of aircraft the development costs will have been covered and further cost reduction is dependent upon improvement of manufacturing techniques. When the cost can be reduced to a level below the price a customer is prepared to pay for the aircraft the manufacturer makes a profit, but often this is not the case for civil aircraft until a large number of the type have been produced.

The prediction of first cost during the project definition phase is fraught with difficulty. Simple cost models are usually based on aircraft volume and/or mass and are derived empirically from past experience. Such models can be very misleading. For example the assumption that first cost is more or less directly proportional to mass is erroneous for a sophisticated design as is illustrated in Figure 1.3. This figure shows that initial refinement of the design results in reduction of both mass and first cost. However, technical demands require the use of sophisticated, expensive, advanced materials and manufacturing processes and some increased expenditure to reduce mass further is always accepted.

Likewise an attempt to reduce cost by limiting overall size may be self-defeating as the resulting design compromises can be expensive to achieve. More elaborate cost models require the use of manufacturing data which may be unreliable or restricted to a particular set of circumstances. On the whole it may be concluded that first cost is not a very suitable criterion for project optimisation, see paragraph 1.4. It is, of course, necessary to avoid design features which cause cost increases unnecessarily, see paragraph 1.2.7.

1.3.3 Operating costs

Operators of civil aircraft usually use hourly operating costs as a basis for financial assessment. In actual fact the real criterion is the annual operating cost since this coincides with the usual accounting period. Clearly the relationship between hourly and annual costs depends upon the utilisation of the aircraft. As many of the contributions to the operating costs are fixed on an annual basis there is pressure to achieve a high utilisation to keep hourly costs as low as possible.

Conventionally the operating costs are split into two categories:

a) Indirect costs which are assumed to be independent of the characteristics of a given aircraft and represent the overall costs of the operation of the airline.

b) Direct operating costs which include:

- i) Write off of the initial purchase price, or equivalent lease payment. This includes the cost of raising the purchase sum where relevant
- ii) Insurance
- iii) Crew costs
- iv) Engineering replacement items
- v) Maintenance
- vi) Operational charges such as landing and *en route* navigation fees
- vii) Fuel and other expendable items

It will be seen that the purchase price, presumably related to the manufacturer's production cost, plays a part in the direct operating costs but items such as the fuel used are likely to be more significant. For project work it may be difficult to calculate the actual direct operating costs because they depend to some extent on the organisation and policy of a given operator as well as aircraft utilisation. Nevertheless used on a relative basis with consistent assumptions they do provide a good measure of the overall performance of the aircraft. In this respect the direct operating costs are a better criterion for optimisation than first cost. Interestingly it is often found that for small changes from a given datum there is a close relationship between direct costs and aircraft mass. This may not be true when different configurations or technology standards are compared.

1.3.4 Life cycle costs

Annual operating costs are not very meaningful in circumstances where the aircraft operation does not result in the generation of income. Examples of this range from the privately owned light aircraft to most military types. In these circumstances a better measure is the life cycle cost. As the description implies this covers the following items:

a) Design and development of the type apportioned to each individual aircraft or fleet as relevant.

b) Procurement of the aircraft its spares and the associated ground support equipment.

c) Cost of operations including manpower, maintenance, fuel, etc., over its whole life.

d) Final cost of disposal of the aircraft.

The first two items account for up to half of the life cycle cost of a typical combat aircraft, and the effective design, development and procurement costs represent a larger proportion of the total than would be the case for a civil transport. This is, in part, due to the relatively low utilisation of this type of aircraft. Life cycle cost is a relevant criterion for optimisation. However, as with direct operating costs accurate evaluation of its magnitude is difficult and recourse may have to be made to comparisons on a relative basis. There is some evidence to suggest that the life cycle cost of a military aircraft is more or less directly proportional to mass and maximum operating speed for Mach numbers up to about 1.3. At higher speeds the costs increase disproportionately.

1.4 Optimisation

1.4.1 The aim of optimisation

The optimisation of the configuration of the aircraft within the constraints imposed by the specification is an essential feature of the project definition process. Optimisation implies that the proposed design concept not only meets the specification, but does so when a target criterion has been imposed. Most commonly the optimisation criterion is the minimisation of mass or some aspect of cost.

1.4.2 Optimisation criteria

1.4.2.1 Mass or size

Although there are a few exceptions it is not usual for size constraints as such to be of significance in aircraft design. Among the exceptions may be mentioned the cases of aircraft designed for operation from ships, where space is always at a premium, and when a relatively large transport aircraft may be required to operate within the limitations of existing terminal facilities.

In practice the size and mass of a given class of aircraft are very closely related. As a general rule the lightest aircraft which can be designed to fulfil a given task is the most efficient and has the greatest potential for development providing this has been built in to the concept from the outset. For these reasons the minimisation of mass as an optimisation criterion is always worthy of consideration.

1.4.2.2 Cost

There are several aspects of cost which may be appropriate to a particular aircraft concept and be relevant for consideration as a basis for optimisation. The most important of these are those covered in paragraph 1.3, that is the first, operating and life cycle costs.

1.4.2.3 Selection of optimisation criteria

From above and paragraph 1.3 it could be concluded that the most appropriate criteria for optimisation are direct operating costs for revenue generating aircraft and life cycle costs for others.

Unfortunately, as has been stated, the accurate prediction of costs is difficult because of their major dependence upon the organisational characteristics of particular manufacturers and operators, although comparisons on a relative basis are valuable. It has also been stated that both the direct operating and life cycle costs are fairly closely related to mass for a given type of aircraft and technology standard. For initial design purposes mass is a much more convenient criterion to use for optimisation since fundamentally it is a function only of the specification. Nevertheless a potential manufacturer or operator will always wish to evaluate the potential of the aircraft in terms of costs derived from its own organisational procedures.

For design convenience mass will be used as the optimisation criterion in this text but it cannot be over-emphasised that it is essential to evaluate cost aspects as soon as sufficient data are available, and it could well lead to the need for a re-optimised concept.

1.4.3 Optimisation procedure and techniques

1.4.3.1 Configuration and parametric studies - trade-off studies

The basis of optimisation is a comparison of different design concepts and configuration variations within a given concept to determine the one which best meets the specification. Broadly the process is undertaken at two levels.

a) Concept/configuration studies. At this level alternative concepts and configurations are investigated to establish the one which would seem best suited to meet the requirements. For example a military combat requirement might possibly be met by aircraft of conventional tail, foreplane or tailless layout. Mostly the configuration of transport aircraft is well established. This phase of the study may be included at the feasibility level.

b) Parametric studies within a given configuration. At this level the dominant parameters are varied to ascertain the best combination of them. These parameters include such items as wing geometry determined by aspect ratio, sweep, taper and thickness as well as variations in fuselage layout, powerplant installation and so on. The benefits of selecting near optimum values of certain parameters are "traded-off" against the implied penalties imposed upon other parameters. The parametric studies may be commenced during feasibility investigations and certainly form a major part of the project definition phase.

The two fundamental design characteristics which drive the parametric studies are:

i) Wing loading, that is wing area as a function of take-off mass or weight
ii) Thrust or power loading, defined as the ratio of the basic powerplant to the aircraft mass or weight

1.4.3.2 Referee designs - sensitivity studies

A considerable number of assumptions have to be made in order to undertake the configuration and parametric variations. As the process develops towards an optimum it is desirable to select a likely final solution for detailed analysis. This is sometimes called a referee design and the purpose of the detailed analysis is to check the validity of the necessary assumptions in the region of the optimised design. The sensitivity of the design solution to variation of various parameters is investigated to identify the more critical ones. Correctly chosen the referee design can be the basis for the detail design phase.

1.4.3.3 Optimisation procedures

Various techniques are available for optimisation. These may be divided into those which rely mainly on graphical comparison and those which are more mathematically based.

a) Graphical techniques. These techniques depend upon comparing the results of the parametric studies in the form of graphs which may be superimposed one upon another. The procedure leads to the definition of a 'design space' or region within which the various performance requirements are met. The design configuration may be selected at an appropriate point in the design space and indeed it may be refined so that the space converges to a point. While the technique is a very useful one for initial synthesis it is limited in the number of parameters which may be conveniently handled.

b) Mathematical techniques. These may be powerful tools which can deal with the variation of a large number of parameters at one time. One such technique is multivariate optimisation, MVO, which makes use of penalty functions and gradient-projection-restoration procedures to minimise the chosen optimisation criterion. Care is necessary in the definition of the parameters to reduce the possibility of sudden changes to the optimising path arising from discontinuities in the formulation. The method can require the use of reasonably powerful computational aids and some experience in the interpretation of the results is desirable, if not essential.

1.5 Synthesis process

1.5.1 Introduction

Schematic outlines of the project definition procedure may be found in numerous references, see paragraph 1.6 and the Appendix to Chapter 9. These outlines correctly cover the whole iterative sequence of synthesis and analysis which is necessary to arrive at a fully authenticated proposal. This complete process is referred to in more detail in Chapter 9, where it will be seen that to make a start on the analysis requires the definition of an initial, or baseline, concept. Some references also suggest how this initial concept may be derived. This usually requires either a read across from an existing aircraft, a very simplified analysis of the requirements or a crude sketch. None of these approaches is really satisfactory, except possibly when the aircraft requirements relate to a very well defined class of aircraft. The emphasis of this work is on this initial synthesis phase with the intention of describing a procedure which enables a reasonably convincing baseline to be derived for virtually all classes of fixed-wing aircraft.

In practice, with experience, the initial synthesis activity consists of a series of inter-related events which are undertaken in parallel. However, for ease of description and application the procedure described here consists of a sequential set of tasks as shown in Figure 1.4. The sequence has been carefully selected to minimise the feedback, or parallel activity, required and yet aims to yield a satisfactory result. The number shown in the various task blocks refers to the relevant chapter in this book. Introductory comments on each of these tasks follow, it being presumed that a fairly complete set of requirements, or a specification, is available as discussed in paragraph 1.2.5. Some care has been taken to ensure consistency of assumptions in the various tasks and it may be inadvisable to combine the procedures outlined here with data from other sources.

1.5.2 Selection of configuration (Chapter 2)

The first task is the selection of one or more basic configurations for examination in the context of the specification. Chapter 2 describes those configurations which have found some application in the past and includes comments on the logic behind them as well as their future relevance. It is necessary to state that:

a) *Technological developments* are likely to result in the introduction of novel configurations, which must be considered where appropriate. For example the advent of fully automated control systems combined with the deflection of powerplant thrust may lead to layouts where conventional control and stabilising surfaces can be dispensed with. It is always desirable to review applied research topics to enable such possibilities to be anticipated.

b) *Unconventional configurations* often appear to possess significant advantages. Experience has shown that in many instances the attempt to optimise these configurations results in a tendency back to the conventional. Therefore, unconventional layouts should only be contemplated when the supposed advantage is relevant to a dominant requirement in the specification. The Wright Brothers chose a foreplane configuration for their first successful Flyer and at first examination this idea does seem to have some major advantages in the efficiency of overall lift production. Nevertheless long experience has shown that secondary effects can more than offset the advantages for most subsonic aircraft. For good reasons the advantages may be realised with aircraft layouts suitable for supersonic flight, or when a naturally unstable concept is acceptable.

c) *Well established conventional layouts* exist for some types of aircraft, as is the case with subsonic transports. In these circumstances the selection of the configuration is straightforward. In other cases there may be two or more configurations which are worthy of investigation. This is likely to be the situation when the requirements introduce a need for advanced conceptual technology, such as referred to above in sub-paragraph (a).

d) *The final configuration* selected for offer to potential customers, and hopefully, detail design and manufacture, may not necessarily be the one which is revealed as the best by the overall optimisation process. A novel, or even somewhat unusual, configuration may be eliminated on the grounds of lack of experience, uncertainty of design data, or even customer reticence to operate it. This latter point is especially relevant when the general public is involved as there is often a tendency to be suspicious when some visible novelty is introduced.

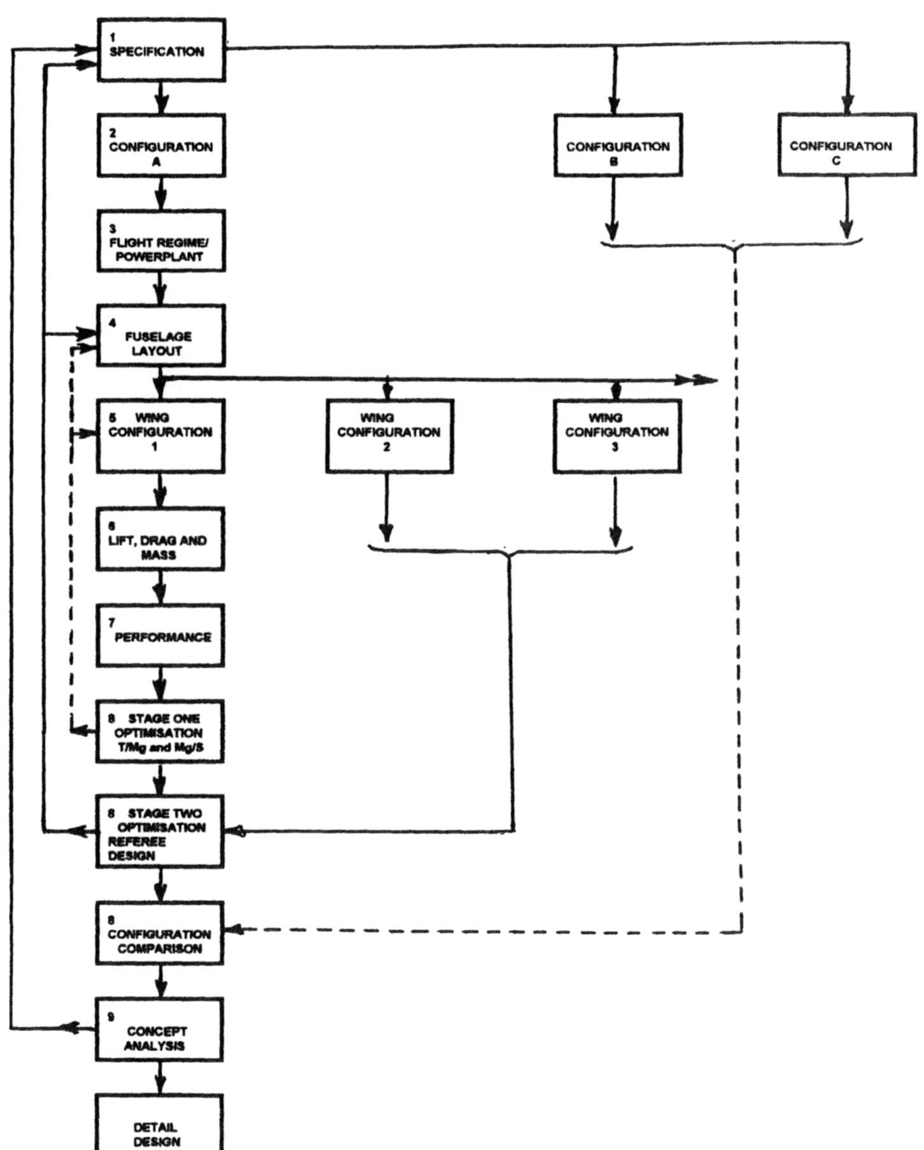

Figure 1.4 Synthesis process

1.5.3 Flight regime and powerplant selection (Chapter 3)

When a basic requirement is expressed in general terms it is necessary to assume a set of operating conditions for use in the aircraft synthesis process. Alternatively the specification may define them. The required operating speed, or speeds, and altitudes effectively define the general class of powerplant appropriate to the concept. Clearly this choice will have some influence on the configuration of the aircraft and hence it may be necessary to consider them together. One example where this situation can arise is that of a requirement to operate at moderate subsonic speed over long distances. In these circumstances both a propeller or ducted fan powerplant are worthy of consideration with the obvious effect upon the choice of aircraft configuration. Fortunately in the great majority of designs the impact of the class of powerplant upon the configuration is a matter of detail.

The actual size and performance of the selected powerplant system are determined subsequently in the synthesis process.

1.5.4 Fuselage layout (Chapter 4)

Figure 1.4 suggests that the more detailed aspects of the synthesis should commence with the layout of the fuselage. There are several reasons for this:

a) Fuselage geometry in many cases, may be established with only secondary reference to the lifting surfaces.

b) Payload definition is often, although not always, a major consideration in the layout of the fuselage and details of this are given in the specification. Even when the payload is not by itself a major issue in fuselage layout other items in the specification can be expected to play a major part, for example crew provision.

c) Only the overall fuselage dimensions are required in order to make the first predictions of the aerodynamic drag and the mass of the concept. Thus the initial layout of the fuselage does not necessarily have to be completed in detail.

d) Geometry and size of the fuselage are primarily derived by a layout procedure since only in a small number of special cases is it possible to use analytical methods. This introduces some difficulties in the sense that there is no obvious single solution, but it does mean that the importance of describing the aircraft geometrically and graphically is present from the outset of the work.

1.5.5 *Wing configuration* (Chapter 5)

The selection of the aerodynamic and geometric properties of the wing is fundamental to the performance of the aircraft and the properties of the auxiliary lifting surfaces follow to some extent. This is a complex matter due to the numerous parameters which must be considered and refined during the optimisation of the design. The wing configuration has a major impact upon the lift, drag and mass characteristics of the whole aircraft and hence it is necessary to give consideration to them when making a first selection of the wing layout. The aim of Chapter 5 is to present typical overall data to enable a reasonable preliminary choice to be made while at the same time minimising the number of parameters which have to be used in the initial synthesis. Nevertheless it is inevitable that the synthesis process returns to reconsider the wing configuration as the design is optimised.

It is not necessary to determine the actual size of the wing at the initial task level. This is determined subsequently in the process, first in terms of a wing loading and finally in absolute terms when the first prediction of aircraft mass is obtained.

1.5.6 *Lift, drag and mass representations* (Chapter 6)

Lift, drag and mass are the three primary characteristics which determine the performance of a fixed-wing aeroplane within the constraints imposed by a given powerplant. It is essential to predict their values as accurately as possible in order to relate the aircraft performance equations to the specified performance requirements. One way of doing this is to use typical values from previous aircraft of a similar type. However, this technique, although commonly used, has two major disadvantages:

a) Relevant data for aircraft of newer technology standards are difficult to find. Manufacturers naturally regard such data as being commercially sensitive. The use of older data can be misleading even when allowances are made for new technological advances.

b) Overall representation of data does not allow the investigation of the effect of differences in aircraft configuration, and hence precludes the use of optimisation until subsequent analysis is undertaken.

The aim of Chapter 6 is to represent the lift, drag and mass by simple analytical expressions which retain only those parameters needed to enable a first optimisation to be undertaken. This aim does require the sacrificing of much of the detail needed to undertake a subsequent, full analysis of the derived concept. Hence more comprehensive methods are necessary once the baseline design has been formulated. Reference to these may be found in Chapter 9.

The interdependence of lift, drag and mass with the wing configuration means that feedback from this task is likely to be needed at some stage of the synthesis process.

1.5.7 Performance representation (Chapter 7)

The interpretation of the specified performance and its expression in terms of the design parameters is a vital part of the synthesis process. This is done by expressing the various aspects of performance in equation form. Many references contain statements of aircraft performance equations but those given in Chapter 7 are unusual in certain respects as they have been formulated specifically for use during synthesis. They have the following characteristics:

a) All the equations have been derived by a theoretical approach, that is by using the basic laws of mechanics. However, in many cases they have been modified by the introduction of empirical data to enable their application to the design process to be simplified as far as is possible.

b) Arrangement of the equations is such that the best values of the two dominant properties of a given concept, the wing loading and the thrust/power loading, may be determined as the first stage of the optimisation.

1.5.8 Parametric analysis and optimisation (Chapter 8)

1.5.8.1 First stage

The first stage of the parametric analysis involves bringing together the results of the previous tasks. The way in which this is done is:

a) The derived fuselage dimensions and wing geometric parameters are used with the lift and drag representations so that these latter may be expressed only in terms of wing loading for the one chosen wing configuration. Other, as yet undefined, features of the design such as the details of the empennage are represented by generalised expressions based on the selected data and a given class of aircraft.

b) The specific lift and drag expressions are used with the relevant powerplant representation in the performance equations. This results in an expression of the variation of wing loading with thrust/power loading for each of the individual performance requirements. Comparison of these performance relationships enables a design space to be established within which all the requirements are met. A suitable combination of wing loading and thrust/power loading is then selected, usually the one which gives maximum acceptable wing loading and/or minimum thrust/power loading.

The importance of evaluating the effect of different wing geometries was mentioned in paragraph 1.5.5. This is done by subjecting each wing geometry to the procedures outlined above. Thus, for example, the alternative wing configurations could consist of a set of three different aspect ratios and three pairs of sweep and aerofoil thickness to chord ratios.

1.5.8.2 Second stage

The selected values of wing loading and thrust/power loading are then used in the mass representation and the corresponding absolute value of mass is calculated. To do this the derived values of these parameters are used in the mass representation to enable the total mass of the aircraft to be expressed in terms of itself, the wing loading and thrust/power loading. If necessary alternative acceptable combinations of wing loading and thrust/power loading may be analysed and the combination which gives the minimum absolute value of mass chosen as the optimum for that particular wing and powerplant configuration.

For each configuration the process yields an aircraft total mass, and the lowest value may be regarded as the optimum from the selected parametric analysis. Paragraph 1.4.2.3 outlines the relationship of aircraft mass to other possible criteria which may be used as a basis for optimisation. The configuration which has the lowest mass is appropriate for selection as the referee design for more detailed analysis and evaluation.

The actual wing size follows directly from this procedure, and it becomes possible to produce a first definitive layout of the chosen configuration. If desired this can include a notional representation of empennage and landing gear (see Chapter 8, paragraph 8.10 and Addenda 1 to 3). Assumptions made during the preparation of the fuselage layout and the wing configuration, such as wing/fuselage intersection, can be checked and if needed the fuselage layout adjusted. It may be necessary to repeat the whole process with revised values if there are any major discrepancies.

At this stage of the synthesis process it is also possible to make comparisons between alternative overall aircraft configurations, should more than one have been chosen for investigation. However, until a referee design for each overall configuration has been analysed and the assumptions made are verified or adjusted, such a comparison may be considered to be premature.

1.5.9 Analysis of the derived design (Chapter 9)

The conclusion of the process outlined above is the derivation of a concept which, within the validity of the assumptions and for a given overall configuration, best meets the specification. Some drastic assumptions will have been made to simplify the procedure. It is essential to check these by using more comprehensive analytical tools, especially those concerned with lift, drag, mass and control and stability. The sequence of undertaking this verification is:

a) Complete the aerodynamic layout of the aircraft by making first predictions of the sizes of control and stabiliser surfaces. This is preferably done by employing a method which uses the required control and stability as inputs, although it may be adequate to base it on past data (see paragraph 1.5.8.2 above and Addenda 2 and 3).

b) Complete the layout of the landing gear. Some aspects of this will already have been considered during the fuselage and wing layout, but the assumptions need to be verified (see Addendum 1).

c) Estimate the lift, drag and mass of the completed configuration using the most comprehensive analytical methods available (see Addendum 4 for mass prediction).

d) Repeat the performance calculations described in paragraph 1.5.7 with the improved predictions of lift, drag and mass and possibly also using data for an actual powerplant. It may be appropriate to use somewhat more elaborate performance estimation methods.

e) Re-optimise the design if necessary to derive a new mass using the process outlined in paragraph 1.5.8. If this new mass differs significantly from that previously estimated it may be necessary to re-visit the wing parametric analysis.

f) If necessary reconsider the control and stability requirements especially if empirical values have been used at sub-paragraph (a) above for simplicity. Where significant changes are needed repeat the previous sequence.

g) Repeat the whole process until mass convergence is achieved.

Chapter 9 discusses possible references for use during this analysis and refinement phase. The satisfactory conclusion of the study should be the definition of an aircraft in a form suitable for detail design and subsequent manufacture.

1.6 Design information

The primary purpose of this volume is to give sufficient information to enable a first synthesis of an aeroplane design to be completed. Some of the data are presented in an unusual form simply to facilitate the synthesis procedure. There are numerous other sources of information which are of value, and as the design proceeds through the iterative analytical phases it is necessary to refer to these sources. A summary of some of the available books on aircraft design and important sources of data is included in a bibliography in the Appendix to Chapter 9.

Chapter 2

Aircraft configuration

2.1 Introduction

Selection of the configuration of the aircraft is fundamental to the synthesis process. When it is not clear which layout is the most appropriate it is necessary to identify two or more possibilities for independent study and subsequent comparison.

As aviation developed during the twentieth century a conventional configuration for fixed-wing aircraft emerged. Numerous variations within this conventional configuration are possible. There are also a number of fundamentally different concepts, one of which may offer particular advantages in special circumstances. Nevertheless, experience has shown that in many instances there is a tendency for these unconventional layouts to tend back towards the conventional as the concept is refined. Such a tendency can be avoided by clearly identifying the potential advantages conferred by a particular unconventional concept and applying constraints to ensure that they are realised.

The configurations described are appropriate to both manned and unmanned aircraft and, with some reservations, to guided weapons.

2.2 Conventional configuration

2.2.1 Basic definition

For the present purposes a basic, conventional, configuration is defined as one having the following layout characteristics:

a) A cantilever monoplane wing

b) Separate vertical and horizontal tail surfaces

c) A discrete fuselage used to provide volume and continuity to the airframe

d) A retractable tricycle landing gear

As would be expected there are numerous examples of this layout, Figure 2.1 being just one such.

Figure 2.1 Conventional configuration - Airbus A319 Airliner
(Courtesy British Aerospace Airbus, B3430)

The number of powerplants used should be the minimum possible needed to meet power and operational requirements. At least two engines are necessary for most commercial operations and some military types. The need for more than two engines for long overwater commercial flights has been largely superseded by the ETOPS (extended twin engine operation) concept and it is less likely that there will be new three-engine airliners of conventional design.

2.2.2 Layout characteristics

2.2.2.1 Powerplant location

The location of the powerplants within a conventional configuration depends to some extent on the role of the aircraft. The ease with which alternative makes of powerplant may be used can be an important consideration. Five usual situations may be identified:

a) Nose location - propeller propulsion, Figure 2.2. Traditionally the location of a single tractor propeller engine is in the nose of the aircraft. There are clear reasons for this of which propeller clearance is the most obvious. Another major advantage is that it enables the powerplant unit to be largely self-contained and independent of the rest of the aircraft. This is beneficial in design, manufacture and maintenance.

b)Fuselage central/aft buried location, Figure 2.3. Jet and fan engines are frequently located within the fuselage of an aircraft. This is especially the case for single- and twin-engine military jet trainer and combat types. A compact overall layout results but access for maintenance and engine removal requires careful consideration if severe structural penalties are to be avoided. In general it is best to minimise the length of the jet pipe by placing the powerplant as far aft as possible within constraints imposed by centre of gravity considerations. Jet efflux should emerge behind all parts of the local airframe to minimise acoustic fatigue. Buried shaft engines driving a rear pusher propeller are unusual and at the very least introduce difficulties with ground clearance in aircraft tail down attitude.

Figure 2.2 Nose powerplant - Scottish Aviation Bulldog
(Courtesy College of Aeronautics)

Figure 2.3 British Aerospace Hawk
(Copyright British Aerospace MAA, 86573)

Figure 2.4 Britten - Norman Islander
(Author)

c) Wing-mounted powerplants, Figures 2.1 and 2.4. The majority of larger aircraft utilise wing mounting for the powerplants. Jet and fan engines are podded and usually mounted on short underwing pylons while propeller engines are most frequently located immediately on the wing. The podded arrangement especially is versatile in that alternative types of engine may be incorporated with minimum change. It is also compact and access for maintenance is convenient. Spanwise location of the powerplant depends upon such matters as propeller diameter or fan burst trajectory. Typically inner engines are at about 30% semi-span while on four-engined aircraft the outer units are at around 55% semi-span. The major restriction on the use of this configuration is ground clearance. If this is a problem it may be overcome by mounting the wing higher on the fuselage and/or by using a longer landing gear, both of which can imply design penalties, see paragraph 2.2.2.3. Alternative solutions in this case are to accept the drag penalty of locating the powerplant over the top of the wing, see Chapter 6, paragraph 6.3.2.1 and Figure 2.5, or to locate the powerplants elsewhere.

Figure 2.5 **Hawker - Siddeley HS 748**
(British Aerospace K1/1/3069)

d) Rear fuselage podded powerplants, Figure 2.6. In the past jet engines have been located in pods mounted from the rear fuselage of moderately large transport aircraft. However more recent designs universally prefer an underwing location for this class of aircraft, and the use of rear fuselage pods is primarily restricted to smaller jet aircraft in the business/feeder line categories. The generally small size of this type of aircraft precludes the use of underwing pods. A major difficulty associated with rear podded layouts is an aft centre of gravity tendency. Fan/turbine burst and acoustic fatigue considerations may also place restrictions on powerplant fore and aft location.

e) Upper or lower podded powerplant. Some unmanned aircraft have single podded engines located above or below the fuselage. This results in a simple fuselage layout.

2.2.2.2 Air intakes

Inevitably podded turbine powerplants have a pitot intake. For subsonic applications the length of the intake can be relatively short, typically about 50% of fan diameter. Supersonic intakes demand a longer diffuser which may incorporate variable geometry. Efficient supersonic flight above about Mach number 1.7 demands precise control of geometry. Some geometry control can be obtained by using a translating centre body of appropriate configuration in an axisymmetric pitot intake, Figure 2.7. An alternative is to have moving ramps in a two-dimensional intake layout, of which the Concorde design is a good example, Figure 2.8. The Boeing (McDonnell Douglas) MD11 third engine arrangement, Figure 2.9, is an interesting example of a podded pitot intake concept.

Figure 2.6 Raytheon Hawker 800XP
(Author)

Figure 2.7 British Aircraft Corporation Lightning
(Copyright British Aerospace MAA, 54937)

A number of air intake configurations are possible when the powerplants are buried in the fuselage:

a) Pitot (nose) intake, Figure 2.7. A nose intake is generally not preferred as it is of excessive length and occupies an undesirably large part of the fuselage volume.

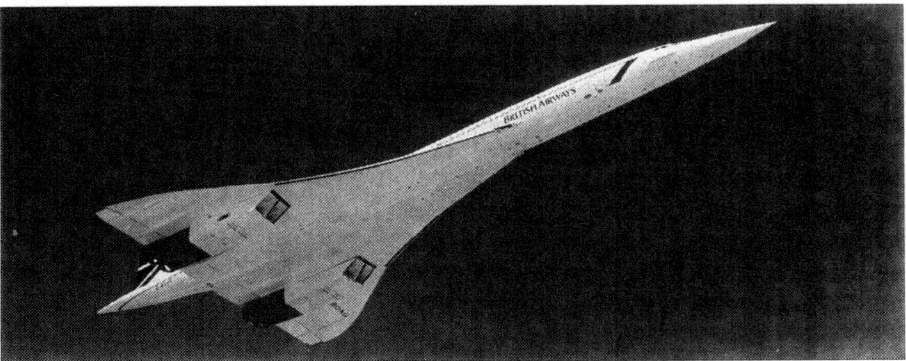

Figure 2.8 British Aircraft Corporation - Aerospatiale Concorde
(Courtesy British Aerospace Airbus, C2018)

b) Side intakes above wing, Figure 2.3. Side intakes generally can be designed to achieve a minimum intake length and hence a low volume requirement. Care must be taken in designing the geometry to avoid unduly sharp bends which could cause flow distortion. Intakes positioned above the wing are usually associated with low-wing trainer/strike aircraft and general experience suggests that this may not always be the best arrangement.

c) Side intakes below wing, Figure 2.10. Side intakes located below a mid to high wing is a more common layout than above-wing intakes. A better overall performance may be anticipated and benefit may be gained by extending the wing leading edge above the intakes to enhance flow at high angles of attack.

d) Ventral intakes, Figures 2.11 and 2.13. Ventral intakes are becoming common for high performance combat aircraft where high angle of attack manoeuvrability is important. Reservations with regard to ventral intakes include concern about the probability of debris ingestion during ground operation.

e) Upper fuselage, dorsal intakes, Figure 2.12. A few airliners have been fitted with three engines, the third being buried in the rear fuselage. A dorsal intake located in the root of the vertical stabiliser supplies air to the centre engine in this arrangement. This has proved to be satisfactory in applications required to operate only at low angles of attack. Intakes mounted above the centre fuselage, aft of the cockpit, appear to confer some layout advantages for military combat types. Although wind tunnel tests have suggested that such an arrangement may function efficiently up to moderate angles of attack, there remains considerable doubt as to whether this would be realised in practice. The adoption of a dorsal intake is not recommended unless low observability is a primary consideration (see paragraph 2.4.3).

Figure 2.9 Boeing (McDonnell Douglas) MD 11
(Licensed - Boeing/McDonnell Douglas MO-00019)

Figure 2.10 Panavia Tornado
(Copyright British Aerospace, MAA, 150723)

2.2.2.3 Vertical location of the wing

The vertical location of the wing has an impact on the layout of the fuselage, and this is discussed in Chapter 4, paragraph 4.4.1. Nevertheless it is appropriate to comment upon the issues involved here in as much as they affect other aspects of the configuration.

Aerodynamic considerations suggest that a mid-wing arrangement is likely to give the lowest interference drag penalty, while a high wing is beneficial in having an efficient spanwise lift distribution and hence low drag due to lift.

Structurally, with only a few exceptions, it is important that the primary spanwise beam should be continuous across the fuselage. A conflict with aerodynamic requirements may therefore arise. A mid- or shoulder wing location can be achieved when the wing is of low aspect ratio, as on a combat aircraft (Figure 2.13 and Chapter 4, Figure 4.7) but is likely to be ruled out on most higher aspect ratio configurations.

In many respects the advantages of a low wing are the disadvantages of a high wing and vice versa. Which is used depends upon the particular requirements. See Chapter 4, paragraph 4.4.1.

A high wing does imply either a very long wing-mounted landing gear, Figure 2.4, or a narrow track fuselage-mounted configuration, Figure 2.14.

Similar considerations apply to other classes of aircraft to a greater or lesser extent. Low wings tend to be most common on general aviation aircraft, Figure 2.2, although there is little relative penalty for a high wing, Figure 2.15. High wings are often preferred for unmanned aircraft.

Combat aircraft most frequently have a mid- to high-wing configuration, Figures 2.10 and 2.16. The choice is determined by such considerations as intake configuration, landing gear length and, importantly, access/clearance below the wing for maintenance and stores.

Figure 2.11 Eurofighter 2000
(Copyright British Aerospace, MAA, 122041)

Figure 2.12 Lockheed Tristar
(Courtesy Lockheed - Martin Aeronautical Systems)

Figure 2.13 Lockheed - Martin F 16
(Courtesy Lockheed - Martin, 31-48305)

Figure 2.14 Lockheed C 5A
(Courtesy Lockheed - Martin Aeronautical Systems, RS6456-26C)

2.2.2.4 Empennage

The basic configuration for the tail surfaces consists of a horizontal fixed tailplane, or stabiliser, and a vertical fixed fin, each having a hinged rear flap acting as an elevator for pitch control and a rudder for yaw control respectively. A dorsal fairing is often incorporated into the base of the fin to preclude the possibility of fin stall.

Variations on the basic arrangement are common and include:

a) Variable incidence tailplane. The forward section of the horizontal surface is capable of rotation through a range of angles of attack. In this way it may be used to adjust the pitch trim, as opposed to control, of the aircraft especially when deployment of the high lift devices introduces significant pitching moment increments.

b) All moving or "flying" tailplane. In this concept the whole surface is used as the primary pitch control with the elevator, if it is retained, being used only for trim. Such an arrangement offers significant advantages at transonic and supersonic speeds when the effectiveness of conventional trailing edge controls is much reduced and fuselage bending can result in unfavourable loads on a fixed tailplane. Some combat aircraft use differential movement of the two sides of the horizontal surface to provide roll control.

c) Vertical position of the horizontal tail. The horizontal tail is within the wing downwash field which has the effect of reducing the effectiveness as a stabiliser. The degree of this reduction is a function of the vertical location of the tail relative to the wing and the effect may be reduced by significant upward movement in tail location (Figures 2.4, 2.5, 2.9, 2.12 and 2.14). In general a horizontal tail mounted at the top of the fin can be smaller than would otherwise be the case. Unfortunately such an arrangement is not without disadvantages. There is a mass penalty on the fin due to higher loading and aeroelastic effects, and there is also the possibility of a deep stall. Essentially this may occur when the aircraft pitches nose up rapidly and reaches an attitude such that the tailplane is virtually ineffective as a stabiliser in the conventional sense. However, the aircraft is in a stable stalled condition from which it may be difficult or impossible to recover. While means are available to resolve this difficulty it is suggested that a high-mounted 'T' tail should only be used when it is really necessary, as may be the case of a high-mounted swept back wing configuration or when an engine intake is placed at the bottom of the fin. A possible alternative to the 'T' tail which does not suffer from the deep stall problem is to mount the tailplane very low. Unfortunately in most cases this is not an option because of tail down ground clearance limitations, but it is worthy of consideration on smaller aircraft, such as combat types, especially when the wing is positioned high on the fuselage, Figure 2.17.

2.2.2.5 Landing gear (see also Addendum 1)

The basic configuration is presumed to have a retractable, tricycle, landing gear but there are many examples of low performance light aircraft where the gear is fixed, Figure 2.15. For these aircraft the drag advantage of a retractable configuration does not offset the mass and complexity penalties which are associated with it. However, the fixed gear is the exception rather than the rule and for the great majority of aircraft it is advantageous to retract the gear. Some unmanned aircraft dispense with a conventional landing gear and use catapult launch with parachute or net recovery, possibly also with an air bag.

Figure 2.15 Auster Aiglet
(Author)

Figure 2.16 British Aerospace Harrier
(Copyright British Aerospace, MAA)

Figure 2.17 British Aircraft Corporation TSR2
(Copyright British Aerospace, Weybridge, MP20801)

The conventional tricycle gear consists of a pair of main legs which are located just behind the centre of gravity of the aircraft, together with a nose leg, Figures 2.2 and 2.6. Each of the three legs incorporates a shock absorber to dissipate vertical energy on landing and carries either a single wheel, or a pair of side-by-side wheels. Only the main wheels are fitted with brakes but the nose wheel can usually be steered for ground manoeuvring. Effective steering requires that the nose leg should support between 6 and 14% of the mass of the aircraft. From this it may be deduced that at a mid centre of gravity position the fore and aft location of the wheels should be such that each main leg supports about 45% of the mass. Lateral disposition of the main wheels must be such as to preclude any tendency to overturn during ground manoeuvring. This is a function of shock absorber characteristics as well as the height of the centre of gravity as a ratio of the track. During the initial definition of the layout of the aircraft it is necessary to make some provision of suitable locations both for attachment of the landing gear legs and stowage of the units when retracted.

Many civil airfields have runways of an accepted minimum strength but there are exceptions, especially in more remote locations and smaller airfields. Military aircraft are often required to operate from unprepared or semi-prepared surfaces of relatively low load carrying capability. This is a complex issue but the consequence is that as aircraft mass increases, operation from a runway of given strength dictates the need to provide a greater number of wheels to spread the load. This can be done in various ways depending on the number of wheels necessary and, possibly, on the overall layout of the aircraft:

a) Two-axle bogie, Figure 2.18. The use of a bogie with an axle at each end enables the number of wheels to be doubled relative to a single-axle arrangement. Usually each axle supports two wheels although there may be four tyres. Such an arrangement is likely to be necessary for civil airliners having a mass in excess of about 90,000 kg. Occasionally a bogie arrangement with only a single wheel on each axle is used to facilitate stowage in a confined location, Figure 2.17.

b) Three-axle bogie. As aircraft mass increases further, above about 210,000 kg for an airliner, further load spreading becomes necessary. One way of achieving this is to add a third axle to the bogie. It may be done either by simply locating another axle in the centre of the bogie, Figure 2.19, or by using a split bogie with a pair of side-by-side axles at the rear, Figure 2.14. In either case it is likely to be necessary to incorporate some bogie steering.

c) Three main gear legs. An additional, centreline-located, main leg can be used to spread the load. The centre leg may not have to carry as much load as the outer legs and the concept has been used where it is necessary to cope with significant mass increase of an aircraft basically equipped with two main legs, Figure 2.18.

d) Four main legs, Figure 2.14. When the gross mass of an airliner exceeds about

300,000 kg it is likely to be necessary to use four main legs. For each of them to be equally loaded a means of balancing the load is needed. Both the mounting of the units on the airframe and stowage is a complex problem. Available evidence suggests that an airliner fitted with four main legs, with a three-axle bogie, could have a mass of about 500,000 kg.

Figure 2.18 Airbus A340
(Courtesy British Aerospace Airbus, B2102)

e) Multiple legs with single axles, Figure 2.20. A further possibility is the use of a number of single-axle legs mounted in a fore and aft line. This concept can be incorporated into a high-wing layout, the landing gear units being mounted along the sides of the fuselage and retracted into blisters. It is applicable to large freight aircraft where there is a need to operate from low-strength surfaces. Some, or all, of the wheels have to be steerable.

Figure 2.19 Boeing 777 Airliner
(Licensed -The Boeing Company K58291)

Figure 2.20 Antonev AN 124-100
(Courtesy AirFoyle)

2.2.3 Variations of basic configuration

2.2.3.1 Braced wing, Figure 2.15

Reduction in structure mass can be achieved by using a bracing strut to support the wing, but the associated drag penalty can be high. This layout is only likely to be used on small aircraft which fly at low speed and where there are operational advantages of having a simple wing/fuselage connection.

2.2.3.2 Sweep

A wing may be swept for a number of reasons, by far the most important of which is to delay the onset of compressibility drag, see Chapter 5, paragraph 5.3.4. Sometimes there is a layout advantage of employing moderate sweep, for example to facilitate landing gear attachment and stowage behind the primary wing structure, or to enhance static longitudinal stability.

Figure 2.21 VFW Hansa Jet
(Courtesy DASA 3609 25a)

Figure 2.22 Boeing (McDonnell-Douglas) F 18B
(Licensed - Boeing/McDonnell Douglas C12-25300-239)

Generally sweep back, Figure 2.18, is used rather than forward sweep, Figure 2.21, because:

a) Overall it gives a better compromise of aerodynamic characteristics, especially in terms of stability and control.

b) Forward swept wings are prone to aeroelastic divergence which can only be overcome by accepting a mass penalty.

c) The loading towards the front of a forward swept wing at the root can be high and this introduces a further mass penalty. An exception where limited forward sweep may confer a layout advantage is when it is helpful to locate the wing carry through structure behind the centre of gravity so that the payload volume is not interrupted by it, Figure 2.21.

d) Unless there are overriding layout requirements the amount of sweep employed should be the minimum required for aerodynamic reasons. Large sweep angles are required to maintain subsonic leading edge flow when an aircraft is flying supersonically. As a consequence the low wing span, delta, configuration becomes advantageous, Figure 2.8. Low aspect ratio wings have a relatively poor lifting capability which may be enhanced by incorporating leading edge strakes, or extensions, along the fuselage at the wing root, Figure 2.22.

2.2.3.3 Wing tip fins (winglets), Figure 2.18

Wing tip end plates or fins may be used to reduce drag during cruise by effectively increasing the wing span. There may also be some contribution to directional stability if the wing is swept back. It is usual to keep tip fins relatively small to avoid severe structural penalties, and their primary use is when an improvement is required to an existing design or when wing span is limited by operational considerations.

2.2.3.4 Twin tail fins, Figure 2.23

At one time it was common for the conventional single vertical tail surface to be replaced by a pair of fins located at the extremities of the horizontal tail, but there are a number of reasons why the configuration is now rarely used. These include the associated mass and aerodynamic interference penalties. With the exception of twin fins used with a twin boom layout, paragraph 2.2.3.6, it is not possible to avoid the possibility of fin stall at high sideslip angles by incorporating a dorsal fin.

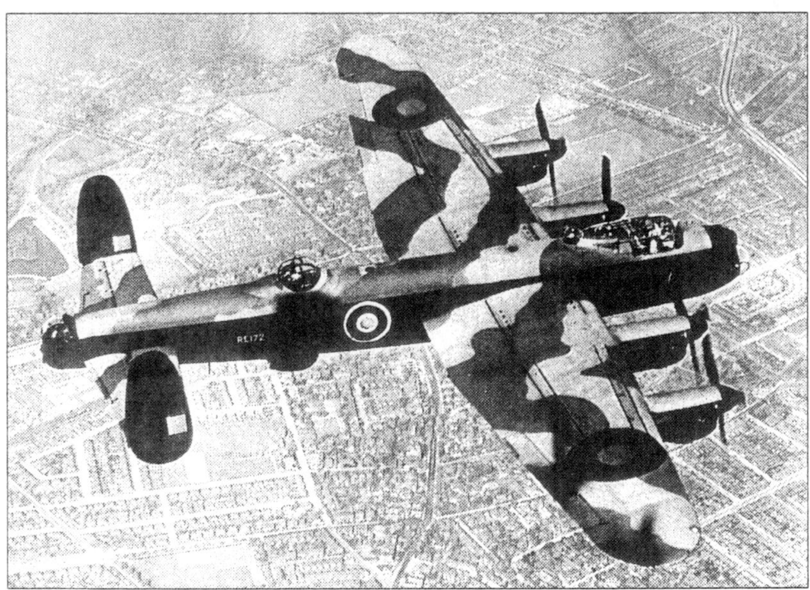

Figure 2.23 Avro Lancaster
(Hawker Siddeley Aviation Ltd A-9-24)

2.2.3.5 All moving fin

A one-piece fin, rather than the conventional fin/rudder arrangement, has been used on rare occasions, Figure 2.17. There seems to be little reason to do this since a conventional rudder may be designed to provide adequate yaw control at all speeds.

2.2.3.6 Twin booms

Twin booms extending aft from the wing can be used to support the tailplane as an alternative to the more conventional rear fuselage arrangement. Twin fins located on the booms are inevitable but dorsal fairings can be incorporated. Two main reasons can be identified for using this layout:

Figure 2.24 Cessna 337
(Author)

Figure 2.25 Armstrong - Whitworth Argosy
(Sir W.G. Armstrong Whitworth Aircraft Ltd DO1230)

a) Provision for a pusher propeller for a general aviation or unmanned aircraft where engine mass is such that it must be located close to the centre of gravity of the aircraft, Figure 2.24. Some early jet-powered aircraft used twin booms for the same purpose.

b) To give unrestricted access for a rear freighthold on an aircraft of small to moderate size, Figure 2.25. The booms may conveniently extend rearwards from engine nacelles and may provide volume for landing gear stowage.

As a generalisation a twin boom layout should only be used if it is essential to enable requirements to be met. Although there is probably little difference in fuselage mass relative to the conventional layout, there is a wing mass penalty, greater interference drag and less usable volume.

2.2.3.7 Landing gear variations

Three variations on the basic tricycle landing gear can be identified:

a) Tailwheel layout, Figure 2.26. In this arrangement the two main legs are located somewhat forward of the centre of gravity and the third support consists of a small tailwheel or skid. Although this configuration is both simpler and lighter than the conventional one it is much less satisfactory from an operational aspect. Ground manoeuvring is more difficult, in part due to poor visibility, and take-off and landing more complex. It is now limited to simple light aircraft where the operational penalties are accepted in exchange for simplicity.

Figure 2.26 Cranfield A1
(Courtesy College of Aeronautics 43821)

b) Single main gear leg, Figure 2.16. On occasions there are layout advantages in concentrating the main load carrying capacity into a single main leg, roll stability on the ground being obtained from a pair of lightweight outriggers. It is essential to arrange the layout so that the outrigger loads are small, as otherwise they tend to become more substantial and revert to a conventional, twin main leg, form.

c) Bicycle layout, Figure 2.27. The bicycle layout may be regarded as a special case of the single main leg configuration, except that the rear wheel is located significantly further aft than normal. Because of this the nose unit carries a similar proportion of the mass as the rear leg. The advantage is an uncluttered wing and a long length of the fuselage free from landing gear components which can be used for payload, such as a weapons bay. The disadvantages are the highly loaded nose leg, which makes ground manoeuvring and rotation at take-off difficult, and the need for a special landing technique to avoid extreme nose leg loads during touch down and braking. A variation of the basic arrangement is where the two legs are replaced by two pairs, located at the sides of the fuselage, but similarly fore and aft, as on the B 52 bomber. The difficulties associated with the bicycle layout are severe and it should not be used unless there is no acceptable alternative.

Figure 2.27 Boeing B 52 Bomber
(Boeing Aerospace Company P12462)

2.3 Alternative configurations

2.3.1 General comments

As well as the variations of the conventional configuration discussed in paragraph 2.2.3, there exist a number of radically different layouts which can offer specific advantages in particular circumstances. As a general rule these alternative configurations should only be considered when there is a clear advantage since it may be anticipated that corresponding disadvantages will be present. In some cases design refinement results in a tendency towards a conventional layout.

2.3.2 Biplane, Figure 2.28

In the earliest days of aeronautics multiplanes, especially biplanes, were normal. At that time aerodynamic technology was such that a relatively large wing area was required, that is the wing loading had to be low. In addition a light wing structure could be achieved by cross-bracing between the upper and lower planes. As technology advanced and speeds increased the drag penalties of the traditional biplane became unacceptable. At the same time the development of high lift systems and improved materials enabled wing loadings to be increased so that the monoplane became, and remains, the preferred solution. Now the only reason for using a biplane configuration is when a low wing loading must be combined with a compact layout on an aircraft of moderate speed performance. An aerobatic aircraft is possibly an example of such a requirement.

Figure 2.28 Pitts Special Aerobatic Biplane
(Courtesy Aerobatic Displays Ltd)

Figure 2.29 Lockheed F 117
(Courtesy Lockheed - Martin Aeronautical Systems)

2.3.3 Butterfly tail, Figure 2.29

A butterfly tail is essentially one where the conventional horizontal and vertical tail surfaces are combined into a pair of inclined surfaces. In terms of stability and control these surfaces combine the separate roles of the tailplane/elevator and fin/rudder. The claimed advantages are a smaller total surface area and less interference drag. The disadvantage is the cross-coupling of stability and control characteristics which introduces handling difficulties unless a fully automatic flight control system is used. Use of a butterfly tail is inadvisable unless there is a compelling reason, such as a stealth requirement, see paragraph 2.4.3.

2.3.4 Tailless layout

It has long been recognised that the only essential airframe component is the wing. From this it may be deduced that the most aerodynamically efficient configuration is an "all wing" aircraft, the tail surfaces being dispensed with and the payload being located within the volume of the wing. In practice the issues are not as clearly defined as this deduction would suggest. Often the wing shape is unsuitable to accommodate the payload and it is not easy to provide directional stability and control from a basically horizontal lifting surface. Tailless designs, Figure 2.30, usually, but not always, incorporate a measure of sweep back since this increases overall length and facilitates the provision of longitudinal stability and control. In the limit this leads to the relatively highly swept, low wing span delta configuration, Figure 2.8. While tailless designs do exhibit relatively low drag the theoretically achievable gains are not realised because:

a) Longitudinal control is limited due to the short moment arm of the elevators. This means manoeuvre capability is limited and, as it may be difficult to use powerful high lift devices, the wing loading may be lower than on a comparable conventional design.

b) A rudimentary fuselage is often required for payload volume, unless the aircraft is very large. Even then the internal shape of a wing section introduces a severe mass penalty when pressurisation is a requirement.

c) The retention of a vertical fin is the most efficient way of providing directional stability and control, unless this is ruled out by low observability demands.

d) The need to keep the centre of gravity movement within a small range may compromise the design. This is a consequence of the low longitudinal control effectiveness.

Stealth requirements may be such that an "all wing" design is the best solution, Figure 2.30, but otherwise the value of the tailless concept must be questioned. Pure tailless delta designs are no longer favoured.

2.3.5 Variable sweep, Figure 2.10

Highly swept wings can possess poor low speed aerodynamic characteristics, both in terms of available lift and handling near stall conditions. Variable sweep back was originally conceived as a means of overcoming this difficulty by matching the sweep angle to the ideal at a given flight speed. It was claimed that using it enabled optimum aerodynamic conditions to be achieved at all speeds. In practice this was not found to be the case, primarily due to the drag associated with the interaction between the fixed and moving parts of the wing. When this penalty is combined with those due to the extra mass and system complexity the extent of the real benefits is less than might be anticipated. Variable sweep does enable an aircraft to possess good performance over a wide speed range and therefore be capable of fulfilling a multirole requirement. However, it is unlikely to be as effective in a given role as an aircraft designed specifically to meet that role.

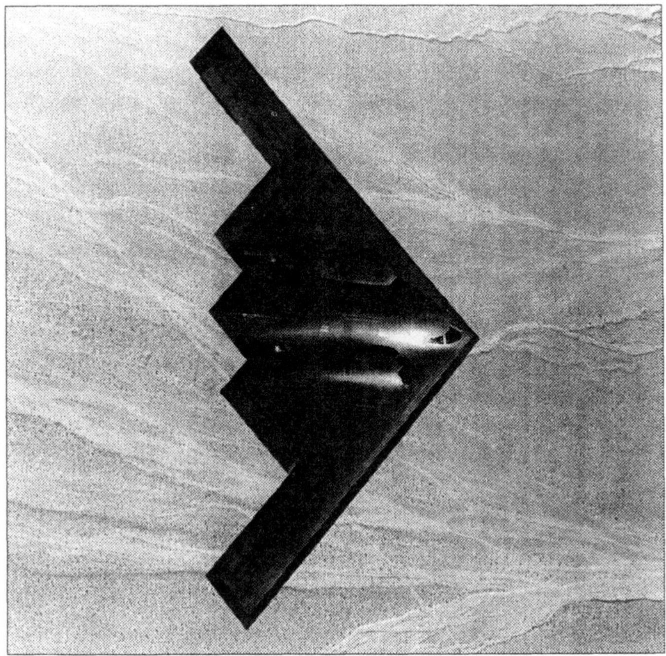

Figure 2.30 Northrop-Grumman B 2 Stealth Bomber
(Courtesy Northrop-Grumman, 89AFE115-18)

2.3.6 Canard layout

A canard layout is one where the conventional aft tailplane is replaced by a foreplane, Figure 2.31. The wing is located further back on the aircraft, effectively behind the centre of gravity, and provides such longitudinal stability as may be present. The foreplane provides control for longitudinal trim and manoeuvre. The merit of this configuration is that foreplane trim forces are generally in the same direction as those on the wing, unlike those on a conventional tailplane. Thus there should be negligible trim drag penalty and manoeuvre response is more rapid. However, airflow from the canard does interfere with the wing and it is essential to match the layout geometry to achieve the best overall result. The use of a foreplane may confer a layout advantage when the aft-located wing structure may pass behind the cabin in much the same way as that of the forward swept wing Hansa Jet, Figure 2.21.

Three variations of the canard configuration may be mentioned:

a) Long coupled with swept back or delta wing, Figure 2.31. In this arrangement a small foreplane is located sufficiently far ahead of the wing that the interference effects are small. It is a layout which is particularly suited to a long range supersonic aircraft, such as an airliner, where the foreplane effect is beneficial both in terms of cruise trim drag and at low speed, especially take-off rotation.

b) Short coupled with delta wing, Figure 2.11. The foreplane is placed just ahead of and, usually, just above the wing in this configuration. Careful location of the two components enables the lift effectiveness of the pair to exceed that of the sum of their isolated values. It is applicable to a highly agile combat aircraft, but is disadvantageous from the stealth aspect.

c) Foreplane with forward swept wing, Figure 2.32. On the whole there does not appear to be any real advantage of forward sweep, as was discussed in paragraph 2.2.3.2. However, when forward sweep is associated with a foreplane some of the airflow problems at the root of the wing may be overcome. Hence there may be some merit in employing this configuration if it also confers an overall layout advantage.

Foreplane lift requirements generally tend to be disproportionately high and it may be necessary to limit then by imposing layout and size restrictions. Otherwise there can be a tendency for the foreplane size to increase to that of a conventional wing, while the wing size decreases towards that of a tailplane. For this reason foreplane area is usually relatively smaller than that of an equivalent tailplane, but there are exceptions.

Figure 2.31 Raytheon (Beech) Starship
(Courtesy J.P.Fielding)

Figure 2.32 Grumman X 29
(Courtesy Northrop - Grumman 851070)

2.3.7 Three-surface configuration, Figure 2.33

As its name implies the three-surface configuration is one which employs both a tailplane and foreplane. The aim is to obtain the advantages of both, that is the stabilising capability of a tailplane with the favourable trim and control function of a foreplane. As is often the case with unconventional layouts it would seem that secondary penalties, such as a fuselage mass penalty, largely offset the advantages.

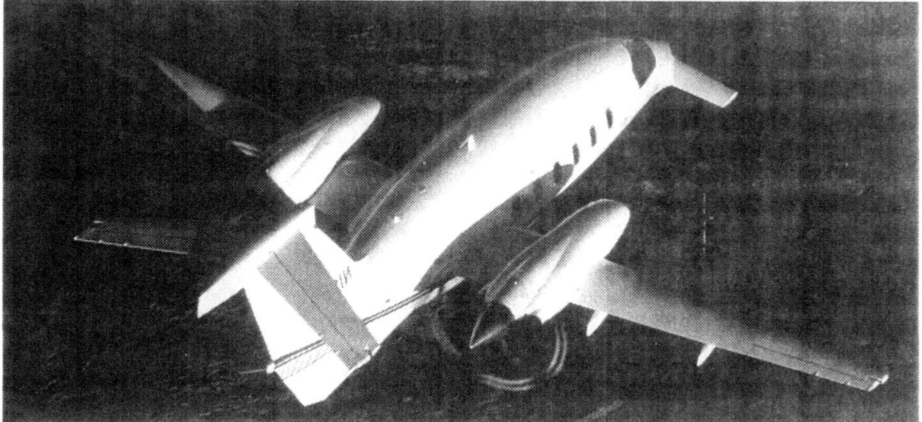

Figure 2.33 Piaggio P180 Avanti
(Courtesy Piaggio Aero Industries)

2.3.8 Twin-fuselage configuration

From time to time twin-fuselage configurations are proposed and indeed have been used on occasion for flying boats, effectively creating a hybrid craft. Although arguments may be advanced in favour of twin fuselages the advantages would not appear to be such as to justify their application in usual circumstances.

2.4 Special considerations

2.4.1 General comments

It is always possible that the requirements for an aircraft will include a special provision which has a dominating impact upon the configuration chosen. Such a condition may well override the comments made in paragraphs 2.2 and 2.3.

2.4.2 Short and vertical take-off and landing

2.4.2.1 General remarks
Some decades ago a considerable effort was expended in the investigation of short and vertical take-off and landing civil transport aircraft. Experience now suggests that there is no requirement for this class of aircraft except for small short range types which typically have low wing loading anyway. There is, however, a continuing requirement for military aircraft in this category.

2.4.2.2 Military freight aircraft
Military freighters are often required to operate to and from airstrips of short length and poor surface strength. Short take-off and landing capability is necessary for this kind of operation, but it is not likely to have any major effect upon the overall configuration of the aircraft The emphasis is on relatively high installed thrust, complex high lift devices, and large diameter, low pressure, tyres on multiple wheels.

2.4.2.3 Combat aircraft (see also Addendum 5, paragraph AD5.5.2)
The usual requirement for military combat aircraft is STOVL - Short Take-Off and Vertical Landing. The vertical landing requirement implies that there must be a vertical thrust component somewhat greater than the landing weight of the aircraft. It is logical to make use of this component for short take-off.

Many ways of providing vertical thrust have been investigated. Most of them are complex and require a large volume. In all the more likely systems some use is made of the forward flight propulsion unit or units, by downward deflection of the main exhaust gases in the take-off and landing modes. Often it is not practical to locate this deflected exhaust immediately below the centre of gravity so some additional provision for vertical thrust is required. Plausible means of providing this are:

a) Vectored bypass flow, Figure 2.16. The Rolls-Royce Pegasus engine used in the Harrier aircraft is provided with four nozzles, each of which may be rotated to vector the efflux as required The rear two nozzles exhaust the hot gases while the front two exhaust bypass air taken directly from behind the fan. The whole system is compact but bulky and has the disadvantage that it must be located at the centre of gravity of the aircraft. Attempts have been made to augment the thrust by burning fuel in the bypass air, so-called plenum chamber burning, but this introduces problems of hot gas ingestion into the air intakes, and possibly ground erosion.

b) Remote location of nozzles. The centre of gravity difficulty of the Pegasus concept may be overcome by ducting the bypass flow forward to a remote nozzle where it may be augmented by fuel injection. The volume requirement is high.

c) Vertical lift engines. Another possibility is the use of one or more dedicated vertical lift engines in addition to the deflected thrust of the cruise engine. The cruise engine may now be located aft in the aircraft and the lift engine forward resulting in much greater flexibility in aircraft layout than is the case with the Pegasus-type engine. Such an arrangement was used in the YAK-141. However, against this there is the mass and volume penalty of the lift engines which are only used in low speed flight.

d) Remote driven lift fan. An alternative to a dedicated lift engine is the use of a lift fan driven remotely from the cruise engine. The drive may either be mechanical, which places some restrictions on fuselage layout, or compressed gas which is bulky and relatively inefficient. The total effective volume of the fuselage occupied by this system is likely to be more than that of the systems previously mentioned except, possibly, the remote nozzle arrangement.

Whichever propulsion system is chosen it is necessary to provide low speed control effect by means of reaction jets supplied with air bled from the compressor of the cruise engine. These are certainly required for roll and yaw control and may also be needed in pitch control.

Clearly the choice and layout of the propulsion system must have a major impact upon the configuration of the aircraft and may well limit the options available.

2.4.3 Low observables (Stealth)

(See for example ASME Papers 90-GT-116 and 117.)

2.4.3.1 General comments

There are two basic ways of approaching the problem of conferring low observability characteristics on a combat aircraft:

a) To specify an aircraft configuration which fundamentally will possess good low observability characteristics and to design the details appropriately. The all wing layout of the Northrop-Grumman B 2 shown in Figure 2.30 is a good example of this approach.

b) To determine a configuration of the aircraft primarily on the basis of performance requirements and then to incorporate low observablity characteristics in the details of the layout and design. In most cases this is likely to be the preferred approach and one which appears to have been adopted in the design of the F-22, Figure 2.34.

Figure 2.34 Lockheed - Boeing F 22 Raptor
(Courtesy Lockheed - Martin Aeronautical Systems)

2.4.3.2 Overall low observablity configuration

In selecting an overall configuration to meet stealth requirements the following general points should be considered:

a) Foreplane surfaces are best avoided when the critical direction is forward and downwards.

b) Powerplants should be buried with long, curved, intake ducts and exhausts shielded from the critical direction, Figure 2.30.

c) Surfaces at right angles to one another should be avoided, for example by using inclined fins, Figure 2.34.

d) Discontinuities in shape and surface must be minimised, but radii of curvature should continually vary, Figure 2.30, or the shape should be faceted, Figure 2.29.

e) Edges of surfaces should be parallel as far as is possible, Figure 2.30.

f) Overall the configuration must be as clean as possible which suggests that weapons should be carried internally.

g) Cockpit transparencies demand special treatment, so an unmanned aircraft has an advantage. Doors, panels, etc. also require special treatment.

2.4.4 Water-borne aircraft

2.4.4.1 General comments

At one time aircraft designed to operate from water were very common. They possessed significant advantages in comparison with landplanes in the earlier period of aviation. In particular they could operate from any location where there was a stretch of reasonably calm water and this was of importance at a time when landing strips were confined to developed areas. The situation changed with the wide use of air power in World War 2 which had the consequence that many airfields were constructed across the world. At the same time developments in technology resulted in a trend toward much higher wing loadings than used previously and this brought with it higher take-off and landing speeds. This trend placed waterborne aircraft at a severe disadvantage since the higher wing loading needed to achieve good cruise performance, and the consequent higher water speeds, was difficult to accept because of the implied high water resistance forces. Because landplanes have a much superior performance the use of waterborne craft is now restricted to relatively small aircraft which operate within costal regions or remote locations having an abundance of lakes and rivers.

Figure 2.35 Cessna 185 Floatplane
(Courtesy Mission Aviation Fellowship)

2.4.4.2 Floatplane, Figure 2.35

As the name suggests floatplanes are aircraft in which the conventional landing gear is replaced by floats. Usually there are two floats, corresponding to the more usual main landing gear. General aviation floatplanes are invariably propeller driven and are nearly always direct conversions of land-based types. Floatplanes of up to around 12,000 kg mass have been used, but they are usually much lighter than this. Small wheels may be incorporated in the floats to facilitate handling when the aircraft is taken from the water. The air drag of floats is significant.

2.4.4.3 Flying boats, Figure 2.36

Flying boats are usually, but not always, larger than floatplanes. In this configuration the fuselage also fulfills the role of a hull for water operation and is shaped accordingly. Wing tip floats, which may be retractable, or fuselage sponsons are needed to provide roll stability while the craft is waterborne. Air drag is inevitably higher than that of a conventional fuselage but less than that of separate floats. The use of jet engines on flying boats has been rare, primarily due to the danger of water ingestion. Hence when they are used it is necessary to locate them in positions which would be considered to be inefficient on a conventional landplane, such as over the top of the wing.

Some flying boats are fitted with a retractable conventional landing gear, Figure 2.36, so that they can operate from either land or water, and they are know as amphibians.

Figure 2.36 Piaggio P136 Amphibian
(Courtesy Piaggio Aero Industries)

Chapter 3

Flight regime and powerplant considerations

3.1 Introduction

The maximum, or maximum operating, speed is one of the more important of the requirements which are specified for a particular aircraft. This has a dominant effect upon the overall configuration of the aircraft and is directly related to the type of powerplant system employed. Aircraft designed for supersonic operation frequently also have an important subsonic role which implies a need for compromise in both the aircraft and powerplant configurations, or possibly some form of variable geometry.

A general knowledge of the characteristics of various powerplants is necessary to enable the correct selection of powerplant type to be made for a given aircraft application.

3.2 Powerplant characteristics

3.2.1 Thrust

The thrust developed by an engine is the rate of change of momentum imposed upon the propelling medium:

$$T = \frac{d}{dt}(\bar{m}v) \tag{3.1}$$

where T is thrust

$\bar{m}$ is mass of propelling medium

v is velocity of propelling medium

The working out of Eq (3.1) depends upon the type of powerplant. For example in the case of a rocket engine in vacuum conditions:

$$T = V_j(\frac{d\bar{m}}{dt}) \tag{3.2a}$$

where V_j is the characteristic exhaust velocity of a particular propellant and $(d\bar{m}/dt)$ is the rate of burning of the propellant.

In the case of an air breathing engine having some measure of jet propulsion:

$$T = m(V_j - V_0) + m_F V_j + (p_j - p_0)A_j \tag{3.2b}$$

where m is now the mass flow through the engine per unit time
m_F is the rate of fuel usage
V_j is the jet velocity
V_0 is the aircraft speed
p_j is the exhaust static pressure
p_0 is the freestream static pressure
A_j is the exhaust area

Usually, for this class of engine, m_F is small in comparison with m and if there is complete expansion in the exhaust p_j is equal to p_0 so that approximately:

$$T = m(V_j - V_0) \tag{3.2c}$$

This equation also applies to any propulsion system which uses an actuating disc, such as a propeller.

It can be seen from Eq (3.2c) that a given thrust may be generated by an infinite number of combinations of mass flow and velocity increment and different types of air breathing engine reflect this possibility. It is also clear that the thrust will vary both with forward speed and altitude as the mass flow is dependent upon density. In practice a further complication is the variation of internal prime mover characteristics with forward speed.

For the purposes of comparing different types of engine it is convenient to use a specific thrust defined as:

$$T_{SP} = \frac{T}{m} = (V_j - V_0) \tag{3.3}$$

which reduces to V_j for static conditions, V_j being the equivalent of the characteristic exhaust velocity of a rocket engine.

3.2.2 Efficiency

The overall efficiency of a powerplant system is the product of a so-called "ideal" efficiency and the various mechanical and thermal efficiencies of the engine, air intake and exhaust/nozzles, as relevant. It is beyond the scope of this discussion to consider the internal engine efficiencies in detail, but some comments concerning the ideal intake and exhaust/nozzle efficiencies are relevant.

3.2.2.1 Ideal propulsion efficiency

The ideal propulsion efficiency, η, is defined as:

$$\eta = 2/(1 + V_j/V_0) \tag{3.4}$$

It can be seen that since V_j must be greater than V_0 to produce thrust, the highest efficiency is realised when V_j is only a small increment above V_0. This implies that the most efficient powerplant, in ideal terms, is one where the thrust is generated by imposing a small velocity increment on to a large mass of propulsive medium. In practice the behaviour of the individual components of a powerplant system modifies the ideal efficiency and introduces practical limitations. However, if these effects are ignored it can be seen from Eq (3.4) that for a given characteristic velocity, V_j, the ideal efficiency increases with forward speed, V_0, from a zero value in the static condition.

3.2.2.2 Air intake pressure recovery

At higher aircraft speeds it is necessary to reduce the velocity of the air entering the intake of a jet engine to a value which the compressor can tolerate. The velocity reduction is accompanied by a change of pressure and a contribution to the net thrust. The ratio of effective pressure recovery can be regarded as an intake efficiency, and clearly it should be as high as possible. The major considerations arise when the airflow entering the intake is greater than a Mach number of 1, resulting in the formation of shocks as the air is slowed down. At higher entry Mach numbers the number of shocks required to achieve acceptable pressure recovery is increased, as indicated in Figure 3.1. In addition the intake performance drops significantly with small changes from the optimum configuration. An essentially fixed geometry intake can be designed to provide acceptable pressure recovery over a wide subsonic Mach number range, although auxiliary doors may be necessary for low speed conditions. At increasing supersonic Mach number the compromises involved in using a fixed geometry intake become more penalising. Above a Mach number of about 1.6 prolonged, efficient, flight demands the use of a variable intake system. Inevitably at these higher speeds an intake solely designed to give satisfactory lower speed performance will have a poor pressure recovery and there will be a reduction in overall thrust. Typically at moderate subsonic speeds the intake system accounts for less than 10% of the total thrust, while at a Mach number of about 2 a correctly designed intake may provide 30% of the thrust.

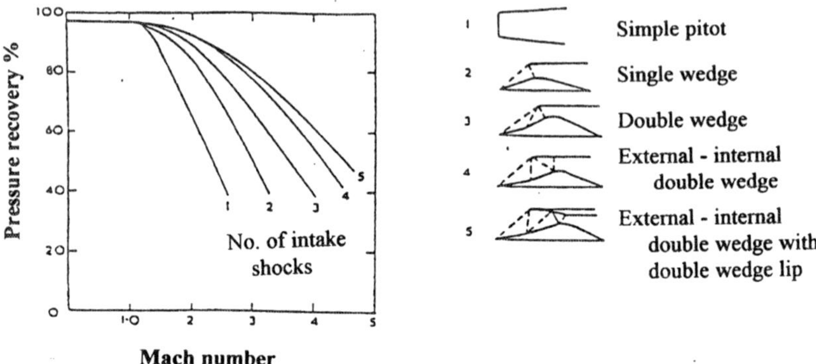

Figure 3.1 Intake efficiency as a function of Mach number

3.2.2.3 Exhaust and nozzle considerations

Although Eq (3.2b) would appear to indicate that additional thrust is produced when the exhaust pressure is greater than the static value, this is not actually the case since this situation means that the gases have not expanded to their full potential velocity and V_j is less than would be the case if complete expansion were to be achieved. Therefore the design of the exhaust/nozzle should be such as to achieve full expansion.

It is found that a fixed geometry convergent nozzle can be designed to give acceptable performance for aircraft which operate up to transonic speed conditions. At higher Mach numbers a convergent-divergent nozzle configuration is required to give complete expansion. The wide range of engine and flight conditions which have to be met almost invariably results in a need for such a nozzle to be of variable geometry. While at subsonic speed the actual contribution of the exhaust/nozzle to the thrust is small, possibly around 5% of the total, a figure of nearer 30% is likely at a Mach number of 2.

3.2.3 Noise

The primary sources of turbine engine noise are the airflow interactions with the rotating components and the exhaust. The noise from both of these sources is critically dependent on their velocities. That due to rotating components is dependent upon velocity to a power of approximately six while in the case of jet mixing the dependence is on velocity to the power of eight. Thus while the actual noise generated is dependent upon many detail considerations the importance of reducing the associated velocities is clear. This is particularly true of the exhaust gases.

3.2.4 Relationship between power and thrust of propeller propulsion

It is usual for the performance of propeller engines to be stated in terms of power, rather than thrust. In forward flight the thrust is given by:

$$T = \frac{\eta P}{V_0} \tag{3.5}$$

where P is the power and V_0 is the flight speed; η is derived from Eq (3.4) for a given propeller design.

Equation (3.5) breaks down under static conditions since in this case η is zero. In order to calculate static thrust it is necessary to derive a static thrust coefficient, which may be limited either by propeller design characteristics or by engine static power. See also paragraphs 3.6.2.3, 3.6.2.4 and 3.6.2.5.

3.3 Types of powerplant

3.3.1 General

Essentially only two basic types of prime mover are employed in manned aircraft applications. These are piston and turbine internal combustion engines.

Rocket engines do find wide application in certain classes of guided weapons and vehicles which operate outside the atmosphere. In the past they have also occasionally been used in manned aircraft, but the developments in turbine powerplants have been such as to render a repetition of this unlikely.

Ramjets, which are air breathing engines which rely upon air intake compression rather than rotating machinery, are also used in missiles. However, ramjet useful thrust is only delivered at supersonic speeds and the only likely application to manned aircraft is in a mixed powerplant configuration for speeds in excess of Mach 3.

Electrically driven propellers have very limited application for small, very high altitude, long endurance types which use solar panels as the power source.

3.3.2 Piston engines

Piston engines driving propellers were used exclusively to power aircraft during the first 40 years of practical flight. Individual units of up to about 2000 kW power were developed. This was achieved by using a large number of cylinders arranged either radially or in-line of flight. In either case two or more rows or "banks" of cylinders were incorporated in what was a mechanically complex unit. Usually, but not exclusively, the radial engines used air cooling while the in-line ones were liquid cooled.

More recently the application of piston engines has been limited to relatively small, low speed, aircraft which use engines of up to about 400 kW power. Most frequently these are air cooled and use two banks, each of from one to three cylinders in-line. Liquid-cooled engines have been employed occasionally, especially where noise is a major consideration.

Virtually all aircraft piston engines use gasoline fuel although diesel units have been used from time to time.

Rotary or 'Wankel'-type engines are related to piston engines and there has been some application in the power range up to about 100 kW.

3.3.3 Gas turbines

3.3.3.1 General

The great majority of aircraft are powered by one form or other of gas turbine engine. The only aircraft where they are not universally employed is the small general aviation class referred to in the previous paragraph. This is because small turbine engines are relatively expensive and scale effects result in them being less efficient than larger versions.

The common feature of all gas turbine engines is an assembly of air compressor, combustion chamber and turbine. The power may be extracted mechanically through a turbine-driven shaft or by expanding the exhaust gases through a nozzle in the jet propulsion principle. Various compressor/turbine shaft configurations are found, depending upon the intended application of the engine. Shaft engines are employed to drive propellers or rotor systems through gearboxes. Illustrations of typical turbine powerplants can be found in "The Jet Engine" (see Chapter 9, Appendix 9.1, Reference D6).

Apart from a few very small units axial compressors are used rather than centrifugal ones to enable the necessary high pressure ratios to be achieved. Advanced engines may have an overall pressure ratio of the order of 40.

Kerosine is the most commonly used fuel, although there are other possibilities.

3.3.3.2 Jet engines

A basic jet engine has a single shaft which connects the turbine to the compressor. The role of the latter is to compress the air needed to provide the exhaust gases which are expanded through the nozzle to produce the thrust. In passing from the combustion chamber to the nozzle some of the gas energy is extracted to drive the compressor through the turbine.

A more efficient engine may be produced by dividing the compressor into two parts, each driven by separate turbine/shaft assemblies. While such engines are heavier and mechanically more complex these disadvantages are usually outweighed by the improved performance, especially in terms of lower fuel consumption.

3.3.3.3 Low bypass ratio engines

In a bypass engine the compressor is divided into two or three separate parts, each of which is driven by individual turbine/shaft assemblies. Some of the compressed air output from the first, low pressure, stage of the compressor is ducted directly into the exhaust nozzle. The rest passes through the subsequent compressor stages, combustion chamber and turbines before joining the ducted air in the nozzle. While such an engine is more complex it is more efficient than a basic jet engine since the thrust is obtained by employing a relatively higher mass flow and lower average exhaust velocity. For the same reason it is also less noisy.

Bypass ratio is defined as the ratio of the mass of air passing directly to the nozzle to that passing through the turbo machinery. Most recent powerplants intended for use on high performance combat aircraft utilise a bypass configuration, with bypass ratios typically between 0.4 and 1.0. Engines in this category may have a basic thrust of up to about 70 kN, or more for special applications such as vertical take-off.

3.3.3.4 Afterburning/reheat engines

When the performance requirements for an aircraft demand very high thrust for a short period of time, such as during transonic acceleration or for a supersonic dash, it is possible to augment the exhaust gases by injecting and burning additional fuel between the turbines and the nozzle. This is known as afterburning or reheat, and is applicable both to basic jet and bypass engines. Depending upon the requirements and circumstances the thrust may be increased by more than 70%, but at the expense of, possibly, a fourfold increase in specific fuel consumption.

3.3.3.5 Turbofan engines

A turbofan engine is a development of the bypass jet engine where the first stage of the compressor is substantially increased in diameter to become a ducted fan. In this configuration much of the thrust is derived from the ducted air and the basic engine, known also as a gas generator, is there primarily to provide gases to drive the fan through its separate shaft/turbine unit.

Bypass ratios range commonly from 4 to 7 but values of 8 or more have been employed. In the great majority of engines the fan is directly driven by the turbine although geared fan systems have been used experimentally with bypass ratios in excess of 10. Fan engines are used for subsonic aircraft except for those powered by propellers. Thrust capability covers a very wide range from a low of around 8 kN to a potential high of the order of 500 kN.

3.3.3.6 Turboprop/turboshaft engines

In this type of engine the turbine stage is normally divided into two parts, each driving separate concentric shafts. The high pressure turbine stage drives the compressor which is essentially employed to supply the air needed to produce the gases to power the turbines. The low pressure turbine drives the power output shaft. The low velocity exhaust gases may be converted into a small residual thrust.

Turboprop engines incorporate a reduction gearbox to connect the low pressure turbine shaft to the propeller. There is no local gearbox in the case of a turboshaft engine, any required reduction in drive speed being achieved remotely, such as by a helicopter main gearbox.

Engines in this category typically range from around 300 to 4000 kW power output although both smaller and larger units have been produced.

3.3.3.7 Unducted fan engines

Unducted fan engines are effectively a hybrid between the ducted fan engine and the turboprop engine. The aim is to raise the Mach number at which open rotor engines may be operated, primarily to achieve better efficiency than the ducted fan type. It is not clear whether the difficulties associated with this concept, such as noise, vibration and mechanical complexity, justify the potential reduction in fuel consumption.

3.4 Typical engine parameters

3.4.1 Specific thrust

Figure 3.2 shows typical static specific thrust values for various types of gas turbine powerplant. As would be expected from the definition of specific thrust, Eq (3.3), the higher values relate to those engines which have lower relative mass flows.

3.4.2 Frontal area/disc loading

The mass flow is effectively a function of the "capture area" employed by the basic powerplant. The capture area ranges from the intake area of a jet or bypass engine to the fan or propeller disc area. In some circumstances a knowledge of the thrust as a function of the capture area is useful. This is known as the disc loading for a ducted fan or open propeller propulser. Typical values are given in Table 3.1.

Table 3.1 Typical effective disc loading
(Thrust/frontal area)

ENGINE TYPE	LOADING (kN/m^2)
Jet with afterburning	up to 200
Jet	95
Bypass ratio; one with afterburning	up to 150
Bypass ratio; one	85
Fan; bypass ratio four	60
Fan; bypass ratio six	50
Fan; bypass ratio ten	35
Unducted fan	15
Advanced turboprop	3
Conventional turboprop	2
Piston propeller	1

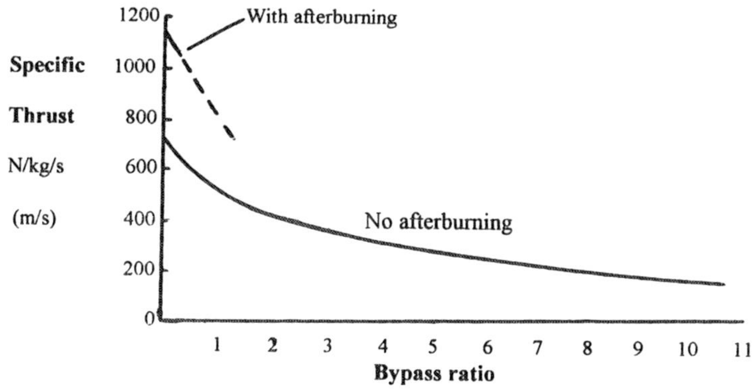

Figure 3.2 Static specific thrust for gas turbines

3.4.3 Propulsive efficiencies

Figure 3.3 shows typical variation of ideal powerplant efficiencies as a function of Mach number. While the trends shown are common to any given class of powerplant, detail variations are possible according to the design requirements in a particular case.

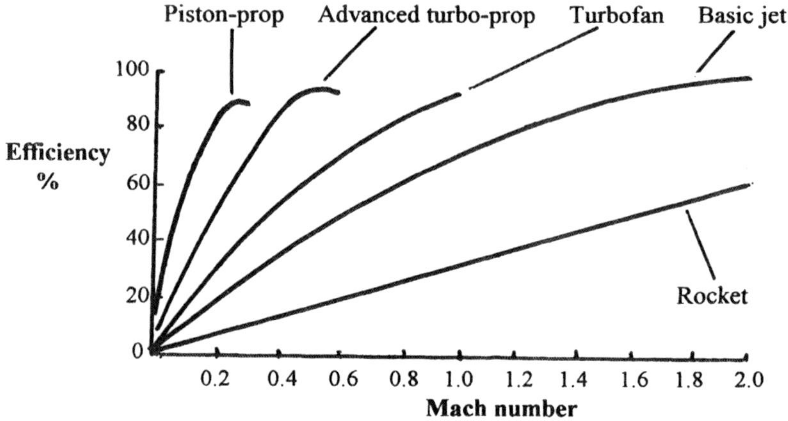

Figure 3.3 Ideal propulsive efficiencies

3.5 Flight regimes of powerplants

3.5.1 General

Simplistically it may be assumed that the internal thermal and mechanical efficiencies of all aircraft air breathing engines are similar. Although this is a drastic assumption it does enable a first order comparison to be made of the various types of powerplant on the sole basis of the ideal efficiency, Eq (3.4). Using this basis Figure 3.3 illustrates how the efficiencies of the various powerplant types determine the Mach number regime appropriate to their application.

3.5.2 Propeller engines

Because the thrust from a propeller is derived by the addition of a small velocity change to a large mass of air, the efficiency increases rapidly with forward speed. However, at higher subsonic Mach numbers the performance of the propeller suffers, mainly as a result of the adverse changes of pressure distribution on the blades due to compressibility effects. There is therefore a limit to the Mach number at which propellers may be efficiently employed. Modern propellers which use wide chord, multiple, swept blades rotating relatively slowly can be usefully used up to a Mach number of about 0.7, although a somewhat lower value, say between 0.6 and 0.65 is a more frequent design point. At low speeds a turboprop engine has an effective bypass ratio which is in excess of 50.

Propellers may be designed to achieve their highest efficiency at various Mach numbers depending upon the required application. The adjustment of blade pitch on an individual propeller enables relatively high efficiency to be achieved over a range of speeds. Propellers designed for small, slow flying aircraft, at say around 40 m/s, may only be able to achieve a peak efficiency of about 70%. At somewhat higher speeds 80% is commonly realised with 85 to 90% efficiency achieved for flight at Mach numbers of between 0.3 and 0.65.

It may be concluded that for flight speeds up to about a Mach number of 0.5 the propeller, whether it be piston or turbine driven, is the only realistic possibility when the efficiency of fuel usage is a dominant consideration. In the Mach number range from 0.5 to, possibly, 0.7 the turboprop engine is generally the most efficient means of propulsion. Other considerations, such as those of aircraft overall layout, begin to become significant at these speeds.

3.5.3 Unducted fan engines

The unducted fan engine is effectively a high disc loading propeller powerplant where useful efficiency is retained up to a Mach number of about 0.8, and it therefore achieves better fuel economy than the ducted fan engines more commonly employed at this speed. While there is an undoubted potential for this lower specific fuel consumption it is achieved at the expense of a complex and heavy powerplant. Local noise aspects suggest that such powerplants should be located at the rear of the aircraft. No clear application has yet arisen where this class of engine is preferable in overall terms to the other possibilities.

3.5.4 Turbofan engines

The ducted turbofan is the obvious choice of powerplant when the maximum normal flight speed is in the high subsonic region, say Mach numbers in the range 0.7 to 0.9. The actual bypass ratio used depends upon the application, and is a compromise between engine diameter and mass on the one hand and fuel economy on the other. As a general rule the higher bypass ratios in the typical range 4 to 8 are used on relatively longer range aircraft.

Turbofan engines also have some application to aircraft which normally operate at Mach numbers in the range 0.5 to 0.7. While their efficiency at these lower speeds is less than that of a propeller engine, their relatively small size can enable a more compact aircraft to be designed which is more efficient in overall terms. Such aircraft are often in the small business/executive class.

At very high subsonic speeds the higher bypass ratio turbofan engine begins to loose efficiency for reasons not dissimilar to those associated with loss of efficiency of propellers.

3.5.5 Low bypass ratio engines

It is clear from Eq (3.2) that as flight speed increases it is necessary to increase the jet velocity to achieve thrust. This implies the use of engines which operate on relatively smaller masses of air. Low bypass ratio and jet engines generally fall into this category.

With just one or two notable exceptions, such as Concorde, aircraft which operate above a Mach number of 0.9 and into the supersonic region are military types. While at one time such aircraft employed basic jet engines, the great majority are now fitted with powerplants having a bypass ratio in the range 0.4 to 1.0. The facility for afterburning is also usual in order to augment the jet velocity at higher Mach numbers and in transonic flight.

Engines in this category have exhaust velocities of up to 700 m/s and this is increased to 1100 m/s when afterburning is used. A bypass ratio of less than 1 is likely for flight at a Mach number of 2.

3.5.6. Basic jet engine

The development of the afterburning low bypass ratio engine suggests that there is little future application for a basic jet engine. A velocity of 1100 m/s is in excess of a Mach number of 3.5 at altitude and in practice the limitation to exhaust velocity is one of engine construction materials and this applies equally to both the basic and low bypass ratio jet engines.

3.6 Powerplant performance representation

3.6.1 Introducton

The synthesis of an aircraft configuration requires a knowledge of certain vital powerplant characteristics. It is important to use the most accurate possible representation of the variation of thrust and specific fuel consumption with speed, altitude and engine conditions. It is always highly desirable to use the characteristics of a known powerplant. When such data are available it must be recognised that there are likely to be some flight conditions where the powerplant has to operate at settings other than the ideal design values and consequently the specific fuel consumption becomes greater than would otherwise be the case.

Sometimes the detailed characteristics of a given powerplant are not available. Even when they are it may be useful to have generalised expressions of their variation with flight conditions. Various literature sources quote generalised expressions for powerplant performance which can be used for initial configuration studies. Typical expressions are given here which are of use when available powerplant data are inadequate to cover all the required conditions. The models can be adapted to represent a given powerplant when the data for several aircraft performance points are known, the critical aircraft performance conditions being used for matching.

3.6.2 Thrust representation

3.6.2.1 General

The maximum thrust or power available for any given engine is primarily dependent upon three main parameters:

a) Flight speed (Mach number).

b) Flight altitude. This is conveniently defined in terms of the relative air density, σ.

c) Engine operating conditions. This covers operation at other than optimum design conditions.

3.6.2.2 Turbojet and bypass engines

a) Flight speed (Mach number). The thrust models are conveniently given as a function of Mach number over three ranges:

- i) Low subsonic speed (Mach number up to 0.4)
- ii) High subsonic speed (Mach number range of 0.4 to 0.9)
- iii) Transonic and supersonic speed (Mach numbers above 0.9)

b) Flight altitude. Up to 11 km altitude the thrust varies approximately in proportion to some power, s, of the relative density. This power varies in the range of around 0.6 for a high bypass ratio engine, up to about 0.85 for an engine of nominally zero bypass ratio. Above 11 km altitude the thrust variation is more or less directly proportional to the relative density. Thus when it is useful to use a simple model to represent the performance of an engine at altitudes significantly above 11 km it is best to base the model on conditions at this transitional altitude. In the absence of better information an assumption of 0.7 for the relative density power gives a reasonable overall average up to around 15 km altitude in the case of fan engines.

c) Engine operating conditions. This covers engine operation with, or without, the use of reheat at engine conditions other than the design value and non-optimum intake and nozzle geometry.

It is convenient to relate operating thrust values to a datum, sea level static dry condition, that is without use of afterburner when this is incorporated in the engine. The values suggested are for the basic engine in typical operating mode with allowance for achievable intake recovery factor and nozzle performance. This implies variable geometry intakes at Mach numbers in excess of about 1.7 and variable convergent-divergent nozzles for supersonic flight generally. There is no specific allowance for other installation losses, see paragraph 3.6.4.

At any given condition it is possible to define:

$$T = \tau T_0 \tag{3.6}$$

where T_0 is the datum sea level static dry thrust

T is the available operating thrust at any given condition

τ is a factor dependent upon (a) to (c) listed above and the bypass ratio, R

The τ factor may be written as:

For $0 < M_N \leq 0.9$:

$$\tau = F_\tau \left[K_{1\tau} + K_{2\tau}R + \left(K_{3\tau} + K_{4\tau}R\right)M_N\right] \sigma^S \tag{3.7a}$$

For $M_N > 0.9$:

$$\tau = F_\tau \left[K_{1\tau} + K_{2\tau}R + \left(K_{3\tau} + K_{4\tau}R\right) \left(M_N - 0.9\right)\right] \sigma^S \tag{3.7b}$$

where s is the altitude factor referred to in subparagraph (b) above and F_τ is a factor to allow for the use of afterburning. F_τ is unity under basic, dry operating conditions.

Typically when afterburning is used:

$$F_\tau = \left(\frac{T_W}{T_D}\right) \Big/ (1.32 + 0.062R) \tag{3.8}$$

where T_W and T_D are the sea level static thrust values in wet and dry operating conditions respectively. When reheat is used F_τ may be taken as unity for engines intended for long range cruise applications, but is likely to be higher, say up to 1.3, for combat aircraft applications.

$K_{1\tau}$, $K_{2\tau}$, $K_{3\tau}$ and $K_{4\tau}$ are assumed to be constant for a given powerplant in a defined Mach number range and operating condition. Typical values are suggested together with those of s in Table 3.2. It should be pointed out that in practice the detail design of the engine, especially the overall pressure ratio, has an important effect upon the actual values of the $K_{i\tau}$

Table 3.2 Powerplant thrust parameters

Bypass ratio R	Mach number range	Operating condition	$K_{1\tau}$	$K_{2\tau}$	$K_{3\tau}$	$K_{4\tau}$	s*
1	0-0.4	Dry	1.0	0	-0.2	0.07	0.8
or		Wet	1.32	0.062	-0.13	-0.27	0.8
lower	0.4-0.9	Dry	0.856	0.062	0.16	-0.23	0.8
		Wet	1.17	-0.12	0.25	-0.17	0.8
	0.9-2.2	Dry	1.0	-0.145	0.5	-0.05	0.8
		Wet	1.4	0.03	0.8	0.4	0.8
3 to 6	0-0.4	Dry	1.0	0	-0.6	-0.04	0.7
	0.4-09	Dry	0.88	-0.016	-0.3	0	0.7
8	0-0.4	Dry	1	0	-0.595	-0.03	0.7
	0.4-0.9	Dry	0.89	-0.014	-0.3	+0.005	0.7

* Strictly s values apply up to 11 km altitude above which the factor is unity based on the 11 km altitude condition as a datum.

3.6.2.3 Propeller characteristics

The efficiency of a propeller, η, is defined in Eq (3.5) and is a function both of the advance ratio, J, and the thrust coefficient, c_T, the latter being influenced by the pitch to diameter ratio.

$$J = V_0 / (nD_p) \tag{3.9a}$$

where V_0 is the forward speed (m/s)

n is the rotational speed (rev/s)

D_p is the propeller diameter (m)

The value of (nD_p) is limited both by tip Mach number and noise considerations. Typical values for unducted propellers lie in the range 60 to 100 m/s. For example:

i) Small directly driven piston engine applications, which typically have two or three blades $(nD_p) = 90$ m/s

ii) Turboprop trainers and related types with, say, four bladed propellers $(nD_p) = 80$ m/s

iii) Small general aviation and regional turboprops, usually with three to five bladed propellers $(nD_p) = 75$ m/s

iv) Large turbo propeller driven transport aircraft with up to six bladed propellers as low as $(nD_p) = 63$ m/s

The maximum achievable efficiency is approximately given by:

For $0.4 \leq J < 1.0$ $$\eta = 0.82J^{0.4} \tag{3.9b}$$

For $J \geq 1.0$ $$\eta = (0.82J^{0.16})/(10^{j}) \tag{3.9c}$$

where $j = 0.3(\log J)^{2.4}$

In some circumstances the thrust, and hence efficiency, may be limited by the achievable thrust coefficient and the efficiency is not likely to exceed:

$$\eta_{LIMIT} = 1.8\rho z^{0.15}(nD_p)^{3.7} J P_0^{0.095} \times 10^{-7} \tag{3.9d}$$

where P_0 is the maximum engine shaft power (kW)
ρ is the air density
z is the number of propeller blades

Current practice suggests that for new aircraft designs z is the nearest whole number to $(0.4P_0^{0.35})$ but not greater than six. Space limitations on developed designs may result in the need to use a greater number of blades than suggested by this relationship.

Equation (3.9d) is most likely to be critical for lower values of (nD_p) and J. There is unlikely to be any significant effect for engines having a shaft power below 1700 kW or when the value of (nD_p) is more than 80 m/s.

The overall efficiency in cruise conditions, and for powers above about 1000 kW is approximately:

$$\eta_{Cr} = 0.59P_0^{0.05} \tag{3.9e}$$

The thrust in any given flight condition, except static, is given by:

$$T = \eta P_0 \times 10^3/V_0 \qquad \text{N}$$

$$T = \eta\, P_0 \times 10^{3}/J\,(nD_p) \qquad \text{N} \tag{3.9f}$$

By definition η is zero in static conditions and in this case the thrust is given by:

$$T_S = (c_T)_S\, \rho (nD_p)^2 D_p^2 \qquad N \tag{3.9g}$$

where $(c_T)_S$ is the static thrust coefficient. A typical maximum value of the static thrust coefficient is:

$$(c_T)_S = 0.0085\; z^{\,0.15} \left(\frac{P_o}{A}\right)^{0.65} \tag{3.9h}$$

where (P_0/A) is the power disc loading $(1.273\, P_0/D_p^2)$
P_0 is the static (max. rated) power, kW

The propeller rotational speed, n, is chosen in conjunction with the diameter to give the best compromise between tip speed and efficiency in a given application in line with the values of (nD_p) suggested above. Typically:

Direct drive piston engines $\qquad n = 45$ rev/s
Geared turboprop engines $\qquad n = 433(P_0)^{-0.4}$ rev/s

Typical power disc loadings are:

a) Direct drive piston engines:

$$\frac{P_0}{A} = 4.7 P_0^{0.5} \qquad kW/m^2 \tag{3.9i}$$

b) Turboprops based on above value of n:

$$\frac{P_0}{A} = \frac{1.45 P_0^{0.27}}{(nD_p)^2} \times 10^5 \qquad kW/m^2 \tag{3.9j}$$

The diameter of the propeller follows from the above equations:

Direct drive piston engines: $D_p = 0.52P_0^{\ 0.25}$ m (3.9k)

Turboprops, approx: $D_p = 3(nD_p)P_0^{\ 0.365} \times 10^{-3}$ m

$$\text{or:}\qquad D_p = 0.1\left(\frac{T_S}{\rho z^{0.15}}\right)^{0.403} \left(nD_p\right)^{-0.089} \quad m \qquad (3.9l)$$

$$\text{where}\qquad P_0 = \frac{1.435}{(nD_p)^3}\left(\frac{T_S}{\rho z^{0.15}}\right)^{1.105} \times 10^4 \quad kW \qquad (3.9m)$$

which follow from Eqs (3.9g), (3.9h) and (3.9k). For values of n other than $433(P_0^{-0.4})$ the product (nD_p) is taken as constant to give a revised value of D_p. If required a revised value of (P_0/A) may be calculated).

3.6.2.4 Turboprop

While it is possible to express the thrust of a turboprop engine as a function of both speed and altitude, it is usually preferable to separate the shaft power from the effect of propeller efficiency. Unless otherwise stated the thrust values are for a single engine.

a) Maximum power take-off conditions

The relationship between climb/cruise power and the sea level static maximum value is complicated by the possibility of "flat rating" the engine to overcome the effects of altitude and temperature on take-off performance. In general maximum take-off power is effectively independent of speed up to a Mach number of about 0.25. However it does fall with altitude.

The take-off performance is conveniently expressed in terms of the static thrust, as given by the inversion of Eq (3.9m):

$$T_S = 1.73\rho z^{0.15}(nD_p)^{2.7}P_0^{0.905} \times 10^{-4} \quad N \qquad (3.10a)$$

The second segment climb performance can be based on the static power rating. Assuming that the engine is flat rated up to about 30°C, or the equivalent, then the equation given in Chapter 7, paragraph 7.3.4 may be used with a correction to allow for the fact that power is not reduced under hot and high conditions. It is likely that the second segment climb speed will correspond to a value of the advance ratio less than unity and when this is so and there is no thrust coefficient limit, from Eqs (3.9b) and (3.9f):

$$T_{SS} = \frac{0.82P_0 \times 10^3}{(nD_p)V^{0.6}} \quad N \tag{3.10b}$$

where T_{SS} is the required thrust to meet the second segment climb condition, as defined in Chapter 7, paragraph 7.3.4. and V is the climb out speed:

$$V = 1.1(V_{US})_{HH}$$

where $(V_{US})_{HH}$ is the true unstick speed in hot and high conditions.

The equivalent static thrust, using Eq (3.10a) and making the above-mentioned correction is:

$$T_S = 3.92N_E\rho z^{0.15}(V_{US})^{0.55}(nD_p)^{3.062}\left(\frac{T_{SS}}{N_E}\right)^{0.905} \times 10^{-7} \quad N \tag{3.10c}$$

where N_E is the number of engines
V_{US} is the unstick speed in standard conditions

When the thrust coefficient limitation applies throughout the second segment climb phase, it may be assumed that:

$$T_S = T_{SS} \text{ approximately} \tag{3.10d}$$

This presumes that the thrust coefficient is effectively constant at the static value, although in fact it may increase slightly at low forward speed.

b) Climb/cruise power and thrust

As far as the climb/cruise power is concerned it is convenient to base the power variation on an equivalent static value $(P_0)_{EQ}$, derived from the performance at maximum continuous rating and 3 km altitude:

i) If the true static maximum power P_0 is known, as is the power, P_1, at 3 km altitude and some Mach number M_{N1}, then approximately:

$$(P)_{3k} = 0.74P_1/(0.74 + 0.58M_{N1}) \tag{3.10e}$$

where $(P)_{3k}$ is a nominal static power at 3 km altitude, and

$$(P_o)_{EQ} = 2.67(P)_{3k} - 1.33P_0 \quad (3.10f)$$

ii) If P_o is not known, but P_I at 3 km altitude and Mach number M_{NI} is known, or can be derived from aircraft performance requirements, then assume:

$$(P_0)_{EQ} = 1.1(P)_{3k} \quad (3.10g)$$

as a typical value and then:

$$P_0 = 1.07\ (P)_{EQ} = 1.17(P)_{3k} \quad (3.10h)$$

iii) The climb/cruise maximum continuous power rating is given by:

$$P = (P_0)_{EQ}(\sigma + 0.75\sigma^{0.85}M_N) \quad (3.10i)$$

iv) For climb/cruise conditions at a Mach number in excess of about 0.25 it may be assumed that $J \geq 1.0$ and using Eq (3.9e):

$$T = 590PP_0^{0.05}/V$$

hence $$T = 1.74PP_0^{0.05}/M_N\sigma^{0.117} \quad N$$

where it is assumed that $M_N = V/340\ \sigma^{0.117}$

Replacing P from Eq (3.10i) and using Eq (3.10h), the maximum cruise thrust is:

$$T = 1.62P_0^{1.05}\ [(\sigma^{0.883}/M_N) + 0.75\sigma^{0.733}] \quad N \quad (3.10j)$$

Equation (3.10j) assumes that there is no thrust coefficient limitation at this value of J.

The thrust given by Eq (3.10j) may be related to the total aircraft static condition by use of Eq (3.9m):

$$T_S = 1.14N_E\,\rho z^{0.15}(nD_p)^{2.7}\left(\frac{T}{N_E}\right)^{0.862}\left\{\frac{\sigma^{0.883}}{M_N} + 0.75\sigma^{0.733}\right\}^{-0.862} \times 10^{-4} \quad N \qquad (3.10k)$$

3.6.2.5 Piston propeller powerplants - general aviation aircraft

a) General

There can be considerable variation of power with altitude dependent upon the induction system, that is whether or not supercharging is employed and if so to what degree. When there is no supercharging the power is approximately proportional to relative density to the power of 1.1. Power is also approximately proportional to engine speed. Supercharging may be used either to :

i) Maintain sea level power up to say, 5 km altitude, or

ii) Increase sea level power by a significant amount, up to a factor of about two. The power then decreases as the relative density decreases, as with an unsupercharged engine.

Variation of power with forward speed is negligible.

The power available in climb/cruise conditions is primarily dependent upon the ratio of the engine revolution speed to its maximum value, for a given altitude and air inlet conditions. Typical climb rating is 90% power and maximum cruise 85%. All values are given for one engine.

b) The static thrust may be taken as:

$$T_S = 42P_0^{0.85} \quad N \quad \text{per engine} \qquad (3.11a)$$

From Eq (3.9k) the propeller diameter in terms of static thrust is approximately:

$$D_p = 0.17T_S^{0.3} \quad m \qquad (3.11b)$$

c) Climb

Piston-engined aircraft usually climb at relatively low forward speed and it is most likely that the advance ratio will be less than unity. Hence from Eq (3.9b):

$$T_{Cl} = \frac{0.82P \times 10^3}{V_{Cl}^{0.6}(nD_p)^{0.4}} \quad N \tag{3.11c}$$

Assuming that the maximum continuous climb rating is 90% of the maximum rated value and that, typically, $(nD_p) = 90$ m/s:

$$T_{Cl} = 0.122\sigma^{1.1}P_0 \times 10^3 / (V_{Cl})^{0.6} \quad N \tag{3.11d}$$

and using Eq (3.11a) the equivalent static thrust is:

$$T_S = 0.71(V_{Cl})^{0.51}\sigma^{-0.94}(T_{Cl})^{0.85} \quad N$$

$$(or: \ \tau_{Cl}T_S = (T_{Cl})^{0.85} = 1.41(V_{Cl})^{-0.51}\sigma^{0.94}T_S) \tag{3.11e}$$

d) Cruise

i) Cruise speed less than about 90 m/s: $J \leq 1.0$
Equation (3.11c) applies to this condition also, and if the engine continuous cruise rating is 85% of the maximum rated value Eqs (3.11d) and (3.11e) become respectively:

$$T_{Cr} = 0.115\sigma^{1.1}P_0 \times 10^3/(V_{Cr})^{0.6} \quad N \tag{3.11f}$$

(This is the maximum available value allowing for propeller efficiency)

The corresponding static thrust is:

$$T_S = 0.74\ (V_{Cr})^{0.51}\sigma^{-0.92}(T_{Cr})^{0.85} \quad N$$

$$(or: \ \tau_{Cr}T_S = (T_{Cr})^{0.85} = 1.35(V_{Cr})^{-0.51}\sigma^{0.92}T_S) \tag{3.11g}$$

ii) Cruise speed greater than 90 m/s: $J > 1.0$
Equation (3.9c) predicts the efficiency for this condition. For the likely range of advance ratio it may be simplified to:

$$\eta = 0.4\ V_{Cr}^{0.16} \tag{3.11h}$$

The maximum cruise thrust is then:

$$T_{Cr} = \frac{0.4\ (V_{Cr})^{0.16} P \times 10^3}{V_{Cr}} \quad N \tag{3.11i}$$

and

$$T_{Cr} = 0.34\sigma^{1.1} P_0 \times 10^3/(V_{Cr})^{0.84} \quad N \tag{3.11j}$$

and the corresponding static thrust is:

$$T_S = 0.3(V_{Cr})^{0.714}\sigma^{-0.92}(T_{Cr})^{0.85} \quad N$$
$$(or: \ \tau_{Cr} T_S = (T_{Cr})^{0.85} = 3.33(V_{Cr})^{-0.714}\sigma^{0.92} T_S\ ) \tag{3.11k}$$

3.6.3 Fuel consumption characteristics

3.6.3.1 General

The specific fuel consumption achieved in a given application depends upon the actual installation details. The values quoted in this paragraph are uninstalled values, and need to be factored appropriately when the fuel consumption is based on actual thrust available, see paragraph 3.6.4.

3.6.3.2 Turbojet and Turbofan

A dominant parameter in the determination of the variation of specific fuel consumption with forward speed is the bypass ratio, *R*. Altitude also has some effect as does the detail of the engine operating cycle and throttle setting.

The following equations attempt to simplify this complex matter.

a) Dry (no reheat)

$$c = c'(1 - 0.15R^{0.65})\left[1 + 0.28(1 + 0.063R^2)M_N\right]\sigma^{0.08} \tag{3.12a}$$

up to 11 km altitude, above which *c* should be assumed to be constant with altitude.

where c is the specific fuel consumption at the design condition

c' is a factor which should be determined by reference to the actual specific fuel consumption of a given powerplant at a critical datum condition

Very approximately for:

i)	Supersonic engine $R \leq 1.0$	$c' = 0.95$ N/N/h (27 mg/N/s)
ii)	Low bypass ratio subsonic engine	$c' = 0.85$ N/N/h (24 mg/N/s)
iii)	Large subsonic turbofans	$c' = 0.70$ N/N/h (20 mg/N/s)

When c is known for a given datum design condition, say for example $c = 0.56$ N/N/h for a long range transport operating at a Mach number of 0.8 and 11 km altitude with bypass ratio of 5, then:

$$0.56 = c'(1 - 0.15 \times 5^{0.65})\left[1 + 0.28(1 + 0.063 \times 5^2)0.8\right] \times 0.907$$

where 0.907 is $\sigma^{0.08}$ at 11 km, $0.56 = 0.819c'$, $c' = 0.684$ N/N/h

When the engine is operated at some thrust less than the design value in a given flight condition there is a tendency for an increase of specific fuel consumption. This is approximately:

$$c_{OD} = c\left[1 + 0.01\left(\frac{T}{T_{OD}} - 1\right)\right] \tag{3.12b}$$

$$for \quad T/T_{OD} < 10$$

where c_{OD} and T_{OD} refer to the off-design conditions.

b) Afterburning (reheat). The specific fuel consumption when afterburning is used is further complicated by the degree of afterburning employed. Approximately:

$$c = 1.05\left(\frac{T_W}{T_D}\right)\left(1 + 0.17M_N\right)\sigma^{0.08} \quad N/N/h \tag{3.13}$$

This applies up to 11 km altitude, above which c should be assumed to be constant with altitude.

3.6.3.3 Turboshaft engines

There is evidence to suggest a small fall in specific fuel consumption with increase of forward speed, but negligible altitude effect. Typically the maximum continuous rating has a specific fuel consumption which is about 95% of the value at the maximum rating. For engines of new design the fuel consumption in terms of power, $(c)_P$, in climb/cruise conditions is:

$$(c)_p = 2.88\left(1 - 0.025P_0 \times 10^{-3}\right)\left(1 - 0.2M_N\right) \qquad N/kW/h \qquad (3.14a)$$

In terms of thrust, $(c)_t$, this becomes:

$$(c)_T = M_N\sigma^{0.117}\left(1 - 0.025P_0 \times 10^{-3}\right)\left(1 - 0.2M_N\right)/\eta \qquad N/N/h \qquad (3.14b)$$

where η is given by Eqs (3.9b), (3.9c) or (3.9e) as appropriate.

It should be noted that while Eq (3.14a) refers to shaft power, not flight power, Eq (3.14b) inevitably includes the effect of propeller efficiency. Equations (3.14a) and (3.14b) are based on the power of a single engine.

3.6.3.4 Piston engines

The specific fuel consumption for a normally aspirated engine is proportional to engine speed and it may increase at lower power settings. Typically for engines operated "economically":

$$(c)_p = c'\left[1 + 0.24\left(\frac{P_0}{P_r}\right)\right]\left[1 + 1.5\left(\frac{P_r}{P_0\sigma^{1.1}}\right)\right] \qquad N/kW/h \qquad (3.15a)$$

where P_0 is the maximum rated power
P_r is the power required in a given flight condition

c' depends upon the actual engine and its mode of operation. It typically has a value of unity, or somewhat less.

In determining P_r allowance must be made for propeller efficiency.

The specific fuel consumption in terms of thrust, making allowance for propeller efficiency becomes:

$$(c)_T = (c)_p \frac{V}{\eta} \times 10^{-3} \qquad N/N/h \tag{3.15b}$$

approximately, where P_o and P_r in Eq (3.15a) are the values for a single engine and η is given by Eqs (3.9b) or (3.9c) as appropriate.

3.6.4 Installation losses

The power and thrust values given in Eqs (3.7) to (3.11) refer to the basic powerplant and make allowance only for propeller, intake and nozzle efficiencies. Allowance must also be made for other installation losses such as power offtakes for the various secondary systems. These may be relatively small during take-off but may be significant under cruise conditions, possibly accounting for 5% or more loss of thrust. Fuel consumption should be based on thrust in the absence of these losses, including allowance for an off-design condition, Eq (3.12b), where appropriate.

3.7 Powerplant mass

The mass of a given type of powerplant is closely related to the static thrust or power. Typical values are quoted in Chapter 6, paragraph 6.4.3.2, in the section dealing with aircraft mass prediction.

3.8 Typical aircraft installed thrust and power

Table 3.3A lists typical values of installed static thrust to weight ratio for various kinds of aircraft equipped with fan or jet engines. Table 3.3B gives similar data for propeller-driven aircraft, and includes typical power to weight ratios.

Table 3.3A Typical static thrust to weight ratios

Jet and fan driven aircraft

TYPE	Thrust to weight ratio	
	DRY	REHEAT
Civil - twin - light executive	0.30-0.40	
- executive/regional jets	0.30-0.37	
- transports	0.29-0.34	
three - executive	0.28-0.31	
- transports	0.27-0.30	
four - regional jets	0.28-0.31	
- transport	0.26-0.28	
- supersonic transport	0.40	0.50
Military - trainers, light attack	0.40-0.50	
- light attack, with heavy payload*	0.24-0.30	
- subsonic attack	0.60	
- supersonic fighter/attack*	0.35	0.60
- air superiority	0.55	0.90
- subsonic bomber, transport and related special missions	0.25-0.31	
- supersonic bomber	0.20-0.25	0.35
- subsonic high altitude survey	0.40-0.60	
- vertical take-off and landing	1.2 - 1.3	
- short take-off and vertical landing	0.7 - 0.9	

* Large variations are possible in these cases, dependent upon the mission requirements

Table 3.3B Typical static power and thrust to weight ratios
Propeller-driven aircraft

TYPE	Power to weight ratio kW/kN (x10^3)	Thrust to weight ratio
Piston engine - single - basic general aviation	14	0.25
- specialist aerobatic	20-30	0.35-0.55
Piston engine - twin - light commuter, etc.	16	0.30
Turbine engine - single - commuter, etc.	18	0.32
- basic trainers	25-38	0.45-0.55
twin - light	18	0.30
- commuter	18	0.30-0.35
- executive	23-28	0.35-0.45
- transport	20-27	0.32-0.38
- special military types, up to	30	0.4
four - transports	18-23	0.22-0.28

Chapter 4

Fuselage layout

4.1 Introduction

The fuselage fulfills several functions. In the majority of aircraft these include the provision of volume for the payload and overall structural integrity. In many designs the fuselage has several other purposes such as the mounting of landing gear units and the housing of powerplant systems. Fuselage layout is simplified if there is no requirement for crew accommodation.

Once the fundamental configuration of the aircraft has been established it is usually possible to propose a layout for the fuselage with only secondary reference to other aspects of the design. Therefore this is often a useful starting point in the overall layout process. Layout of the fuselage can be complex as a result of the numerous detail requirements and alternative possibilities which must be considered.

Certain fuselage layouts, such as those of passenger and freight aircraft, may be derived by using a semi-analytical approach. Others, such as combat aircraft with a fuselage-located powerplant and internal weapons bay, have to be laid out. This latter process may be approached by various means but an effective technique is to work with modules which can be combined together in different layouts. The modules can include such items as crew accommodation, powerplant system, payload configurations, fuel volume, landing gear stowage, wing carry through structure, empennage and so on.

4.2 Primary considerations

4.2.1 General

There are some primary considerations which have a major impact upon the layout. It is necessary to consider these before proceeding to detailed aspects of the configuration.

4.2.2 Payload and related items

In many aircraft a large part of the fuselage volume is occupied by the payload. The main exceptions are:

a) Single- /two-seat light aircraft, where the occupants are considered to be the payload.

b) Trainer/light strike aircraft, where the weapon payload is usually carried on the wings.

c) High performance combat aircraft, where the weapons are carried on the outside of the fuselage or on wing pylons. Nevertheless many aircraft in this category are also equipped with internal guns which do have a major impact upon some parts of the fuselage layout.

Payload and related items carried within the fuselage of various kinds of aircraft include:

i) Passengers and baggage
ii) Freight
iii) Internal weapons, which as well as fuselage-located guns may include such items as free fall bombs and guided weapons located in a dedicated bay.
iv) Crew, which in some cases such as antisubmarine and early warning aircraft may occupy a large part of the fuselage volume.
v) Fuel, which is frequently exchangeable for payload on a mass basis. While fuselage fuel tanks as such are rarely used in passenger aircraft they are inevitable, and hence common, in many trainer and combat aircraft.
vi) Avionics, which for the class of aircraft mentioned in (iv) above is a major part of the payload. Flight test instrumentation may be regarded as the equivalent for experimental aircraft.

Since many of these types of payload constitute a major part of fuselage volume they are considered in more detail in paragraph 4.5.

4.2.3 Pressurisation

Operational requirements may dictate that the fuselage of a manned aircraft is pressurised. When this is the case it is likely to have a major effect upon the overall shape of the fuselage, especially the cross-section.

Two distinctly different levels of pressurisation may be identified:

a) Low differential pressure. This is mainly associated with combat aircraft where the crew are also equipped with pressure suits, but it may also be relevant to some general aviation aircraft. In the former case the cockpit pressurisation is primarily intended to provide a safe environment should the suit system fail while the aircraft is flying at high altitude. In the latter case it is to confer greater comfort for the occupants of a propeller-engined aircraft when it is operating at moderately high altitude. For the purpose of definition a low level of differential pressure may be defined as one no greater than about 0.27 bar (4 lb/in^2) although this is somewhat arbitrary in the case of a general aviation aircraft. As a general guideline the effect of the pressurisation requirement is to necessitate careful detail design of the pressure compartment and the avoidance of flat surfaces where possible. This is less significant in the case of small, highly manoeuvrable, combat aircraft which inevitably have a substantial structure than it is for lightly built general aviation aircraft where some of the considerations of the next paragraph may apply. See Figure 4.1.

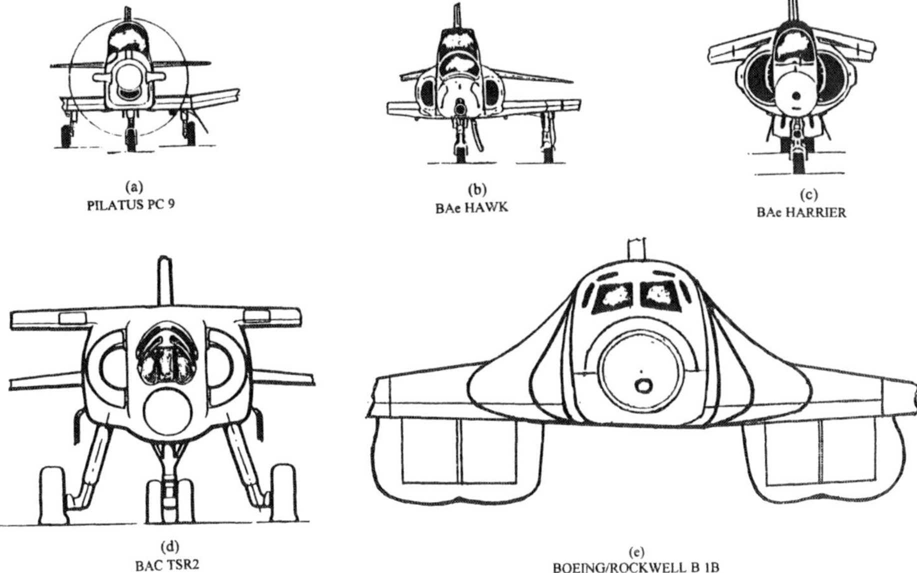

Figure 4.1 Military aircraft cross-sections

b) Normal (high) differential pressure. More usually in the case of turbine-powered transport aircraft the differential pressure requirement is established to ensure that at no phase of operation does the effective altitude of the cabin exceed 2.44 km (8000 ft) and preferably not more than about 1.83 km (6000 ft). These values are appropriate for usual passenger transport operations. The implied pressure differential for subsonic aircraft covers the range from about 0.37 bar (5.5 lb/in^2) for aircraft designed to operate at up to 7.6 km (25,000 ft) to 0.58 bar (8.5 lb/in^2) which is appropriate to 13.1 km (43,000 ft) operating altitude. A supersonic airliner flying at, say, 19.8 km (65,000 ft) altitude requires a differential pressure of 0.65 bar (9.4 lb/in^2). Because the higher order of pressure differential is mainly associated with passenger aircraft the greater part of the fuselage is subject to it and it constitutes a major and possibly overriding fuselage structural design requirement. In particular the cross-section of the outer shell must be based on circular arc cross-sections if significant mass penalties are to be avoided. Figures 4.2 to 4.4 illustrate typical cross-section shapes and show how non-circular, double bubble, configurations may be used providing that a tie, for example in the form of a floor, is incorporated to join points of change of radius of curvature. While such an arrangement can confer layout flexibility in some cases, see also paragraph 4.5.3.5, there is no doubt that a true circular shape is preferable when it can be achieved as it eliminates difficult joints. Non-circular shapes such as ellipses imply bending of the cross-section of the shell and a correspondingly high mass penalty. It is not structurally essential for the fuselage cross-section to be constant along its length. However, in transport aircraft it is usual for the greater part of the fuselage to be of constant cross-section to facilitate manufacture and provide versatility of internal layout.

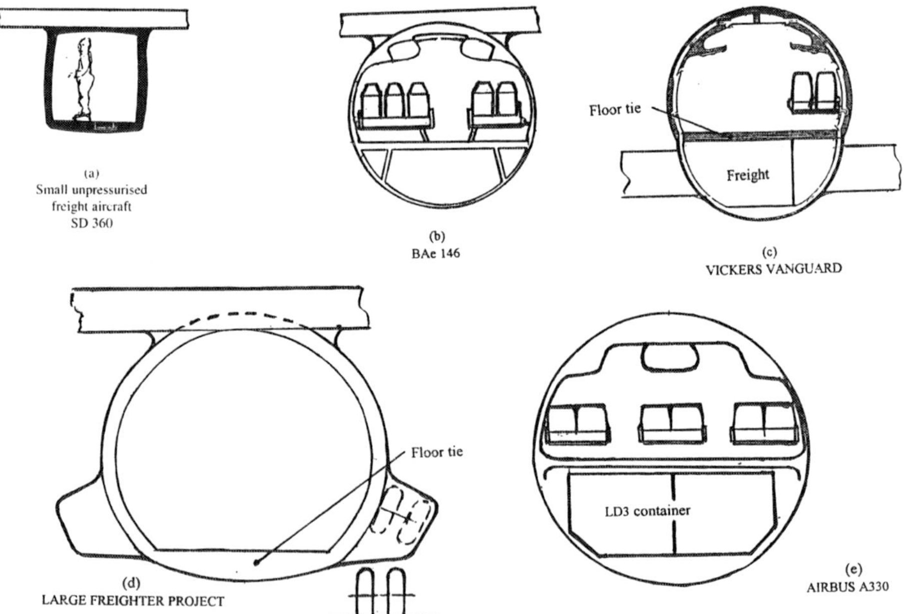

Figure 4.2 Transport aircraft fuselage cross-sections

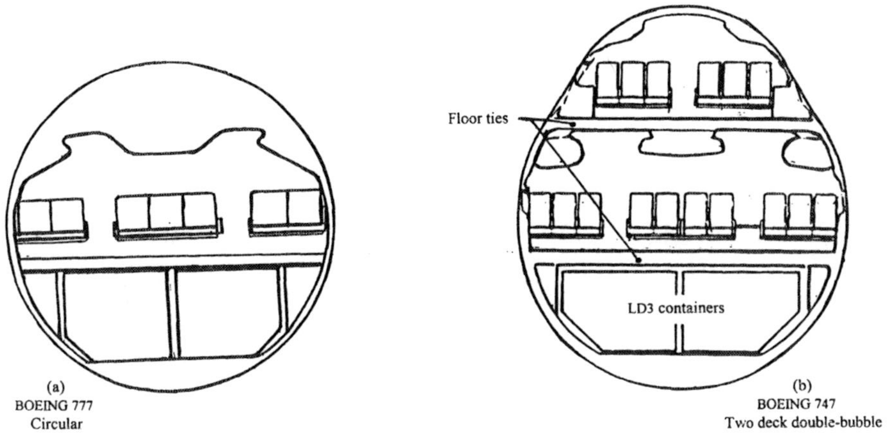

Figure 4.3 Fuselage cross-sections of large transport aircraft

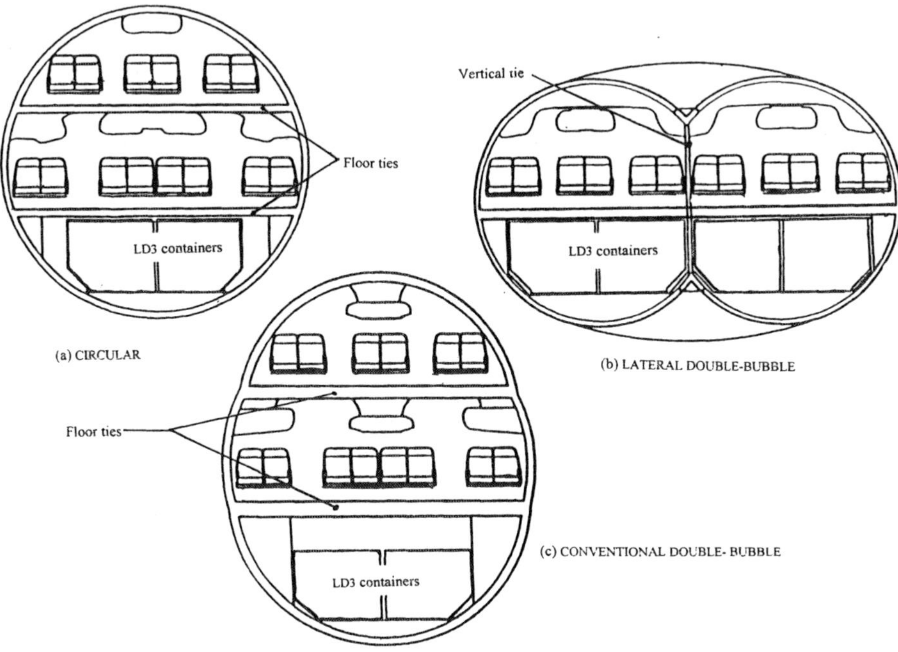

Figure 4.4 Fuselage cross-sections for ultra-high capacity aircraft

4.2.4 Powerplant location

In many designs the powerplant system is located within or on the fuselage and when this is the case it requires primary consideration and may provide the starting point for the layout. As was discussed in Chapter 2, five basic fuselage-located powerplant arrangements may be identified:

a) Nose mounting of engine, see Chapter 2, Figure 2.2. This arrangement is appropriate to both piston- and turbine-driven propeller engines. The powerplant determines the geometry of the front fuselage, including influencing the cross-section, but has little other effect on the rest of the fuselage layout. An exception to this generalisation is when the exhaust gases from a nose-mounted turbine engine are passed rearward through the fuselage rather than being ejected locally, but this is unlikely to be necessary in view of the relatively low exhaust gas temperature and velocity associated with this class of engine.

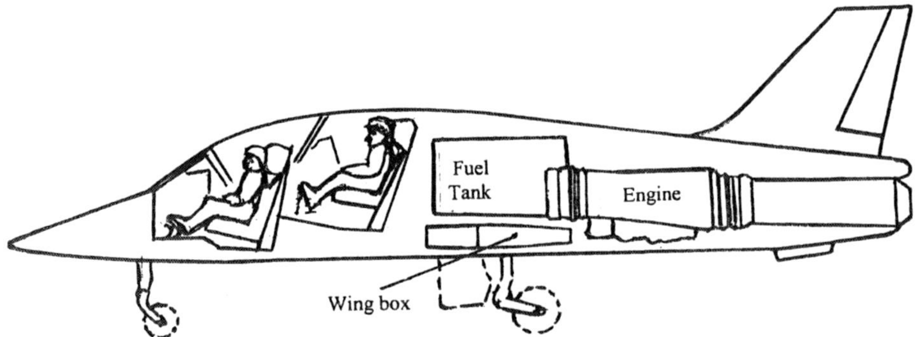

Figure 4.5 Central engine - BAe Hawk

b) Central or central/rear location, Figures 4.5 and 4.6. Location of the powerplant system in the centre of the fuselage can be advantageous in some circumstances, particularly for jet-powered military trainer/strike aircraft having wings of moderate aspect ratio. Positioning of the engine just aft of the main wing structure implies that its relatively high mass is near to the centre of gravity of the aircraft. The intake system usually employs side or ventral fuselage intakes and may pass through the region of the wing centre structure. A major consideration is the means of engine removal. While there are other possibilities it is usually considered that the best approach is to provide sufficient ground clearance for the engine to be removed downwards by removal of a lower surface access panel or through doors. It is also usually preferable for the exhaust gases to be ejected out of the rear of the fuselage. The alternative of fuselage side exhausts is likely to give rise to acoustic fatigue problems at the rear of the aircraft.

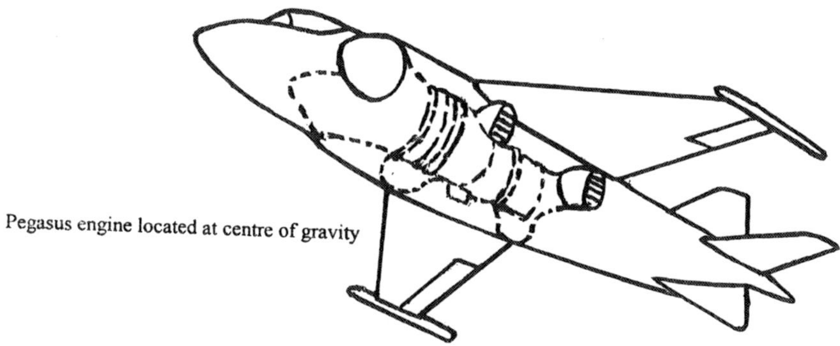

Figure 4.6 Central engine - BAe Harrier (V/STOL)

A special case of a centrally mounted powerplant is that of a V/STOL aircraft, Figure 4.6. Whatever vertical lift system is used it is often found that the main cruise/lift powerplant has to be mounted at, or close to, the centre of gravity. As a consequence the engine and centre wing structure occupy a similar fore and aft location in the fuselage. It would appear that location of the wing above the powerplant is the only practical solution, removal of the engine preferably being downwards. Should downwards engine removal be precluded for reasons of inadequate ground clearance, as with the British Aerospace Harrier, the apparently drastic solution of removal of the wing to enable upwards engine withdrawal may have to be considered.

It is clear that when the engine is located in the centre fuselage the total powerplant system of air intake, engine and jet pipes occupies a large part of the total fuselage volume and has a major effect upon the overall layout.

c) Rear fuselage location, Figure 4.7. It is usual for the powerplants to be mounted at the rear of the fuselage in supersonic combat types which have wings of relatively low aspect ratio. One major advantage of this arrangement is that the high velocity exhaust gases are emitted aft of all major structure without the need for a long exhaust pipe. It also means that because of the wide root chord the wing structure can pass round the fuselage forward of the powerplant, greatly facilitating engine removal. This is usually downwards or downward and aft. Against this it is necessary to consider how to arrange the attachment structure for the empennage but this difficulty is lessened when the aircraft has a canard configuration.

Occasionally there may be a requirement to locate a third engine in the rear fuselage of a transport aircraft, although this is less likely than was once the case due to improved engine reliability. Dorsal intakes are usually used but side intakes are a possibility. The influence of the powerplant in this layout is limited to the rear of the fuselage.

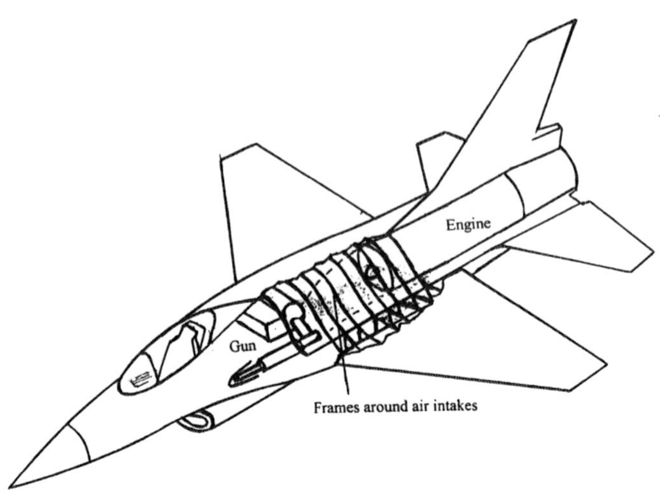

Figure 4.7 Rear fuselage located engine - Lockheed F 16

d) Rear podded powerplant, see Chapter 2, Figure 2.6. This arrangement has only a secondary effect upon the fuselage layout. It is mainly restricted to the need to provide internal supporting structure and to avoid the location of flight critical components in the fan/compressor/turbine burst zones.

e) Podded powerplant located above or below the fuselage. Again this arrangement has only a small effect upon fuselage layout.

4.2.5 Twin boom layout

While most fuselages are complete from the nose to the tail of the aircraft there are sometimes good reasons for considering a twin boom layout. In this configuration the primary fuselage ends after the wing and the empennage is supported from a pair of slender booms mounted off the wing. Instances where such an arrangement may be worth consideration are:

a) Single propeller engine aircraft where there are good reasons to use a pusher arrangement possibly to enhance the view of the occupants.

b) Small to moderate size freight aircraft having a requirement for a rear loading door.

c) Some high performance combat aircraft, especially V/STOL types, although in this case in reality the booms are likely to be formed from lateral extensions of the fuselage.

d) Propeller-driven unmanned aircraft where it is desirable to locate the payload in the nose.

4.3 Overall layout

4.3.1 Aerodynamics - external shape (Figures 4.8 to 4.12)

4.3.1.1 General

In establishing the outside shape of the fuselage the aerodynamic aim is to achieve a reasonably streamlined form together with the minimum surface area consistent with the required volume. Both the drag and mass of the fuselage are significantly influenced by the surface area. The interpretation of what is a reasonable streamlined form does depend upon the class of aircraft although there are some general considerations such as the absence of steps and a minimum number of excrescences. The main considerations follow.

4.3.1.2 Cross-section shape, Figures 4.1 to 4.4

The shape of the cross-section may be determined by primary considerations such as a requirement for pressurisation as discussed in the previous section. Apart from such matters the cross-section shape used is not too critical aerodynamically although sharp corners are best avoided. Where it is acceptable a near rectangular cross-section often enables efficient utilisation of the internal volume and facilitates the fairing of lifting surfaces to fuselage intersections. Changes in cross-section should occur gradually and any unavoidable protuberances should be carefully faired. It is sometimes convenient to define an effective maximum cross-section diameter as that of the circle having the same area as the actual maximum cross-section of the fuselage.

4.3.1.3 Nose shape

As a general rule the nose shape should not be unduly bluff. In some cases it is possible to base the shape on an ellipsoid, but with the major axis in the side elevation curved downward to improve the view from the cockpit of a manned aircraft. It is usually necessary to accept local changes in cross-section in the region of the cockpit to provide satisfactory layout of windscreen panels, but the associated drag penalty may be minimised by careful shaping. The canopy design depends upon the size of the aircraft. A faired in, semi-submerged, canopy is preferred but may be difficult to incorporate in small aircraft where a semi-blister type may be used. In any case the angle of the windscreen requires a careful compromise between aerodynamic, birdstrike and internal reflection requirements. Individual windscreen panels should not be more than about $0.5\ m^2$ in area. The drag associated with a blister canopy may be minimised by careful shaping over the rear section as it blends into the fuselage proper. The satisfactory location of the pilot's eye position is a starting point in nose fuselage layout, see also paragraph 4.5.1.

Although in an ideal streamline shape the length of the fuselage up to the position of the maximum cross-section is about one-third of the total fuselage length, in practice considerations of volume utilisation and, where relevant, wave drag due to compressibility often change this. For most subsonic aircraft the nose portion is relatively short. In the case of a subsonic transport aircraft it has a length in the range of one to two effective diameters, the higher values being for faster aircraft. The nose length is likely to be about four effective diameters for a supersonic aircraft, (see Table 4.1).

When a nose-located powerplant is used the nose shape and the length of the nose are largely determined by the dimensions of the engine and its associated items, possibly in conjunction with the position of a pilot.

Table 4.1 Passenger aircraft - fuselage length proportions

Class of aircraft	Nose length to diameter ratio	Tail length to diameter ratio	Cabin to overall length ratio		Cabin to parallel section ratio
			Basic	Stretched	
Small commuter	1.5 to 2.0	2.5 to 3.0	0.4	-	0.8
Executive	1.2 to 1.8	2.5 to 3.0	0.35	-	0.7
Smaller narrow body	1.1 to 1.6	2.5 to 3.0	0.5	0.65	1.0
Larger narrow body	1.2 to 1.6	2.5 to 3.0	0.65	0.7	1.1
Single deck wide body	1.2 to 1.6	2.5 to 3.0	0.65	-	1.2
Multiple deck wide body	1.2 to 1.6	3.0 to 3.5	0.7	-	1.5
Supersonic	4	6 to 7	0.55	-	1.1

4.3.1.4 Centre fuselage shape and overall length

It is convenient to distinguish between subsonic and supersonic aircraft.

a) Subsonic aircraft. While the fuselage drag may be a theoretical minimum for a streamlined shape having a length to diameter ratio of less than four, this is rarely a practical proposition. Consideration of volume utilisation, overall aircraft centre of gravity location, and the moment arm required for stabilising/control surfaces usually results in the overall length to effective diameter ratio being at least six. A value nearer ten is more typical for many subsonic aircraft with a practical maximum of about 14 for a "stretched" design. There is usually negligible penalty in employing a constant cross-section along the length of the centre fuselage and it is preferable for volume utilisation and manufacture.

b) Supersonic aircraft. A major consideration in the design of high transonic speed and supersonic aircraft is the distribution of the total volume along the length of the aircraft, see Chapter 6, paragraph 6.3.3. The fuselage is normally the major contributor to this volume. Ideally the total volume distribution along the length should be smooth and, somewhat simplistically, close to a sinusoidal shape in order to minimise volume wave drag. The overall length to effective diameter ratio is also of fundamental importance, the higher it is the lower the wave drag. Hence in the case of a supersonic aircraft the overall fuselage length to diameter ratio is likely to be significantly higher than that for a subsonic aircraft, possibly being as high as 20. Where possible the local cross-section area of the fuselage should be matched to that of the other volume contributions from the wing etc., to give the required overall smooth distribution. This cannot be done precisely until the overall layout of the aircraft has been determined but consideration should be given to the requirement from the outset. For initial work it is adequate to assume that the local cross-section does not include the intake area over the length of the powerplant installation.

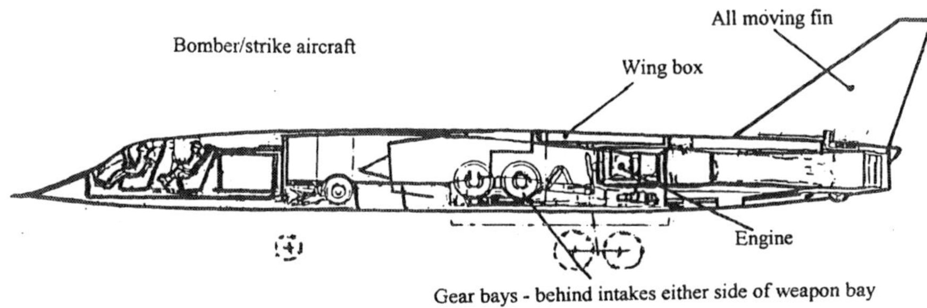

Figure 4.8 Fuselage layout of a military strike/bomber - BAC TSR 2

4.3.1.5 Tail shape

As in the case of the nose shape the tail configuration should change smoothly from the maximum cross-section to nominally zero, excluding any area contribution from the engine exhaust. If the final cross-section area cannot be brought to zero, the base area should be minimised. This is particularly important for transonic and supersonic aircraft.

An important parameter in determining the lower line of the tail fuselage is the ground clearance while the aircraft is at high angle of attack during rotation for take-off and at landing. Allowance should be made for ground effect on the lift characteristics, a typical clearance angle between the main landing gear, extended, and the tail being 12 to 15°, see Addendum 1. In practice the start of the rear fuselage taper depends upon this angle and the landing gear length. A cosine or parabolic shaped region continuing to a straight taper may sometimes be employed. Typically the whole tail section has a length of 2.5 to 3.5 effective diameters on a subsonic design, but it may be in excess of 6 for a supersonic aircraft, (see Table 4.1).

When an aircraft is provided with a rear ramp-type loading door, as in Figure 4.12, and especially when air dropping of supplies is envisaged, the tail up-sweep angle is determined by the clearance for loading and dropping with the door in a horizontal position. The matter is aggravated by the fact that this class of aircraft must also have a low floor line, which implies a nearly flat bottom to the fuselage. Large up-sweep angles may well introduce a large drag penalty and so the angle should be kept as low as possible. If at all possible the use of up-sweep on the top surface of the fuselage should be avoided. This aspect of tail shape is one that requires careful attention as the design is refined.

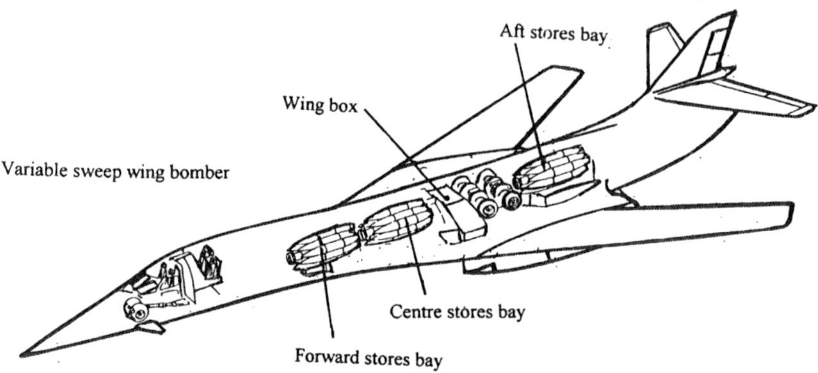

Figure 4.9 Fuselage layout of a bomber aircraft - Boeing/Rockwell B 1

4.3.1.6 Wing-fuselage junction

The region where the wing intersects the fuselage is always critical and is another area which demands careful detail design to keep drag to a minimum. At the initial layout stage it is necessary to ensure that no major difficulty is introduced. For example a wing which passes totally above or below the fuselage requires a very large fairing to ensure satisfactory airflow. This implies a large, undesirable, increase in surface area unless the volume within it can be utilised, say for fuel tankage, landing gear, weapons, equipment or baggage stowage. The situation is made worse for a low wing configuration because the wing is usually set at a positive angle of attack on the fuselage. It is undesirable for the trailing edge to protrude below the fuselage line as this makes fairing difficult. The wing setting on the fuselage should be established such that the fuselage is nominally horizontal in cruising flight and it is usually geometrically between 0 and 4°.

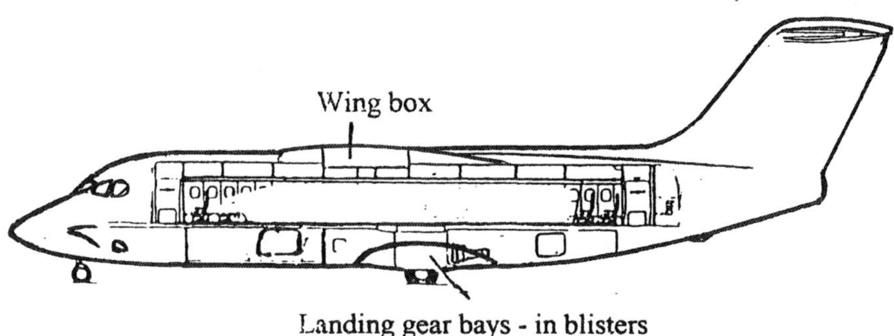

Figure 4.10 Fuselage layout of a small transport aircraft - BAe 146

4.3.2 Structure

4.3.2.1 General

Fundamentally a fuselage is a good shape from the structural point of view. The main structural considerations in fuselage layout are related to circumstances which introduce discontinuities in the load carrying capacity of the outer shell. The two major problems are cutouts and the input of concentrated loads.

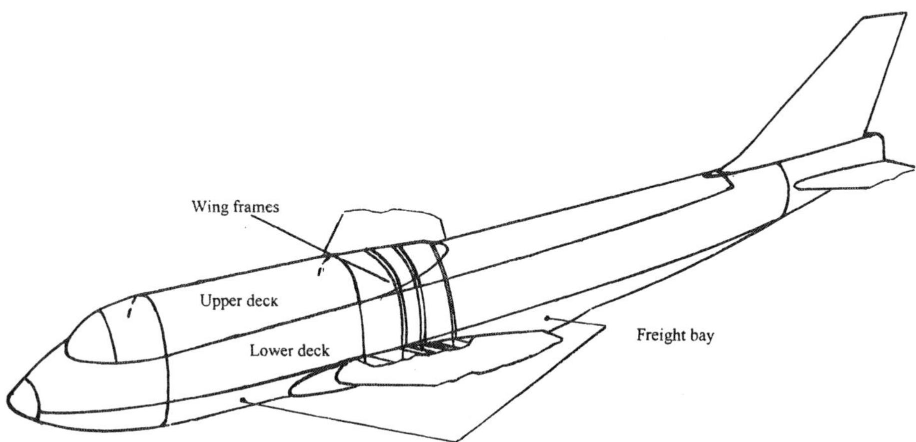

Figure 4.11 Fuselage layout of a large transport aircraft - Boeing 747

4.3.2.2 Cutouts

Cutouts may be loosely divided into those which have a minor effect upon the ability of the structure to transmit loads and those which have a major impact.

a) Minor cutouts. Minor cutouts can be defined as those whose dimensions are an order of magnitude less than those of the overall fuselage cross-section. Thus for example cutouts for passenger windows may be regarded as minor. Generally minor cutouts can be dealt with by providing local reinforcing which effectively replaces the removed material and they may be considered as having negligible impact upon initial fuselage layout.

b) Major cutouts. Major cutouts have dimensions which are similar in magnitude to those of the fuselage cross-section. They cannot be entirely avoided but their location and the ability to provide alternative load paths must be considered at the initial layout stage. Requirements which introduce cutouts in this more serious category are:

i) Entry doors for crew and passengers and doors for servicing. Sometimes the impact of these may be minimised by careful positioning fore and aft and vertically to avoid coincidence of the doors on alternative sides of the fuselage. Pressure loads are taken by the doors but other loads are reacted along vertical and horizontal edge members with local reinforcing, especially at the corners.

ii) Emergency exits. While these are smaller than main entrance doors the same remarks apply. Some are located in the heavily loaded region over a low wing.

iii) Windscreen and canopy. It is necessary to provide continuity of fore and aft bending material by introducing members along the lower edges of the cutout. Some windscreen frames may be designed to carry overall fuselage loads but often they are not.

iv) Access panels and doors for equipment and powerplant. The latter is especially relevant to training and combat aircraft. When the removable components are large it is often necessary to design them to contribute to the overall load carrying capacity. However, access is easiest when they are primarily fairings which may readily be opened or removed, and this may be achieved if alternative loads paths can be provided.

v) Cutouts for landing gear stowage. Landing gear bays are not normally pressurised and so the structure surrounding the bay, which often consists of flat surfaces, must react the local pressures. Nose gear bays are usually located in the relatively lightly loaded forward fuselage and do not present a major overall problem. On the other hand accommodation for main landing gear units may have to be provided adjacent to the centre of the aircraft and when the fuselage is used for this purpose it inevitably has a significant impact on overall structural layout. To minimise the problem provision should be made for the replacement of removed fore and aft bending material by the introduction of substantial edge members which can continue a large distance on either side of the landing gear bay, possibly together with a central keel member. It is also necessary to provide material to complete the cross-section of the fuselage for torsion load carrying capability usually by incorporating a decking above the bay and closing bulkheads at its ends.

vi) Internal weapons bays. The issues involved are very similar to those of main landing gear bays. However, when the weapons bay occupies a large part of total fuselage length the overall layout should take note of this so that the primary structure is above the bay.

vii) Rear ramp loading doors, especially those which have to be open in flight. The problems are not unlike those of main landing gear bays, but it may be beneficial for the doors to contribute to the reaction of some of the overall loading. If possible there should be provision for a continuous roof over the door region to provide a closed structural box, even if it is small.

4.3.2.3 Concentrated load inputs

The input of concentrated loads occurs at attachments for the lifting surfaces, landing gear, powerplant mounting and so on. These loads are usually input at frames or bulkheads which lie in the plane of the cross-section. As far as is possible these structural members should be continuous right round the fuselage so that loads from them can be reacted directly into the shell. This implies that cutouts are undesirable in regions where large concentrated loads have to be dealt with. An area of particular difficulty occurs when a landing gear or weapon bay is adjacent to the wing attachments.

In many designs the main wing structure passes across the fuselage and this considerably eases the attachment design. However in some aircraft, especially high performance combat types, the need for minimum cross-section area results in the wing structure as such terminating at the sides of the fuselage. Wing bending loads are then reacted in a number of frames which pass round the fuselage, often around the air intakes. In these circumstance it is important to ensure the complete integrity of these frames. See also Figure 4.7 and paragraph 4.4.1.3.

4.4 Local layout aspects

4.4.1 The vertical location of the wing

4.4.1.1 General

The choice of vertical wing location relative to the fuselage is a compromise between aerodynamic, structural and operational considerations. In some cases there may be overriding issues such as propeller ground clearance on a multi-engined type or powerplant removal on a V/STOL combat aircraft, both of which may well determine the use of a high wing.

4.4.1.2 Aerodynamic issues

Aerodynamically a mid wing position is attractive in that it is likely to have the lowest interference drag, which is particularly advantageous for a supersonic aircraft. On the other hand a low wing does enable the flaps to be continuous across the lower fuselage or it may be faired in such a way as to provide volume for landing gear stowage. As it is undesirable for the wing trailing edge to protrude below the fuselage the fairing should take note of this.

A wing mounted across the top of the fuselage should enable the drag due to lift to be reduced to a minimum and undue drag rise at critical Mach numbers avoided.

4.4.1.3 Structural issues

Structurally the only loads which must be transmitted from the wing to the fuselage are the shear forces associated with lift, drag and side load, asymmetric bending and torques. There is every reason to avoid passing the main symmetric bending loads into the fuselage and therefore the primary wing structure should be continuous across the fuselage wherever possible. This will usually mean that the whole wing structural box passes through the fuselage, although for some lightly loaded aircraft it may only be one or two main spar members. A large mass penalty may be anticipated should the passage of wing bending loads round the fuselage shell be inevitable.

The desirability of having an uninterrupted wing structure has a substantial impact upon the vertical location of the wing and rules out a mid-position when a single payload volume occupying most of the fuselage is a requirement. On the other hand it can be a good solution when the aircraft has to incorporate a long internal weapons bay.

4.4.1.4 Operational issues

There are a number of operational issues which can influence the choice of wing vertical location:

a) Clearance. A high wing is obviously advantageous when wing-mounted propeller engines, or even large diameter fan engines, are used on relatively small aircraft. It also enables good all round access to the aircraft, which is of particular importance for a freight aircraft. When wing stores are carried a high wing may be advantageous when the aircraft is generally of small dimensions. However, if the lower wing surface is more than about 1.5 m above the ground stores handling becomes difficult.

b) Passenger appeal. It is sometimes argued that a high wing is advantageous from the point of view of occupant appeal as it does not obstruct the view of the ground from the cabin windows. This is really only likely to be an issue on aircraft which fly at relative low altitudes and are used for ground observation.

c) Crashworthiness. A low wing has the advantage that it provides a convenient platform for occupant escape in the very rare event of an aircraft having to force land on to water, and over-wing emergency exits are convenient. Against these advantages is the fact that the wing fuel volume is vulnerable in the event of a wheels-up landing, this is not the case for a high wing.

4.4.1.5 Overall layout issues

The use of a high wing on a small passenger aircraft may introduce cabin headroom difficulties especially when used with a circular cross-section, pressurised, fuselage. The cabin floor width cannot be much less than the full cabin width if passenger foot room is to be adequate. This implies a floor which is located vertically about 0.35 diameters up from the lower surface and that the depth of fuselage below the floor does enable a low wing structure to be passed through as well as providing volume for freight. See also paragraph 4.5.3.6 (b). A high wing implies a long and heavy landing gear if it is located on the wing, the alternative being a narrow track arrangement mounted off the lower fuselage and retracted into blister fairings, see Figures 4.10 and 4.12. The combination of a narrow track and high wing may give rise to an overturning tendency.

4.4.1.6 Summary

Consideration of the issues discussed above enables some general conclusions to be drawn regarding the vertical location of the wing for various types of aircraft. Preferred applications are likely to be:-

a) High wing.
- Freight aircraft
- Smaller propeller-powered transport aircraft
- Some light aircraft, especially single-engine types
- Some combat aircraft, including V/STOL types
- Unmanned aircraft

b) Mid wing.
- Some high performance combat types
- Weapons systems aircraft with a long internal weapons bay
- Possibly multi-deck transport aircraft.

c) *Low wing.* Majority of passenger transport aircraft
Some light single- and twin-engine trainers
Some combat aircraft, including those which use a foreplane configuration. See paragraph 4.4.2.4.

4.4.2 Control and stabilising surfaces

4.4.2.1 General

The configuration of the horizontal and vertical control and stability surfaces has some impact upon the layout of the fuselage.

4.4.2.2 Vertical surface

A single centrally mounted rear fin is by far the most common arrangement with the surface located as far aft on the fuselage as possible in order to maximise its moment arm to the centre of gravity of the aircraft. A requirement for good spin recovery may sometimes result in a fin location partially ahead of the horizontal tail, especially on combat and trainer aircraft.

While twin fin arrangements were once common their use is now usually limited to aircraft having a twin boom fuselage layout and combat aircraft with a stealth requirement. One other application where twin fins may have an overall advantage is on a freight aircraft with a large rear ramp loading door. Here the use of twin fins can result in a significant reduction of rear fuselage torsion loading during asymmetric flight.

4.4.2.3 Horizontal surface, tail

The efficiency of a tailplane is normally critically dependent upon its vertical location relative to the wing because of the effect of wing downwash. Generally speaking on conventional aircraft layouts having a wing of moderate to high aspect ratio and a long tail arm there is no difficulty in locating the tailplane on the rear fuselage, preferably higher than the wing if this is possible. When the tail arm is relatively short or the wing is high mounted it is advantageous to use a lower mounted tailplane, but ground clearance often rules this out.

The alternative is to locate the tailplane on the fin, usually at its upper extremity. Rear fuselage podded powerplants may also dictate the use of this arrangement. While high-located horizontal tail surfaces have major advantages in terms of improved efficiency in normal flight and a longer tail arm which results in a smaller surface area, there are potential disadvantages. One of these is the possibility of the aircraft being prone to a "deep stall". This latter problem can arise if the aircraft pitches rapidly to a high angle of attack causing the horizontal tail to enter the most intense wing downwash region and become ineffective in restoring the aircraft toward level flight. A way of overcoming this problem is to design the control system so as to prevent the situation arising. Other disadvantages of a high tailplane are a fin mass penalty which is caused by the additional loads to which it is subjected and a proneness of 'T' tail layouts to flutter.

When a trimming or "all moving" tail is used it is desirable to design the local fuselage or fin surface to be flat to avoid gaps when the tailplane angle is adjusted.

4.4.2.4 Canard surfaces

The location of a canard surface may cause some difficulties. Generally the canard should be higher than the wing in the side elevation, the actual spacing being a critical detail design feature. On some aircraft it may be convenient to locate the canard just behind the cockpit region, relatively high on the fuselage. In the case of small combat types it is often necessary to locate the surface forward of the cockpit, in which position it obstructs some aspects of the view of the pilot. As with "all moving" horizontal tails it is necessary to arrange the fuselage shape to minimise gaps as the surface moves.

4.4.3 Landing gear

4.4.3.1 General

Although it implies a structural penalty on the fuselage design it is often necessary to mount and/or stow the main landing gear in the fuselage. A number of configurations may be identified.

4.4.3.2 Low wing aircraft of small to moderate size

A possible configuration for this class of aircraft is to mount the landing gear on the wing and to retract it inwards with the wheels stowed within the lower fuselage and the wing to fuselage fairing. Paragraph 4.3.2.2 (b)(v) refers to the structural issues involved in this layout.

4.4.3.3 Low wing aircraft of large size

When the aircraft has a take-off mass of greater than about 210,000 kg, it may be necessary to use more than two main landing gear units. The additional units are usually fuselage mounted and this further complicates the structural layout referred to in the previous paragraph.

4.4.3.4 High performance combat aircraft with thin wings

For this class of aircraft it is not usually feasible to stow any landing gear components in the wing. Hence fuselage location is inevitable. Often the landing gear is of substantial proportions relative to the fuselage cross-section and the volume required becomes a major fuselage layout issue. Ensuring that there is adequate track when the gear is extended demands careful consideration of the mechanical design of the gear, (see Addendum 1).

4.4.3.5 High wing aircraft of moderate to large size

As aircraft size becomes larger it becomes inefficient to employ a wing-mounted landing gear when at the same time the wing is located high on the fuselage. Hence for high wing aircraft of above about 40,000 kg take-off mass it becomes attractive to use fuselage bulges to mount and stow the landing gear. These introduce fuselage mass and drag penalties but they are less than those associated with long wing-mounted units. This is especially true when the aircraft has a requirement to operate off poorly prepared surfaces since this implies large diameter tyres and multi-wheel arrangements.

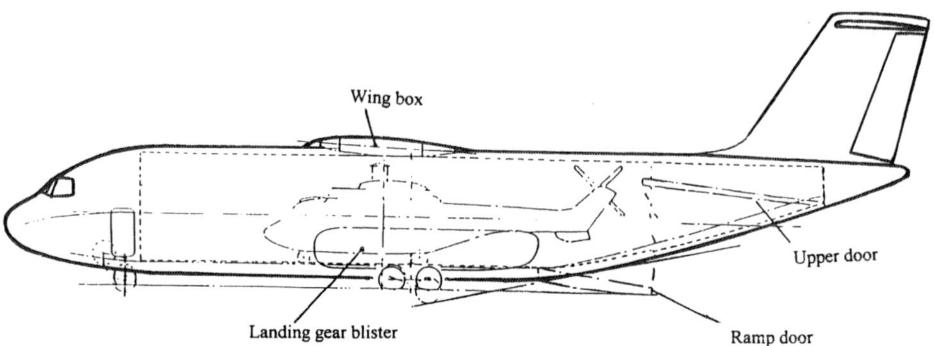

Figure 4.12 Fuselage layout of a proposed large freight aircraft

4.4.4 Systems, fuel and equipment

4.4.4.1 General

The fuselage is required to provide volume for the accommodation of the various items needed for the satisfactory operation of the aircraft. While many of the components are small and do not demand specific consideration at the initial stage of fuselage layout, some items, such as air conditioning units, must be allowed for from the outset.

4.4.4.2 Avionics

With the exception of small general aviation aircraft and basic trainers, aircraft are required to carry complex and extensive navigation, communication and flight control/management equipment. Two aspects of these installations must be allowed for at the initial design stage. Firstly, provision must be made for adequate volume in the correct location, together with ease of access to it. Secondly, the location of aerials, radar scanners, etc. must be considered. Scanners and other sensors are often mounted in the nose of the aircraft facing forward, or forward and down, and can dictate the nose shape. Long range search and early warning radar scanners are sometimes located in the lower fuselage or, depending upon their purpose, above the fuselage on a short pylon.

4.4.4.3 Fuel

With the exception of the centre wing box it is not normal to locate fuel tanks within the fuselage of passenger transport aircraft. There are exceptions, such as the installation of extended range tanks within the freight bay, or centre of gravity balancing tanks.

However, for many other classes of aircraft, ranging from small general aviation types to large flight refuelling aircraft, much of the volume required has to be found within the fuselage. High performance combat aircraft present a special problem as the need to minimise the overall volume results in much of the fuel having to be located in fuselage tanks of awkward size and shape. At the initial layout phase it is important to consider this and to ensure that the space allocated for fuel is both adequate and located appropriately for aircraft centre of gravity control.

4.4.4.4 Auxiliary power unit

An auxiliary power unit is of sufficient size to require location in the fuselage as the layout is developed. Location in the extreme rear of the fuselage is common on transport types.

4.5 Crew and payload

4.5.1 Crew location

4.5.1.1 General

The pilot, or pilots, must have adequate view to enable the aircraft to be controlled satisfactorily at all times. Usually there is no difficulty in the cruise phase, the more critical aspects being ground manoeuvring, landing and combat when it is relevant. The location of a suitable position for the eye of the pilot may be regarded as the starting point for cockpit/windscreen layout. See Figures 4.5 and 4.8.

When two-seat combat and training aircraft are under consideration it is necessary to decide between side-by-side and tandem seating. Side-by-side seating has some advantages in ease of communication between the occupants and in cockpit layout, but it does lead to a relatively wide fuselage. For this reason tandem seating is usually preferred on smaller, high performance, combat aircraft and military trainers usually follow this trend for operational compatibility. A further point is concerned with crew escape since when ejection seats are provided it is necessary to ensure that there is adequate clearance from airframe parts during escape. This may influence the shape of the canopy and, as a general rule, escape procedures are simpler when sequenced from a tandem seating arrangement.

In transport and related aircraft the two flight crew are located side by side, with the duty pilot in the left-hand seat because circling to the left is the usual procedure. The duty pilot must have adequate view forwards and downwards without having to resort to undue head movement Although requirements differ in some ways the view of the pilot should be about 15° down and ahead, or slightly to one side may be acceptable. Large supersonic aircraft are a particular problem in this respect and variable geometry nose/windscreen arrangements have been used to improve the situation.

Apart from special cases the location of other crew members is not critical but where possible the opportunity for direct contact between all the crew should be provided. This is especially true of a military combat aircraft where it is desirable to keep any pressurised volume as small as possible.

4.5.1.2 Size of crew compartment

As a guide for initial layout purposes the minimum size of the space required for a single pilot is about 0.7 m wide by 1.1 m high and 1.3 m long. If an ejector seat is fitted the length increases to about 1.45 m. When two crew are seated side by side the minimum width increases to about 1.25 m. If a crew member is not required to operate flying controls, the length may be reduced by about 0.2 m. These volumes may be allocated within the fuselage with some discretion applied on the corners in the cross-section.

4.5.2 Weapon bays, missiles and guns

4.5.2.1 Weapon bays

Any disposable load must be carried such that its overall centre of gravity is close to that of the aircraft as a whole. Thus a single weapons bay should be more or less equally disposed either side of the centre of gravity. If it is not essential to have one long bay greater flexibility is conferred in the layout by having two or more separate bays which may facilitate landing gear stowage and assist in overcoming structural cutout problems. Where there is a requirement for a long, single, weapons bay then the implication is that the wing will pass over it, and often a high or mid wing is the best solution. See Figures 4.8 and 4.9. The layout in the region of the weapons bay demands careful consideration. Points to be considered are:

a) Avoidance of undue air turbulence when the weapons bay is open. This is particularly a problem at higher subsonic speed and above and may introduce a requirement for a special configuration, such as a rotating weapon carriage to minimise the steps in the lower fuselage when the weapons are to be dropped. This matter may be more difficult to resolve when the weapons bay is short but deep.

b) The airflow into the open weapon bay should be such that the stores fall when released, and do not 'hang up' in the bay. This requirement is likely to be in conflict with that of paragraph (a) since it may well occur when the length to depth ratio exceeds about five or six. The problem may be overcome by fitting downwards ejector units into the stores carriers. Adequate spacing between stores must be allowed to ensure that when released a store does not contact any other, or any part of the airframe.

c) The rear end of the weapons bay should include a rearwards sloping bulkhead, rather than an upright one. This improves the airflow and gives better nose down pitching characteristics should the aircraft force land on water and the weapons bay doors collapse. Free fall weapons may be stored and released vertically, but a horizontal arrangement is more common for all types of store including missiles. When considering fuselage layout it is helpful to prepare one or more weapons configurations for integration with other requirements. These configurations may have different length to width ratios and include various combinations of stores. In determining the width of the weapons bay one consideration is the ability to open the doors adequately while the aircraft is on the ground.

Generally the difficulties of weapons bays on high performance aircraft are such that it is preferable to avoid them if at all possible. Examples of where they are inevitable are stealth aircraft and longer range supersonic types where the drag penalty of external stores is considerable.

4.5.2.2 Missiles

Guided missiles are sometimes carried within weapon bays but may also be located externally on the fuselage. If this is so local fuselage layout details may be affected.

4.5.2.3 Gun installations

It is common for internal gun installations to be required for combat aircraft. By definition of its purpose the gun has to be mounted in the forward region of the aircraft, adjacent to the crew compartment unless a wing podded layout is used. It can therefore be a major factor in the layout of the nose of the aircraft, especially when allowance is made for ammunition stowage and the collection of spent cartridges, etc. (see Figure 4.7).

4.5.3 Passengers and associated freight

4.5.3.1 General

While passenger accommodation is the primary consideration regarding fuselage layout on airliners and related types, the need for adequate baggage and freight volume must not be overlooked. In the majority of cases the freight is carried in under floor baggage holds, often in standard containers. Thus while the cabin length and width are mainly determined by the passenger seating and associated services, the need to accommodate standard containers may influence the cross-section depth. This is discussed in paragraphs 4.5.3.4 and 4.5.3.5, (see Figures 4.10 and 4.11).

For a typical single-deck subsonic airliner it is found that the passenger cabin occupies up to about 70% of the overall length, the remainder being taken up by the nose fuselage and the tapering tail region, but see Table 4.1 for other types. The inevitably longer nose and tail of a slender supersonic configuration reduce the usable cabin length to no more than 55% of the total. Thus for a given typical overall length to effective diameter ratio in a single-deck configuration it is possible to derive an appropriate combination of number of seat rows and number of seats across the width to give a specified passenger capacity. Multi-deck arrangements obviously introduce more flexibility and it is generally necessary to consider various proportions of passengers to be carried on the alternative decks.

It must be emphasised that the choice of the cross-section dimensions is of critical importance, as it becomes a fixed feature of the design. Different passenger requirements may be met by lengthening or shortening the fuselage during development of the type and this is a both usual procedure and an important design consideration.

4.5.3.2 Seating arrangements and associated items

There are numerous points which must be considered in planning the layout of seating for a datum configuration. However, it is important to ensure that the cabin layout is flexible in the sense that alternative requirements may be easily met. Seats are located on fore and aft rails so that the pitch between the rows may be changed easily. Correctly positioned across the width the seat rails can also enable different numbers of seats to be placed across the fuselage. Seats are usually adjustable and include fold-away tables.

The following points should be considered:

a) *The seating should be arranged such that the passengers need to move as little as possible in the fore and aft sense* in order to minimise likely variations in centre of gravity during a flight. This is also true of associated facilities including baggage stowage.

b) *As far as possible the passenger seats and provision for freight and baggage should be equally disposed about the nominal centre of gravity position.*

c) *Seating in line with the plane of propellers should be avoided both for psychological and noise reasons.*

d) *If a seat directly faces a bulkhead additional space is required together with protection in the event of a crash.*

e) *As a general rule staggered seating is not helpful* as it has a tendency to result in loss of useful space. In any cases it is not feasible to stagger seats in a set attached to one pair of seat rails.

f) *Backward facing seats* are only safer in the event of a crash when their integrity and that of the floor is maintained, otherwise forward facing seats are better. Even when adjustment is allowed for there is a tendency for backward facing seats to place the occupant in an undesirable attitude during steep climbs, etc.

g) *While it is desirable to provide a window for each row of seats* this cannot generally be achieved due to the flexibility required in seating layout. It is usually possible to arrange for each row of seats to be close to a window by providing increased numbers of small windows. Small windows are preferable in meeting cabin decompression requirements and careful trim design can give the appearance of a larger transparency. First class seating should not look out on the wing.

h) Access doors are required for:

- i) Passengers; as a rough guide at least one for each 100 passengers, but at least two on smaller aircraft.
- ii) Access to the galley for servicing and replenishment.
- iii) Servicing for toilets.
- iv) Freight/baggage stowage.

i) Passenger emergency exits are required in addition to the doors. The number and type of these depend upon the number of normal access doors as well as the number of passengers. Reference should be made to Table 4.2, but the relevant requirements should be checked since this is an issue which changes from time to time. On a low or a low/mid wing aircraft at least one emergency exit on each side should be located over the wing. Normal entrance doors are included in the number of emergency exits.

j) Typically at least one galley is needed for each 120 passengers.

k) At least one toilet for each 50 passengers is required with a greater number desirable on longer flights.

l) Special stowage is usually provided for mail, bullion, duty-free items and passengers' coats.

m) Overhead lockers are standard for passengers' light baggage. These are built into the ceiling over the outermost seats in a single aisle layout, but also over the central seats when two aisles are used.

n) Headroom should be adequate at all locations within the cabin, preferably 1.8 m minimum and ideally at least 2 m in the aisles.

4.5.3.3 Cabin width

The internal width of the cabin is determined by a combination of:

a) Width of individual seats and the number of them across the cabin. Seats are usually used in sets of two to four. No passenger should have to cross more than two others to reach an aisle. Individual economy class seats are just over 0.5 m wide, but there are variations. Business and first class seats are wider.

b) Aisle width should be at least 0.4 m with 0.5 m desirable in economy arrangements and 0.6 m or more in first class. Up to six seats across, in two pairs of three, can be used with a single aisle. Theoretically up to twelve seats across can be used with two aisles, although a maximum of eleven is more usual.

c) The internal width of the cabin for economy seating is approximately given by the relationship ($0.5p + 0.55a$), where p is the number of seats across the cabin and a is the number of aisles. Typical first class seating layout requires between 125 and 140% of the width of economy seating.

Table 4.2 Passenger aircraft - emergency exits
(FAR/JAR 25 - 803/807)

A TYPES OF EXIT

Type	Size, m width x depth	Notes
A	1.07 x 1.83	Floor level
I	0.62 x 1.22	Floor level
II	0.51 x 1.12	Overwing: 0.25 m step up inside, 0.43 m down
III	0.51 x 0.91	Overwing: 0.5 m up, 0.68 m down outside
IV	0.49 x 0.66	Overwing: 0.73 m up, 0.91 m down
Ventral		Same egress rate as Type I if not obstructed
Tail cone	0.51 x 1.52	(If only Type III size, capacity is 15)

B EXIT CAPACITY

(Numbers for each pair of exits, one on each side of aircraft, except for ventral and tail case)

Type	Capacity
A	110
I	45
II	40
III	35
IV	9*
Ventral	
Tail cone	25

* Only use on aircraft of up to 9 seats capacity

There are specified combinations of exit types for up to 179 seats capacity

(For aircraft with more than 44 passengers complete evacuation must be achieved within 90 s)

4.5.3.4 Cabin length

The length of the cabin is determined by the seat pitch, the number of rows and provision of galleys, toilets, etc.:

a) Seat pitch which varies from as low as an uncomfortable 0.7 m for high density short flights to above 1 m in first class accommodation. The actual pitch used is a decision made by the operator but the designer must make acceptable provision when proposing the capacity of a given design. As a guide the economy seating pitch on long haul flights is typically about 0.83 to 0.85 m with business class at about 1.0 m and first class up to 1.5 m.

b) A typical galley floor area of some 2 m by 0.65 m may be assumed for initial purposes. Location of the 0.65 m dimension along the length of the cabin enables a galley to be accommodated within a typical seat pitch or opposite an entrance door. See paragraph 4.5.3.2(j)

c) Toilets have a typical floor area of 1 m^2, but it is acceptable to reduce the width to about 0.7 m with a corresponding increase in length. See paragraph 4.5.3.2(k).

d) The total length of cabin in any given unit of accommodation is approximately:

$$\left[\left(\frac{P}{p}+g\right)s+t+0.8w\right] \quad m \tag{4.1}$$

where
P is the total number of passengers in that unit of accommodation
g is the number of galleys along the length
p is the number of seats across the cabin width
s is the seat pitch (m)
t is the number of toilets along the length
w is the number of cross aisles

The addition of the provision for different classes of accommodation gives the total length of the cabin on any one deck. Galleys, toilets and cross aisles located at changes of accommodation are only counted once. Note that the equation is based on a constant cross-section and additional length allowance may be required for the usual situation where the front and rear compartments taper. When the passengers are accommodated on two decks, each deck must be treated separately. It is possible that the cabin widths will be different, see paragraph 4.5.3.6.

4.5.3.5 Baggage and freight

A typical average baggage provision is 16 kg per passenger, preferably at a density of no more than 160 kg/m^3. This is in addition to personal items placed in overhead lockers. There is often a requirement to carry freight as well as the baggage of the passengers. The baggage and freight are normally loaded into standard containers or pallets, usually located in underfloor freight holds. Adequate volume for baggage and freight is a major consideration for ultra-high capacity aircraft.

There are a number of container standards. For example the LD series used on wide-body airliners typically have a length of 1.54 m and a height of 1.63 m. The width and cross-section shape vary. LD3 has an upper width of 2 m which is maintained 1.14 m down before tapering to 1.56 m. The corner removed facilitates stowage within the lower, circular arc cross section of the freighthold. LD1 has a maximum width of 2.34 m and a 1.56 m minimum width, tapering in from 0.96 m down. See Figures 4.2 to 4.4. Containers used on narrow body transports are typically about 1.1 m high and 2 m wide.

4.5.3.6 External cross-section dimensions

The external cross-section dimensions follow from the provision for freight containers and the seating arrangements across the width, assuming that the freight is carried below the passenger deck. When two passenger decks are used there may be several feasible solutions.

The maximum external width can be deduced by adding to the dimensions given in paragraph 4.5.3.3 an allowance for trim and structure. This is typically 0.2 to 0.3 m with the larger value for large aircraft.

In the case of a single-deck aircraft the depth of the fuselage is derived from consideration of the location of the floor and ceiling lines. Depth below the floor can initially be estimated from the height needed to accommodate the chosen standard container, with an additional allowance of 0.3 to 0.4 m for structure. A check should be made to ensure that there is adequate depth for the wing carry through structure when a low wing is envisaged. Above the floor, allowance must be made for ceiling trim and structure of say, 0.3 to 0.35 m in addition to nominally 2 m aisle height. If a high wing layout is proposed the space for wing carry through structure above the aisle must be considered. This first estimate of depth may be used with the width to propose a cross-section shape. When the fuselage is pressurised this should ideally be circular. However, on smaller aircraft it may be difficult to achieve a circular cross-section without using an unduly large diameter because of:

a) Provision of adequate depth below the floor for freight and/or wing structure.

b) Adequate floor width to allow for passenger foot room. As a guide the floor width should not be more than 0.4 m less than maximum width. This usually means that the floor has to be located 0.25 to 0.45 diameters up from the lower surface in a circular fuselage.

A common solution to these difficulties is to use a fuselage cross-section consisting of two different radii of curvature above and below floor level, to give additional overall depth as shown in Figure 4.2. There is a structural penalty which should be balanced against the possible aerodynamic penalty of a somewhat larger diameter.

A similar process can be used to determine the cross-section when a double-deck configuration is used. The additional parameter to be considered is the division of passengers between the two decks. It is generally better to locate the majority of the passengers on the lower deck since this reduces the problem of emergency escape from the upper deck. If there are a large number of passengers on the upper deck it is necessary to provide additional emergency exits on that deck and the associated escape chutes may be long and difficult to position.

4.5.3.7 Overall fuselage length

Reference to paragraph 4.5.3.1 and Table 4.1 in conjunction with paragraph 4.5.3.4 enables the overall fuselage length to be predicted.

4.5.4 Dedicated freight

4.5.4.1 General

In addition to the freight carried on passenger aircraft there is the possibility of allocating payload volume on the main deck for freight. A number of concepts are feasible. Some main deck freight may be carried together with passengers on a mixed payload aircraft, or in a basically passenger type aircraft modified to carry freight exclusively. Alternatively some aircraft are designed as freighters from the outset. The majority of these latter result from military requirements although some are subsequently adapted to civil needs.

4.5.4.2 Converted passenger types

Where it is possible nose loading doors are favoured for civil freighters derived from passenger types. They imply considerably less structural penalty than is the case with rear ramp or side doors. Large side loading doors are necessary on converted passenger types if ready access to the payload volume through the nose is prevented by the location of the flight crew. Some passenger/freight aircraft are converted as tankers for flight re-fuelling.

4.5.4.3 Military freight aircraft

The dimensions and total volume of a dedicated military freight payload are dependent upon specific requirements such as the need to carry a particular vehicle or weapons system as well as more general freight. The specific payloads frequently determine the cross-section of the freighthold and possibly also its length. See Figure 4.12.

The freighthold should be of constant section along its length without obstruction and as near to rectangular as is feasible. Ideally the floor line should be about 1 m above the ground line for ease of loading from ground vehicles, although on large pressurised aircraft some relaxation of this is usually inevitable. Width of the freighthold is generally more significant than height except for special loads, since general freight carried on pallets is usually limited to about 2 m height. The density of general freight may be as low as 130 to 160 kg/m^3.

Large loading doors are required so that full advantage may be taken of the freighthold cross-section. The need to airdrop military supplies implies that the door should be of the rear ramp type having a clearance height of at least 2 m when the ramp is horizontal for air dropping.

The low floor line and need for easy general accessibility suggests the use of a high wing, possibly with the landing gear stowed in fuselage blisters.

Pressurisation is usually required, except for small, slow aircraft, and the cross-section shape follows from the required freighthold dimension. The low floor line results in the use of a large radius of curvature on the lower surface to minimise structure depth below the floor, (see Figure 4.2(d)).

4.5.5 Uninhabited aircraft

The payload is usually some form of sensing system and the remarks made at paragraph 4.4.4.2 concerning avionics apply to this case also. Of course there is no need for provision of a cockpit and windscreen.

4.6 Fuselage layout procedure

4.6.1 General

As with all aspects of initial design it is likely to be necessary to consider several different configurations before the most appropriate fuselage layout is derived. Thus some of the decisions required to complete the procedure outlined here are not necessarily final, but must be considered as applying to one of what may be a number of alternatives.

4.6.2 Primary decisions

The primary decisions relate to the matters considered in paragraph 4.2.

a) Is pressurisation required, and if so to what level and extent? If the greater part of the fuselage is to be pressurised the cross-section should be based on circular arc combinations and preferably be circular, see paragraph 4.2.3.

b) Is the powerplant system, or part of it, to be located within the fuselage? If the answer is yes then this may be expected to be a dominant consideration in overall fuselage layout. An exception is when an engine is podded, as it sometimes is on the rear fuselage of a transport type or uninhabited aircraft, where the effect is more local, see paragraph 4.2.4.

c) Will the payload occupy the greater part of the fuselage, as on a transport type? If this is the case the payload layout is the starting point for fuselage layout, see paragraphs 4.2.2 and 4.5.

d) Will the fuselage be of conventional layout or will a twin boom arrangement need to be considered? See paragraph 4.2.5.

4.6.3 Local layout decisions

These secondary decisions relate to the topics covered in paragraph 4.4:

a) What will be the vertical location of the wing? See paragraph 4.4.1, especially 4.4.1.6.

b) Will the horizontal tail be placed on the fuselage, or on the fin? See paragraph 4.4.2. Alternatively will the design use a canard surface or be tailless?

c) Is the main landing gear to be mounted off the fuselage? If not will any parts of the main gear be stowed in the fuselage? See paragraph 4.4.3.

d) Are fuel tanks required in the fuselage? If fuel is to be located in the fuselage this must be born in mind to allow for the volume necessary as the layout is developed, see paragraph 4.4.4.3.

e) Is an auxiliary power unit to be located within the fuselage? See paragraph 4.4.4.4

4.6.4 Layout modules

Once the primary considerations outlined in paragraph 4.6.2 have been determined, together with the secondary decisions of paragraph 4.6.3, it is possible to identify those items of the layout which will have a dominating effect. It is also possible to derive individual modules for these, and other items which may be matched together to form the fuselage layout. Depending upon the outcome of paragraphs 4.6.2 and 4.6.3 these modules will include some of the following:

a) Powerplant installation, i.e. air intake, engine and exhaust. There may be more than one possibility here, especially in respect of the location of the air intake. For each possibility a tentative layout of the installation should be prepared. A fuselage-located powerplant system may well provide the starting point of fuselage layout.

b) Crew. When required the layout of the crew compartment may be determined assuming initially that the nose shape can be arranged to give the required forwards and downward view, paragraphs 4.3.1.3 and 4.5.1.

c) Wing carry through box geometry, paragraphs 4.4.1 and 4.6.3(a).

d) Volume for avionics, etc. Initially it is not usually possible to determine the shape of this volume since it can, to some extent, be fitted in where convenient. There are some exceptions to this, for example scanners and extensive installations for special roles, such as airborne early warning systems, where the avionics is the prime payload.

e) Auxiliary power unit, paragraph 4.6.3(e), and air conditioning equipment which may require provision of significant volume.

f) Landing gear mounting and stowage volume. A prior requirement for this is a definition of the type of landing gear, see Chapter 2, paragraphs 2.2.2.5 and 2.2.3.7, together with an indication of landing gear length and wheel size. See also Addendum 1. However, it is usually possible to make an initial estimate of the size and shape by comparison with similar types of aircraft.

g) Payload. The volume and shape of the payload module may be determined by reference to paragraphs 4.5.2, 4.5.3 or 4.5.4 as appropriate. For passenger and freight aircraft the payload configuration provides a major module which forms the basis of the layout of the whole of the fuselage.

h) Fuel. A recognition of the volume needed for fuselage fuel is necessary, see paragraph 4.6.3(a)

4.6.5 Integration of layout modules

The layout modules may be fitted together in various ways taking note of the aerodynamic and structural considerations outlined in paragraphs 4.3.1 and 4.3.2 respectively. The overall aim is to derive a fuselage configuration which makes the maximum use of the total internal volume with appropriate aerodynamic form and with the minimum of structural difficulties. Parameters of particular importance in the overall synthesis of the aircraft are the overall length and effective diameter, together with the cross-section area distribution for supersonic aircraft.

A simplified indication of the interaction of the modules is shown in Figure 4.13.

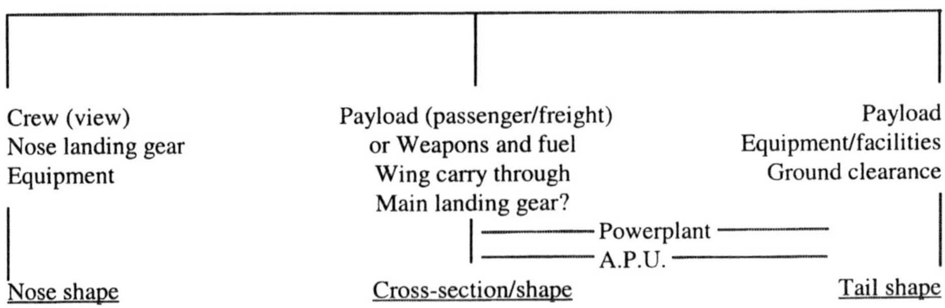

Figure 4.13 Derivation of fuselage layout

Chapter 5

Configuration of the wing

5.1 Introduction

The configuration of the wing is fundamental to the design of the aircraft. In order to discuss the interaction of the many parameters involved in wing design it is helpful to consider them under three separate, but inevitably related, items:

i) Aerofoil section, including the use of high lift devices, see paragraph 5.2.
ii) Planform shape and geometry, see paragraph 5.3.
iii) Overall size, that is the wing area, see paragraph 5.5.

Simplistically the first item follows from the need to obtain the best compromise between all the aerodynamic, structural and operating requirements. The second item is to a great extent determined by the operating Mach number of the aircraft, but is considerably influenced by the aerofoil shape. The last item is determined by the operational requirements for given values of the first two items.

When comparing aerofoil characteristics it is useful to do so on the basis of non-dimensional quantities defined as:

a) Lift coefficient: $C_L = 2L/\rho SV^2$

b) Drag coefficient: $C_D = 2D/\rho SV^2$

c) Pitching moment coefficient: $C_M = 2M/\rho S\,\bar{c}\,V^2$

where L, D and M are the actual lift, drag and moment (positive nose up) acting on the aerofoil respectively, S is aerofoil reference area and $\bar{c}$ the mean chord (S divided by the span, b), V is flight velocity and ρ is local air density.

5.2 Aerofoil section and high lift devices

5.2.1 General comments

As a broad generalisation the choice of the basic aerofoil section is determined by the need to obtain the best aerodynamic efficiency in the primary operating mode of the aircraft. This is frequently the cruising flight mode. Once the basic aerofoil has been selected, with due consideration of the planform shape, it is usual to introduce some form of variable geometry to enable the somewhat different low speed requirements to be met efficiently. Combat aircraft manoeuvring requirements may also require the use of variable geometry.

High lift variable geometry devices, as opposed to drag producing devices such as spoilers, function in three ways:

i) Deflection of the trailing edge and, possibly, leading edge of the aerofoil to increase the chordwise curvature or camber. Greater lift results at the expense of more drag and pitching moment.

ii) Extension of the trailing edge and, possibly, leading edge to increase the chord. This effectively increases the wing area and gives higher lift with relatively small drag penalty.

iii) Introduction of slots between the lower and upper aerofoil surfaces. This enhances upper surface flow, delays flow separation, and again results in more lift potential, but with a drag penalty.

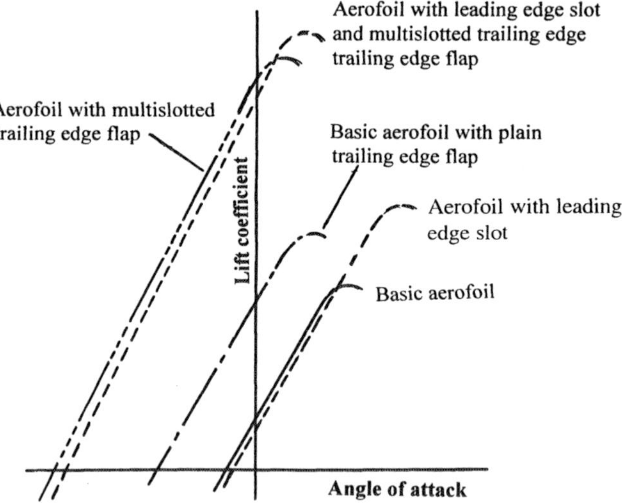

Figure 5.1 Variation of lift coefficient with angle of attack

The effects of chordwise variable geometry may be summarised by reference to the way in which the lift coefficient, C_L, varies with angle of attack, α, as shown in the (C_L-α) curves of Figure 5.1.

a) Increase of camber causes a displacement of the (C_L-α) curve to the left which implies that the aerofoil angle at which there is zero lift, α_o, becomes more negative. Although the angle of attack at which the aerofoil stalls is slightly reduced, the maximum lift coefficient is increased.

b) Increase of chord results in more lift at a given angle of attack due to the effectively increased wing area. Thus relative to the clean wing reference area there is an increase of the slope of the (C_L-α) curve.

c) Slots, especially those in the leading edge region, delay the onset of stall. The effect appears as an upward extension of the (C_L-α) curve along its initial slope.

5.2.2 Basic aerofoil

5.2.2.1 Selection of aerofoil section

In the selection of the basic aerofoil section consideration should be given to the following characteristics:

a) The maximum lift coefficient both at low and higher Mach numbers.

b) The stalling characteristics where a gentle loss of lift is preferable, especially for light aircraft.

c) The aerofoil drag especially in aircraft climb and cruise conditions, when the lift to drag ratio should be as high as possible, and at higher Mach numbers. While most aerofoils operate with the greater part of the chordwise flow in a turbulent state some sections are suitable for laminar flow application. This may be very difficult to achieve in practice but potentially gives much lower drag.

d) The aerofoil pitching moment characteristics which may be particularly important at higher speeds. If it is unduly large there may be a significant trim drag penalty.

e) The depth and shape of the aerofoil in as much as it effects the structural design and the potential volume for fuel, etc.

f) The slope of the lift curve as a function of incidence in that it effects overall aircraft attitude, especially at high values of lift coefficient, such as are required at landing.

The characteristics of the aerofoil are determined by several shape parameters of which the most significant are:

i) the maximum thickness to chord ratio and its chordwise location.

ii) the nose radius, which should be relatively large to give good maximum lift coefficient.

iii) the degree and distribution of camber, if any is used. The aerofoils used for stabilising and control surfaces are usually uncambered. Some degree of camber is normal for a wing section as it gives better lift characteristics in upright, that is usual, flight as opposed to inverted flight and is a recognition of the way most aircraft are flown and manoeuvred.

iv) trailing edge angle, which is often best made as small as is feasible.

Typical early monoplane aerofoils had the maximum depth at 30% of the chord aft of the nose. Maximum lift coefficients of around 1.6 could be achieved associated with zero lift pitching moment coefficients of the order of -0.05. The NACA 23 series are notable in that somewhat higher maximum lift is possible with pitching moment coefficients of only about -0.01, but at the possible penalty of an abrupt stall. Later developments, typified by the NACA 6 series, were aimed at reducing aerofoil drag. They were characterised by an aft movement of the maximum depth to about 0.35 chord in the 62 series, 0.375 in the 64 series, 0.40 in the 65 series and 0.49 in the 67 series. For many purposes where the aircraft operates at speeds below which compressibility effects are significant, the NACA 6 series aerofoils represent a good basis for selection, but see also the next paragraph.

However, the advent of powerful computational aids has resulted in significant advances in aerofoil section design. This is especially beneficial for aircraft intended to fly at higher Mach numbers and it is now usual for a special section to be designed for each aircraft in this category. Some details of advanced or "supercritical" aerofoils have been published. Computational techniques have also been applied to sections intended for use at lower speeds, usually with the emphasis on achievement of higher lift coefficients and, in some cases, laminar flow. Examples of these are the NASA LS and Wortman series intended for use where compressibility effects are negligible, and the NASA MS series for use up to a Mach number of about 0.7. Whilst maximum lift coefficients of the order of two are quoted, they are associated with high zero lift pitching moment coefficients in excess of -0.1 and thin, reflexed, trailing edges which introduce difficulties in the structural design of trailing edge high lift devices. Further, some of these sections may be very sensitive to roughness. When it is necessary to use high lift devices in any case the justification for the use of this class of aerofoil requires careful consideration.

Aerofoils for supersonic aircraft are often adapted from basic biconvex sections to which a small nose radius and possibly some degree of camber has been added.

For preliminary design purposes the most critical aerofoil parameters are the maximum lift coefficient, the thickness to chord ratio and, where appropriate, the related high speed drag characteristics, and lift curve slope. These are discussed further in subsequent paragraphs.

5.2.2.2 Maximum lift coefficient (C_{LMAX})
The maximum lift coefficient of a basic, two-dimensional, aerofoil can vary over a wide range. It is considerably influenced by the nose radius of the aerofoil, decreasing as radius decreases. In the case of a low speed aerofoil and an advanced one for use at high subsonic Mach number a maximum lift coefficient of about 1.6 is typical. For older so-called "high speed" sections and thin aerofoils for supersonic application the typical maximum lift coefficient is around unity. In cruising flight buffet margin considerations may limit the usable lift coefficient to no more than about 40% of the maximum value.

5.2.2.3 Thickness to chord ratio (*t/c*)
This important parameter has some effect upon the maximum lift coefficient and the value chosen is influenced by structural design requirements, see paragraph 5.4.

In incompressible flow conditions relatively high thickness to chord ratios of up to 0.2 are acceptable at the root of the wing and give a good structural depth with a small profile drag penalty. The value at the tip is typically about two-thirds of that at the root.

At higher Mach numbers, where compressibility effects become important, it is usual to use somewhat thinner aerofoils and root values in the range 0.10 to 0.15 are typical. Again the tip value is usually about two-thirds of that at the root, but the spanwise variation is not necessarily linear especially if the wing trailing edge is cranked.

The need to reduce wave drag at supersonic speed dictates the use of thin aerofoils and the thickness to chord ratio is rarely more than 0.06 and may be as low as 0.02 to 0.03. Any spanwise variation is likely to be small.

5.2.2.4 Critical Mach number (M_{NCRIT}) (Two-dimensional aerofoil)
Aircraft which operate in the high subsonic flight regime are designed to fly at about, or just below, a so-called "critical Mach number". There is no generally accepted definition of critical Mach number, but it is the Mach number at which the rate of drag increase due to compressibility becomes unacceptable. This may be stated in terms of the slope of the curve of drag coefficient variation with Mach number or as a specified increment in drag coefficient. A simple definition, following the latter approach, is that it is the Mach number at which the wave drag due to compressibility results in an increment of 20 drag counts (0.002) to the zero lift drag coefficient (see Chapter 6, paragraph 6.3.1).

It is possible to design an aerofoil such that the critical Mach number is unchanged, or even increases, as lift coefficient is increased. However, this may be achieved at the expense of a lower zero lift value of critical Mach number and it is more usual to expect that critical Mach number will reduce with increased lift. Increase of thickness to chord ratio also results in a reduction of critical Mach number. Various formulae and data sources have been derived to enable critical Mach number to be evaluated. A simple formula, originally due to Korn (see, for example, Boppe, C. W., AGARD-FDP Special Course notes, May 1991, Eq (25)), takes the form:

$$M_{NCRIT} = A_F - 0.1C_L - (t/c) = A_f - (t/c) \qquad (5.1a)$$

where M_{NCRIT} is the critical Mach number for a given form of two-dimensional aerofoil
C_L is lift coefficient
(t/c) is thickness to chord ratio

A_F is a number which depends upon the design standard of the aerofoil section. For older aerofoils A_F was around 0.8 to 0.85 but a value of 0.95 should be possible with an optimised advanced aerofoil. Note A_f is equal to $(A_F - 0.1C_L)$, see Chapter 6, Eq (6.13a). Thus for design purposes:

$$M_{NCRIT} = 0.95 - 0.1C_L - (t/c) \tag{5.1b}$$

In the case of a subsonic airliner the cruise lift coefficient is around 0.5 while for a highly manoeuvrable combat aircraft it is somewhat less, say about 0.3, in subsonic cruise. Hence, for an unswept two-dimensional aerofoil, typically:

$$\left.\begin{array}{ll}\text{Subsonic airliner:} & M_{NCRIT} = 0.9 - (t/c) \text{ approx.} \\ \text{Combat aircraft:} & M_{NCRIT} = 0.92 - (t/c) \text{ approx.}\end{array}\right\} \tag{5.1c}$$

For the effect of sweep see paragraph 5.3.4.2.

5.2.2.5 Lift curve slope

The theoretical value of the lift curve slope for a thin aerofoil is:-

$$dC_L/d\alpha = 2\pi \quad \textit{per radian} \tag{5.1d}$$

Practical two-dimensional aerofoils have a somewhat higher value but it falls with reduction in aspect ratio, and the introduction of sweep, see paragraphs 5.3.2 and 5.3.4. An approximate value in these circumstances is given by:

$$\frac{dC_L}{d\alpha} = A\Big/\left[(0.32 + 0.16A/\cos\Lambda_{1/4})\{1 - (M_N\cos\Lambda_{1/4})^2\}^{1/2}\right] \tag{5.1e}$$

where A is aspect ratio, see paragraph 5.3.2
M_N is the flight Mach number
$\Lambda_{1/4}$ is the sweep of the quarter chord line

Deployment of high lift devices has only a small effect on the lift curve slope, unless there is a considerable increase of wing area, see paragraph 5.2.3 and Figure 5.1.

5.2.3 High lift devices

5.2.3.1 General comments

A great variety of high lift devices exists, many of which are mechanically complex. As aerodynamic design has progressed there has been a tendency to use mechanically simpler systems in spite of their lower lift capability, both to save mass and to improve reliability. A summary of the more common types is shown in Figure 5.2 and Table 5.1 gives an indication of their lift potential in practical applications, see paragraph 5.2.6.2.

Trailing edge high lift devices are almost universally used but leading edge devices are best avoided where possible. Apart from the additional mechanical complexity they are located in a sensitive position on the aerofoil where small gaps and steps can cause large drag increments. As a guide current design practice suggests that leading edge high lift devices are likely to be necessary when:

$$\frac{(Mg/S)_0}{\cos\Lambda_{1/4}} \geq F_{LE} \tag{5.2}$$

where $\left(\frac{Mg}{S}\right)_0$ is the take-off wing loading in N/m^2, Mg being the weight and S the wing area.

F_{LE} is approximately 5500 N/m^2 for airliners and but may be less than 4000 N/m^2 for combat aircraft depending upon specific conditions, such as a requirement for short take-off and landing.

5.2.3.2 Trailing edge high lift devices

The simplest systems, such as plain and split flaps, change only the camber of the aerofoil. More complex concepts, such as multi-slotted or Fowler flaps, not only change camber but also extend the chord, opening up slots as they do so.

Trailing edge devices rarely occupy less than the aft 20% of the chord or more than the aft 40%. A mean between these two values is typical.

The maximum angle through which a flap is deflected depends to some extent on the type used. However, maximum effective angles of 35 to 45° are typical with higher values only being employed when there is a particular requirement for high drag.

Trailing edge flap operating systems are numerous but may be summarised by:

i) Simple hinges within or on the aerofoil for plain and spilt flaps, but below the aerofoil for slotted flaps to give chord extension.

ii) Linkage systems, usually arranged to produce significant aft movement of slotted and Fowler flaps before the main rotation occurs.

iii) Track systems, where the track shape can be used to control flap movement as is required. Recent devices tend, however, to use simple track shapes, such as circular arcs and straight portions.

iv) Combination of the above, especially for multi-slotted systems. For example the forward segment of a double-slotted flap may move out on a track, while the other segment may be controlled by a linkage or be simply pivoted to the main segment.

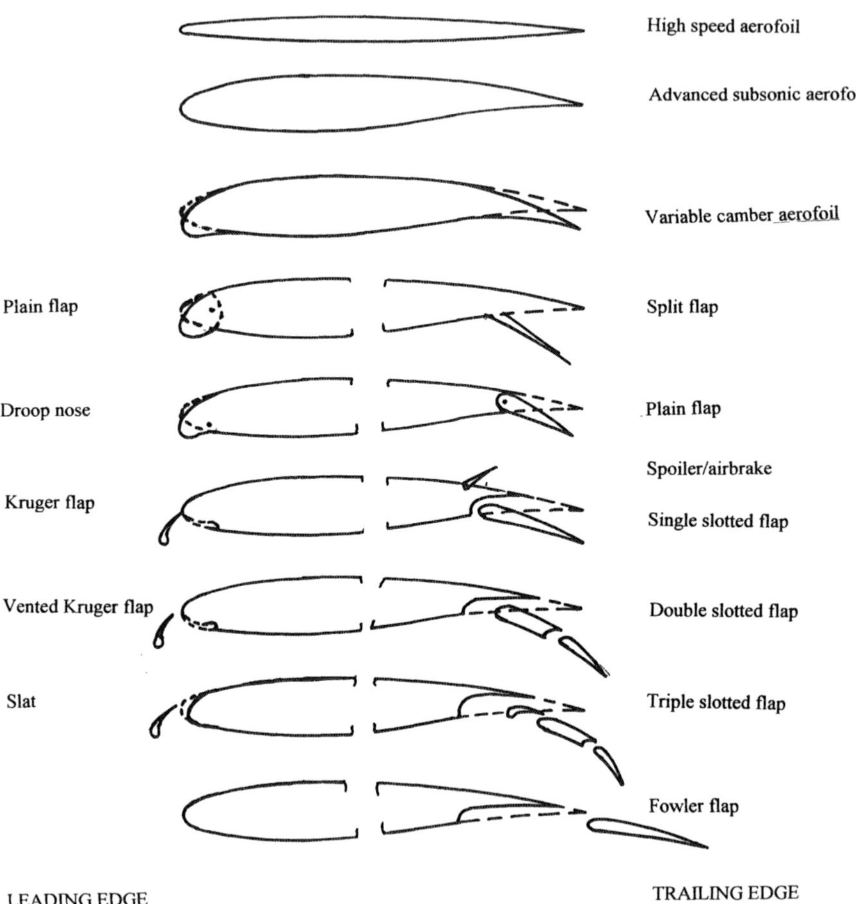

Figure 5.2 Aerofoils and high lift devices

5.2.3.3 Leading edge high lift devices

The variation of geometry obtained from leading edge devices may take one of the basic forms illustrated in Figure 5.2.

a) Hinged nose - plain flap or drooped nose. A plain nose flap is hinged at a point near to the mid-depth of the aerofoil while a drooped nose has a pivot point near to the lower surface so that a smoother upper surface may be achieved. The concept is mechanically simple, and it may be used to locally increase the angle at which the aerofoil stalls. A typical flap deflection angle is 10°.

b) Kruger flap. A nose flap is formed by moving the front lower surface of the aerofoil outwards and forwards about a hinge point located on the aerofoil surface at the leading edge, or slightly behind and above it. When extended the flap forms a forward extension of the upper surface thereby increasing chord as well as camber. The shape of the lower extremity of the Kruger flap is important and the radius required results in a depression being formed in the lower wing surface when the flap is closed. The extended surface shape is also important and sometimes a flexible arrangement is employed to enable better performance to be achieved than would otherwise be possible by using the fixed lower aerofoil shape. It is also possible to locate the pivot point within the aerofoil so that a gap is opened between the Kruger flap and the nose of the basic aerofoil as it extends. This produces a vented Kruger flap.

c) Slat/slot systems. Some aircraft have been designed with a fixed wing slot located just behind the leading edge, but this gives rise to a significant drag penalty in cruise conditions. The usual slat system consists of a section of the leading edge, mainly forming the upper surface of the aerofoil, which is arranged to move forwards and downwards thereby extending the chord, changing the camber and forming a slot. The movement is often based on a circular track. The choice of the centre of rotation is influenced by the need to achieve the optimum gap as the slat opens and in some cases an additional device may be used to rotate the slat relative to the track to facilitate this. The maximum slat rotation angle is likely to be around 30°.

Leading edges devices usually occupy at least 10% of the leading edge of the aerofoil while slat and Kruger flap concepts may extend up to 20%. However, something rather less than this, around 16%, is more typical.

5.2.3.4 High lift device extended positions

Some high lift devices use only one extended position but the majority employ two or more. The main considerations are:

a) Take-off. At take-off the aim is to achieve enhanced lift with the minimum associated drag penalty, so that high lift/drag ratio is the criterion. The emphasis tends to be on chordwise extension associated with small changes in camber. The use of slots requires care as they have a tendency to introduce significant drag penalty. Leading edge devices, such as slats and vented Kruger flaps, may be arranged so that they are sealed in the take-off position for this reason. Typical trailing edge flap deflections for take-off are up to about half of the maximum landing values, whilst in the case of slats the take-off rotation is likely to be around two-thirds of the maximum.

b) Landing. High lift is the overriding consideration in the landing configuration. The associated drag can be tolerated and may be advantageous. However, the case of a baulked landing must be considered either in terms of available thrust or partial retraction of the high lift devices. The limit on the landing flap deflection, referred to in paragraph 5.2.3.2, is likely to be determined either by the need to keep pitching moment to an acceptable value or by mechanical considerations.

5.2.3.5 Variable camber concepts

The aim of so-called "variable camber" or "mission adaptive" wings is to achieve good high lift performance by smooth changes of aerofoil contour as shown in Figure 5.2, rather than by the more conventional use of movable segments. The variable camber is often achieved by combining a flexible drooped nose with a flexible extending trailing edge flap. The latter may, or may not, incorporate a slot system. Experience suggests that the mechanical complexity and mass penalties rarely justify the potential performance gains of the concept although there may be viable applications in special cases.

5.2.4 Wing control surfaces

Conventionally roll control is provided by ailerons which effectively are a pair of differentially moving plain flaps located over the rear 20 to 30% of the chord of the aerofoil. Usually ailerons occupy the outer 25 to 30% of the wing span, although in some designs they are supplemented by short span controls located inboard in the region of power plants and between flap segments. The inboard ailerons are used in high speed flight to avoid possible aeroelastic difficulties associated with aileron reversal. See also Chapter 8, paragraph 8.10.3. Ailerons are sometimes "drooped" symmetrically to augment lift in association with the flaps, and the droop hinge may be offset to introduce a slot.

5.2.5 Spoilers and airbrakes

Hinged flaps located on the upper surface of the aerofoil above the front of the trailing edge flaps are used for various purposes such as:

- i) Roll control at high and low speed by differential operation
- ii) Air brakes at high and low speed
- iii) Lift dumping after touchdown

While their role is important, in some respects they are secondary in terms of determination of wing configuration.

Table 5.1 Summary of high lift device effectiveness

Device	Max. increment in lift coefficient*	
	2- dim. potential	Typical 3- dim. value*
Basic aerofoil - subsonic	1.6	1.50
Basic aerofoil - sharp nose	1.0	0.95
Plain trailing edge flap: 20% chord	0.80	0.55
40% chord	1.10	0.75
Split flap (no gap) (t/c) = 0.15; 20% chord	0.9	0.60
40% chord	1.4	0.95
Single-slotted flap: 20% chord	1.2	0.80
40% chord	1.8	1.20
Double-slotted flap: 40% (+26%) chord	2.5	1.65
Triple-slotted flap: 40% chord overall	2.9	1.90
Fowler flaps: 20% chord	1.2	0.80
40% chord	1.8	1.2
Fowler plus split flap: 40% chord	2.2	1.45
Plain leading edge flap: 15% chord	0.5	0.4
Vented slat: 18% chord	1.0	0.85
Kruger flap: 20% chord	0.8	0.65
Vented Kruger flap: 20% chord	1.0	0.85

* Notes: 1) Typical 3-dimension values are for moderate to high aspect ratio unswept wings, allowing for part span effects. Multiply by ($cos\Lambda_{1/4}$) for swept case. A typical corresponding value for plain/single-slotted flaps on a low aspect ratio wing is 0.25.

2) Take-off lift coefficient values are usually 50 to 60% of maximum (landing) values quoted, see Chapter 6, paragraph 6.2.4.4.

5.2.6 Three-dimensional and part span effects

5.2.6.1 Three-dimensional flow

Two-dimensional lift values are not obtained on a practical wing of finite span especially when it is swept. The combination of finite aspect ratio, sweep and taper of the planform causes spanwise flow interactions which modify the effective angle of attacks of local chordwise sections. On unswept and aft swept wings this gives rise to a tendency to higher lift coefficients outboard resulting in the possibility of tip stall generally and nose-up pitch when sweep is present. Incorporation of some means of reducing the local angles of attack outboard relative to the root can overcome this problem. This may be done by a leading edge device, such as a droop nose, or by built-in geometric properties. If the latter approach is used it is called "wash out" and is typically equivalent to about 2° nose down twist at the tip.

When the wing is swept forward the spanwise flow effect is reversed and tends to give rise to the possibility of root stall. However, the pitch-up tendency is still present and must be dealt with.

The consequence of a finite wing, therefore, is that the achieved maximum lift coefficient is less than the two-dimensional value, both due to tip losses and spanwise angle of attack variations.

5.2.6.2 Part span effects

Layout considerations imply that leading and trailing edge high lift devices cannot occupy all of the actual wing span. There are therefore further reductions of lift relative to the two-dimensional case.

Leading edge devices are not full span because of:

i) Presence of fuselage.

ii) Shape of wing tip required for good cruise performance which restricts the outboard extremity of the slat.

iii) Possible limitations in the region of engine pylons, although careful design can minimise the losses here.

Overall it is likely that the effective three-dimensional maximum lift coefficient of a leading edge high lift device will be no more than:

$$(\Delta C_L)_{3D} / (\Delta C_L)_{2D} = 0.85 \cos \Lambda_{1/4} \tag{5.3a}$$

where $\Lambda_{1/4}$ is the wing quarter chord sweep

As far as trailing edge devices are concerned the limitations are:

1) Presence of fuselage.
2) Provision for ailerons for roll control.
3) Possible interruptions in regions of powerplants.

Unless special provisions are made for three-dimensional drooped ailerons the effectiveness of a trailing edge high lift device can be about:

$$(\Delta C_L)_{3D} / (\Delta C_L)_{2D} = 0.67 \cos\Lambda_{1/4} \tag{5.3b}$$

These practical three-dimensional limitations for high lift devices are included in the numerical values quoted in Table 5.1.

5.3 Planform shape and geometry

5.3.1 General comments

The performance of the high lift devices discussed in paragraph 5.2 is influenced by the geometry of the wing planform but in practice that geometry is primarily determined by high speed rather than low speed flight considerations. The three primary planform parameters are aspect ratio, taper ratio and sweep. Cranked trailing edges and, less frequently, cranked leading edges may be used for detail layout reasons but these are secondary to the primary parameters.

5.3.2 Aspect ratio, A, and wing span, b

Aspect ratio is defined as the square of the wing span, b, divided by the wing reference area, S. The aerodynamic trend is towards high aspect ratio since this is most efficient in reducing the inevitable drag due to lift of a finite wing. However, higher aspect ratio implies higher structural mass so that a compromise is necessary with due consideration given to aerofoil and other geometric parameters. In practice there is a wide variation of aspect ratio across different classes of aircraft. It may be as low as 1.5 for a highly swept, supersonic wing to more than 20 for a high performance sailplane. Typical values for subsonic design are generally in the range 5 to 10.

There may be a limitation on the maximum value of wing span in some cases, for example naval aircraft. Airlines have to be able to operate into airport gates of limited width and typical values of span for different classes of transport aircraft are given in Table 5.2.

Table 5.2 Typical airliner wing span limits

Range category	Limiting wing span	
	m	ft
Commuter/Regional	20-21.5	66-70
Narrow body:		
Short haul	28.5	93
Short/med. haul*	34	112
Wide body:		
Med./long haul	50	165
Long haul	61	200
Ultra high capacity	77-80	250-262

*Boeing B757 is an exception with a wing span of 38.1 m (125 ft)

5.3.3 Taper ratio, λ

The taper ratio is the ratio of a nominal tip chord to that at the centreline, or possibly the fuselage side. Both chords are measured in the streamwise direction. Taper ratio is chosen primarily to give a near to semi-elliptical spanwise airload distribution although a high taper (low value of taper ratio) is structurally beneficial. Some rudimentary designs use an untapered wing for simplicity of construction but usual taper ratio values to give spanwise semi-elliptic loading lie in the range of 0.1 to 0.6 with the higher values being found on higher aspect ratio, unswept configurations. An unduly highly tapered wing may result in high local lift loading at the tip and a consequent tip stalling tendency, see paragraph 5.2.6.1. Reduction of tip chord may also be limited by the requirement of adequate chord for the roll controls (ailerons). As a guide it is suggested that for initial studies the taper ratio, λ, should not be less than:

$$0.2\, A^{1/4} \cos^2\Lambda_{1/4} \tag{5.4}$$

5.3.4 Sweep

5.3.4.1 General remarks (see also Chapter 2, paragraph 2.2.3.2)
Sweep may be forward or back and is most commonly defined along the 25% chord line. Increase of sweep raises the critical Mach number and also reduces the value of the peak wave or compressibility drag. Sweep back is much more common, primarily for reasons of layout and stability. Swept forward wings are prone to aeroelastic divergence. As a general rule sweep angles should be as low as possible for a given design flight condition and aerofoil configuration, since sweep implies both structural and possible handling penalties. Sometimes a moderate sweep is incorporated in the wing for layout reasons. For example some sweepback associated with an unswept trailing edge over the wing root region can be used to provide stowage volume for a retracted landing gear.

5.3.4.2 Effect of sweep at high subsonic Mach number

Sweep angles vary from nominally zero for relatively slow flying aircraft to 60° or more for delta and variable sweep wing configurations. The effect of sweep in raising the critical Mach number at high subsonic speed is approximately:

$$(M_{NCRIT})_{3D} / (M_{NCRIT})_{2D} = \frac{1}{(\cos\Lambda_{1/4})^{1/2}} \quad (5.5a)$$

$$for\ 0° \le \Lambda_{1/4} \le 35°$$

where $(M_{CRIT})_{2D}$ is the two-dimensional value derived from Eq (5.1).

At sweep angles greater than 35° there is an increasing benefit from sweep so that:

$$(M_{NCRIT})_{3D} / (M_{NCRIT})_{2D} = \frac{1}{(\cos\Lambda_{1/4})^{0.6}} \quad (5.5b)$$

$$for\ \Lambda_{1/4} > 45°$$

with appropriate variation for $35° < \Lambda_{1/4} < 45°$.

Figure 5.3 summarises Eq (5.5) while Figures 5.4 and 5.5 show the result of combining Eq (5.5) and Eq (5.1c).

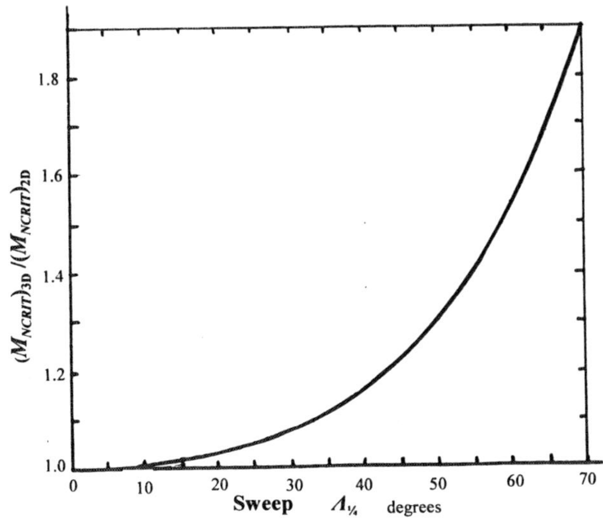

Figure 5.3 Effect of sweep on critical subsonic Mach number

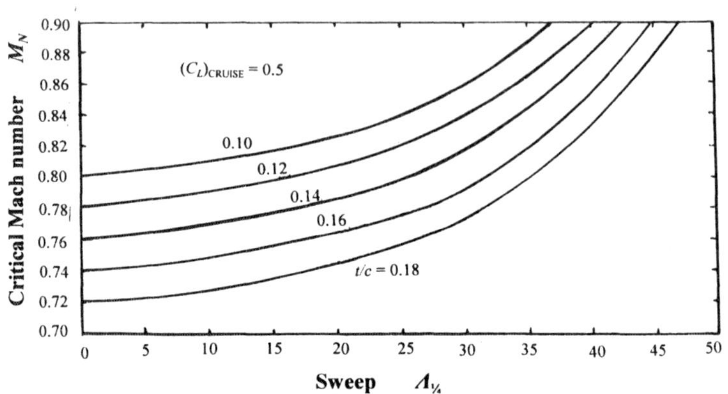

Figure 5.4 Typical subsonic airliner sweep requirement

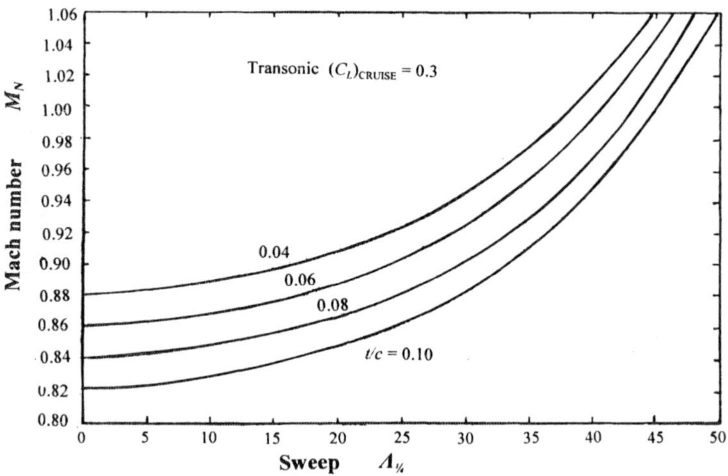

Figure 5.5 Combat aircraft sweep requirement - transonic cruise

5.3.4.3 Effect of sweep at supersonic speeds

a) Subsonic leading edge. There is advantage in retaining a subsonic leading edge even when an aircraft is flying at supersonic speed. As a first approximation the increment in zero lift drag of a wing in these conditions is close to a minimum when:

$$\left(M_N^2 - 1\right)^{1/2} \cot\Lambda_{LE} = 0.8$$
$$for\ 1 \le A\left(M_N^2 - 1\right)^{1/2} \le 4\ and\ (t/c) \le 0.06 \tag{5.6a}$$

where Λ_{LE} is leading edge sweep.
This condition, but not the actual drag, is thus approximately independent of thickness to chord ratio. Equation (5.6a) may be used to predict the leading edge sweep necessary to achieve near to minimum zero lift drag:

$$\Lambda_{LE} = \cos^{-1}(1/M_N) + \Delta\Lambda_{LE} \tag{5.6b}$$

where in the range $1.1 < M_N < 3.0$, $\Delta\Lambda_{LE}$ is found to vary from about 5.2 to 6.4°. Thus approximately for usual supersonic flight Mach number:

$$\Lambda_{LE} = \cos^{-1}(1/M_N) + 6° \tag{5.6c}$$

required Λ_{LE} is shown in Figure 5.6.

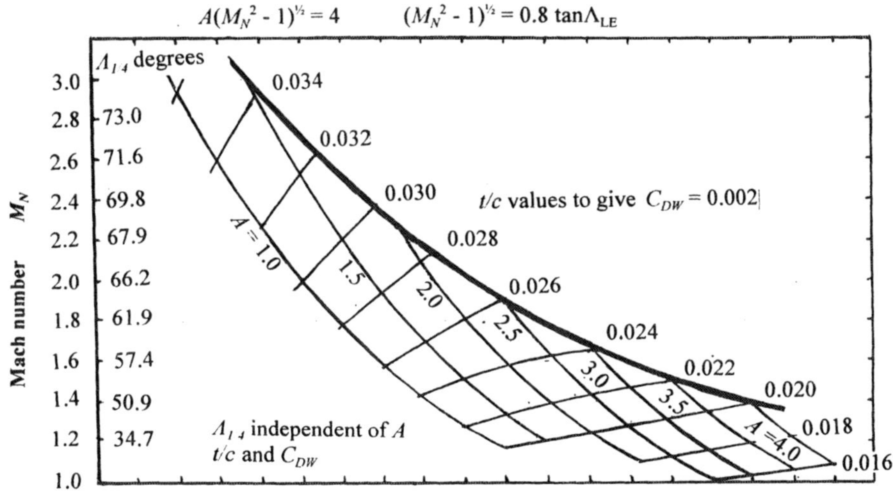

Figure 5.6 Conditions for subsonic leading edge in supersonic flight

b) Interaction of thickness to chord and aspect ratio. Also shown in Figure 5.6 is the implied aspect ratio limitation for Eq (5.6a) to be valid and combinations of parameters required to give a wave drag coefficient, C_{DW}, of 0.002. The aerofoil wave drag is proportional to $(t/c)^2$ so there is considerable merit in keeping thickness/chord ratio as low as possible within structural considerations. Increase of aspect ratio also increases aerofoil wave drag, but the need to keep drag due to lift to a low value offsets this. Generally it is suggested that the aspect ratio is likely to be close to the limiting value given in Figure 5.6.

c) Effect of sweep at supersonic speed - supersonic leading edge. High compressibility drag is associated with a just sonic leading edge. For this reason the sweep of the leading edge should be somewhat less than $cos^{-1}(1/M_N)$, where M_N is the supersonic design Mach number. There is likely to be aerodynamic advantage in employing a sweep near to the maximum tolerable value to reduce the compressibility drag penalty in transonic conditions, although a sweep angle significantly less than $cos^{-1}(1/M_N)$ is a possibility.

5.3.5 Typical wing planform geometry

Table 5.3 summarises typical wing geometry parameters over a range of operating flight regimes.

5.3.6 Dihedral - (lateral stability of aircraft)

Dihedral, or anhedral, of the wing may be determined by one, or both, of two primary considerations:

a) Natural lateral static stability which requires that the rolling tendency due to sideslip should be negative. However, an unduly large negative value may well adversely affect dynamic stability characteristics. This is discussed more fully in Addendum 3.

b) Layout requirements, such as the need for ground clearance for wing-mounted powerplants and stores, or fuel system design.

The following are the major contributions to positive lateral static stability:

- i) Wing position vertically on the fuselage, which is beneficial when the wing is located above the centre of gravity
- ii) Sweep back of the wing.
- iii) Dihedral.

From the point of view of stability the following may be used as an initial guide to the desirable dihedral/anhedral:

A) Low wing, unswept: Dihedral 3 to 5°
B) Low wing, swept back: Dihedral about 3° at 30° aft sweep
C) High wing, unswept: No dihedral
D) High wing, swept back: Anhedral of about minus 3° at 30° aft sweep (increasing somewhat at higher aft sweep)

Dihedral or anhedral of a horizontal tail is sometimes used. This may contribute to directional stability but the main reasons are usually layout or longitudinal stability.

Table 5.3 Typical wing geometry parameters

	Flight regime			
Parameter	$M_N \leq 0.65$	$0.65 \geq M_N \leq 0.95$	$M_N \geq 0.95$ subsonic LE	Supersonic LE
Sweep; $\Lambda_{1/4}$ (Greater sweep may be used for layout reasons)	0	$\cos^{-1}\left(\frac{0.95-0.1C_L-{}^t/_c}{M_N}\right)^2$ $\Lambda_{1/4} \leq 35°$	$\cos^{-1}\left(\frac{1}{M_N}\right)+6°$	Not more than $\cos^{-1}\left(\frac{1}{M_N}\right)-6°$
Aspect ratio, A	Short range 5-7 Long range 10-12	Combat type 4-6 Transport, etc. 7-10	1.5-3.0	2-4
Taper ratio, λ	0.5-0.6	0.2-0.3	0.1	0.2-0.4
Thickness/chord ratio (Root) $(t/c)_R$	0.15-0.20	0.10-0.15	>0.06	0.02-0.03
Thickness/chord ratio (Tip) $(t/c)_T$	65% root value	65% root value	Root value	Root value

5.3.7 Winglets

Wing tip fins, or winglets, have the effect of reducing drag due to lift and hence effectively increasing aspect ratio. Approximately the effective increase of aspect ratio is about half that of adding the tip fin height directly to the span. Winglets are of use when an existing design is being developed, or when there is an operational wing span limit.

5.4 Interaction between aerodynamic, structural and wing volume considerations

5.4.1 Spar positions and fuel volume

Apart from some light aircraft the structure of a wing consists of a spanwise box which is defined in the chordwise sense by front and rear spars. The chordwise positions of these spars follow from the definition of the leading and trailing edge high lift devices, controls and possibly other considerations such as the retracted landing gear. The front spar is located just aft of the rear edge of the leading edge devices when they are used. In any case it has a typical position at 10 to 20% of the chord. Likewise the rear spar is located just ahead of the trailing edge devices, typically around 60 to 70% of the chord. In most cases the volume within the spanwise box is used for fuel tankage. To give development potential this volume should always be maximised, even when the predicted fuel mass to meet the initial specification can be readily accommodated. In the case of a long range airliner the fuel volume required may suggest that it is the parameter which determines overall wing area. This may not be an optimum solution and the volume required may be better achieved by increase of thickness/chord ratio with corresponding increase of sweep.

The actual volume available for fuel does depend upon numerous geometric and system details, but for a typical airliner the maximum available wing fuel mass is approximately given by:

$$420bS(t/c)(1 - 0.89\lambda + 0.49\lambda^2)/A \quad kg \tag{5.7}$$

5.4.2 Structural mass

Low wing mass is associated with low values of aspect ratio and sweep and high thickness to chord ratio and taper (low taper ratio). There is thus a direct conflict between aerodynamic and structural requirements. Design practice suggests that for the majority of aircraft a structural parameter, SP, can be established to indicate likely structural limitations.

$$SP \leq sec\Lambda_E \left[\frac{NA^{1.25}}{(t/c)^{0.5}}\right]^{0.5} \tag{5.8a}$$

where Λ_E is the effective structural sweep, usually approximately $\Lambda_{1/4}$ but it may be less on a highly swept, low aspect ratio wing

N is the ultimate (factored) normal acceleration factor, being the greater of that due to manoeuvre and discrete gust encounter. See Addendum 4.

Typical values of SP are:

Executive jets	SP = 12 to13
Subsonic military strike and trainer aircraft	SP = 18
Supersonic military strike/interception aircraft	SP = 18 to 20
Long range supersonic, excluding strike, aircraft	SP = 10
High performance sailplanes	SP > 30
All other types	SP = 15 to 16

The value of SP should not be used as an absolute design constraint, but it is useful in establishing a set of parameters for detailed investigation. Inevitably there are special cases where the value of SP is greater or less than typical values. This possibility is illustrated by long range supersonic aircraft where the need for a large fuel fraction is reflected in a comparatively low value of wing mass, and hence unusually low value of SP.

The value of N can usually be assumed within reasonably close limits for a given design so that Eq (5.8a) may be used to correlate probable values of sweep, thickness/chord ratio and aspect ratio. Taper ratio is less significant and the assumption of a typical value is adequate for initial purposes.

It is sometimes convenient to rearrange Eq (5.8a) in the form:

$$A = \frac{(SP\cos\Lambda_{1/4})^{1.6}}{N^{0.8}}\left(\frac{t}{c}\right)^{0.4} \tag{5.8b}$$

5.5 Wing area (wing loading)

5.5.1 General comments

The overall size of the wing defined as the reference area, S, is to some extent dependent upon the geometric parameters. However, it is primarily determined by the performance requirements of the aircraft and the corresponding available thrust. The latter is conveniently defined non-dimensionally as thrust/weight ratio, (T/Mg), which for a given powerplant configuration can be related to a datum sea level static value, $(T/Mg)_0$, see Chapter 3. Although the powerplant potential of propeller driven aircraft is usually expressed in terms of power, the thrust available at any given flight condition follows directly from the power as a function of forward speed and propeller characteristics. The relationship between power and thrust is also covered in Chapter 3, paragraphs 3.6.2.4 and 3.6.2.5.

It is convenient in performance analysis to express the wing area in terms of wing loading, (Mg/S), which may also be related to an initial take-off value, $(Mg/S)_0$.

An important factor in the performance of an aircraft is the lift/drag ratio, (L/D). For a given geometry and high lift configuration the lift /drag ratio may be taken as the sum of two terms one of which is directly proportional to wing loading and the other inversely proportional to it, see Chapter 6, paragraph 6.3.1. All other things being constant, the maximum lift/drag ratio corresponds to the condition where the two terms have equal value.

5.5.2 Influence of performance requirements

5.5.2.1 General comments

The performance requirements are partly defined by the operating conditions given in the particular specification for the aircraft and partly by more general airworthiness requirements. For a given selection of high lift devices, wing planform geometry, fuselage layout and general aircraft configuration the performance criteria can be discussed in relation to the flight profile. Some, or all, of the following general flight modes, which are treated more fully in Chapter 7, may be relevant.

5.5.2.2 Take-off: runway length to achieve screen height

The take-off distance is primarily determined by the static thrust/weight ratio divided by the product of wing loading and the lift coefficient at rotation. There is thus a direct relationship between thrust/weight ratio and wing loading for a given unstick lift coefficient which may be assumed constant within the assumptions stated above. The further parameter of the braking deceleration is involved in the determination of the accelerate-stop distance when an engine failure occurs. For simplicity it is adequate to assume a balanced field length for initial synthesis, the justification for which must be checked subsequently.

5.5.2.3 Critical engine failed climb out: second segment climb

Experience suggests that the second segment climb is most likely to be the critical engine failed climb condition, where this requirement is applicable. The specified minimum climb gradient may be expressed in terms of the relevant thrust/weight ratio and lift/drag ratio with the landing gear retracted and high lift devices in the take-off setting. The speed is defined as a factor of the stalling speed which is also a function of wing loading for a given lift characteristic.

5.5.2.4 Normal climb to altitude: ceiling

While the high lift devices are usually retracted the rate of climb to operational attitude depends upon the thrust/weight ratio and wing loading in the same way as the second segment climb. It is convenient to express the rate of climb in a form which directly relates thrust to weight ratio, wing loading and the assumed drag characteristics. Fuel used in climb is established directly by powerplant performance. There must alway be some residual climb rate at operating attitude, this being related to the service ceiling.

5.5.2.5 Cruise

The distance covered in cruise is the product of three items:

a) A propulsion term which is the ratio of flight speed (or Mach number) to specific fuel consumption, and is thus defined for a given requirement, powerplant and flight condition.
b) The aerodynamic term which is the lift to drag ratio.

c) A mass term which effectively relates the fuel mass available for cruise to the mass of the aircraft at the end of the cruise.

Thus for a given cruise speed and powerplant the cruise performance depends upon the wing loading as it influences the lift/drag ratio and mass of the aircraft. Diversion and stand-off flight which give rise to reserve fuel requirements are a form of cruise which is often off-design.

5.5.2.6 Maximum level speed
Achievement of a maximum level speed may be a requirement for some aircraft, especially combat types. It is simply the level flight case of thrust to weight ratio being equal to the drag to weight ratio, the latter being a function of wing loading.

5.5.2.7 Sustained manoeuvre
Sustained manoeuvre is another thrust equal to drag case except here the drag is greater than in steady level flight due to the higher lift demanded by the manoeuvre. As with climb and maximum level speed there is a relation between thrust to drag ratio and wing loading for a given drag characteristic.

5.5.2.8 Instantaneous manoeuvre
The instantaneous manoeuvre is determined either by structural limits or by the maximum available lift coefficient. It is thus independent of thrust but does depend upon wing loading.

5.5.2.9 Specific excess power
A given point performance specific excess power is identical to the potential rate of climb at that point and so the comments made in paragraph 5.5.2.4 apply. Essentially it is a function of the difference between thrust and drag as a ratio of the weight.

5.5.2.10 Transonic acceleration
The time taken and fuel used during the transonic acceleration of a supersonic aircraft may be an important design requirement. The case is analogous to that of specific excess power in that the critical factor is the difference between thrust and drag. However, the analysis covers a range of Mach numbers over which the drag varies considerably. For a given drag the important criteria are still thrust/weight ratio and wing loading.

5.5.2.11 Descent
The time and fuel used during descent from operating altitude to landing approach are not usually of great concern at the initial design stage. When a high rate of descent is required, as may sometimes be the case, secondary devices such as air brakes must be provided.

5.5.2.12 Landing approach and touch down: landing runway length

For a given high lift configuration the wing loading is the only criterion which determines the approach and touch down speeds. Apart from a vertical landing concept it always places a limit on the maximum acceptable wing loading, and in many cases it indirectly determines the take-off wing loading.

The rest of the landing distance is determined by the achievable braking deceleration.

5.5.2.13 Baulked landing

It is usual to consider the performance requirements when a landing is rejected at a late stage in the approach. This condition may have an impact upon the thrust/weight ratio as the aircraft has to accelerate with an initially high drag setting of the high lift devices.

5.5.3 Sensitivity to turbulence in cruise

In the case of passenger transport, executive aircraft and some other types it is important to ensure that the aircraft is not unduly sensitive to atmospheric turbulence. Based on simple discrete gust analysis it may be assumed that the sensitivity to turbulence is more or less directly proportional to:

$$\frac{Va_1}{(Mg/S)} \tag{5.9a}$$

where V is the cruise speed
a_1 is the lift curve slope at the cruise condition
(Mg/S) is the appropriate wing loading

Analysis of aircraft considered to be satisfactory in this respect suggests that:

$$(Mg/S)_0 \geq \frac{2.7V_D A}{(0.32 + 0.16A/\cos\Lambda_{1/4})\{1 - (M_N \cos\Lambda_{1/4})^2\}^{1/2}} \tag{5.9b}$$

where $(Mg/S)_0$ is take-off wing loading (N/m^2)
M_N is maximum cruise Mach number
A is aspect ratio and $\Lambda_{1/4}$ is sweep of 0.25 chordline, as previously
V_D is the structural design speed, see Chapter 7, paragraph 7.2.2

This condition may provide a lower limit to the wing loading for a given wing geometry. A ride control system can be used to overcome this problem.

5.5.4 Practical range of wing loading

Wing loadings used for aircraft design vary over a wide range from as low as 200 N/m to as high as 8000 N/m. In spite of this wide overall variation there is a close correlation between individual aircraft of a given type. Table 5.4 illustrates this.

5.5.5 Determination of wing loading and thrust to weight ratio

For a given overall configuration the various performance requirements as defined in paragraph 5.5.2 may be analysed to give a number of curves, which individually show the wing loadings and corresponding thrust/weight ratios to meet a given requirement. The process may be repeated for other configurations, such as alternative wing geometry or high lift devices, in order to obtain other matches of the two primary criteria. It must be noted, however, that the changes in geometric and other parameters imply mass differences which are not accounted for except in as much as the structural parameter of Eq. (5.7) may have been used. A preferable approach is to complete the evaluation of aircraft mass for a given configuration and make comparison between different configurations at this level. The procedure is outlined in Chapter 8.

Table 5.4 Typical wing loading

Class of aircraft	Range of wing loading N/m ^{2}
Ultra light aircraft	200-400
Light single piston engine aircraft	500-800
General aviation single turboprop aircraft	1000-1800
General aviation twin piston engine aircraft	1000-2000
Small turboprop commuter aircraft	1500-2000
Large turboprop commuter aircraft	2000-3000
Small executive jets	2200
Medium executive jets	3000
Large executive jets	4000
Military jet trainers	2500-3000
Turbopropeller transports	3000-4000
Naval strike/interceptor aircraft*	3500-4000
Land based strike/interceptor aircraft*	4000-5000
Supersonic long range bomber and transport aircraft	5000
Subsonic long range bomber aircraft	5000-6000
Short/medium haul jet transport aircraft	5500-6500+
Long haul jet transport aircraft	6200-7000+
Naval strike/interceptor aircraft*	3500-4000
Land based strike/interceptor aircraft*	4000-5000
Supersonic long range bomber and transport aircraft	5000
Subsonic long range bomber aircraft	5000-6000
Short/medium haul jet transport aircraft	5500-6500+
Long haul jet transport aircraft	6200-7000+

* Swing wing aircraft have a value of about 1000 N/m^2 higher than comparable fixed wing aircraft while incorporation of stealth features on a major scale reduces the value by around 1000 N/m^2.

Chapter 6

Basic lift, drag and mass representations

6.1 Introduction

The representations of lift, drag and mass are of fundamental importance to the design synthesis process. Inevitably an accurate evaluation of them demands a knowledge of many details of a design which are not available at the initial synthesis stage. For these reasons the need to make some assumptions is unavoidable. Certain significant characteristics must be retained as parameters to be optimised during the process. However, it is desirable to restrict the number of these to an absolute minimum by utilising representations which make the fullest possible use of past experience.

Aerodynamic characteristics may be critically dependent upon the Reynolds number which is defined as the product of a characteristic length and the true velocity divided by the kinematic viscosity. In the following representations for lift and drag there are certain implied allowances for Reynolds number effects

6.2 Lift

6.2.1 Introduction

Some aspects of the lift characteristics associated with various wing configurations are discussed in Chapter 5. The assumptions which follow are consistent with these characteristics.

As stated above for the purposes of preliminary design, intended to lead to an initial synthesis of an aircraft configuration, it is desirable to reduce the input information to as little as possible. With this in mind the estimation of lift characteristics is limited to the following phases of the flight:

a) Take-off, unstick lift coefficient and value appropriate to initial climb out.

b) Cruise, lift coefficient as limited by buffet, Mach number effects and climb ceiling.

c) Manoeuvre, maximum available lift coefficient for combat aircraft and related types.

d) Landing, approach lift coefficient based on the maximum achievable value with the high lift devices fully deployed.

6.2.2 Aircraft configurations

It is convenient to simplify the lift representation by relating it to just two basic aircraft configurations:

a) Subsonic and transonic aircraft having moderate to high aspect ratio wings. Usually the sweep is less than 45° and the mean thickness to chord ratio is greater than 0.08.

b) Transonic and supersonic aircraft having wings of aspect ratio less than about 4.5. In most cases the wing is swept and the thickness to chord ratio less than 0.075.

6.2.3 Initial assumptions

In addition to the selection of the basic aircraft configuration the prediction of lift coefficients requires the knowledge of, or the assumption of, some or all of the following:

i) Sweep
ii) Type of high lift system
iii) Aspect ratio
iv) Mach number

Clearly these are all important parameters in the refinement of the design. While Chapter 5 gives some indication of typical values which may be used for initial work, once the basic layout has been established it is necessary to vary them in combination with others to derive an optimum solution. This procedure is outlined in Chapter 8.

6.2.4 Moderate to high aspect ratio wing configurations ($A \geq 5$)

6.2.4.1 General comments

Aircraft in this category almost invariably use some form of trailing edge high lift system. Leading edge high lift devices are likely to be used on transport and related types when:

$$\frac{M_0 g}{S \cos\Lambda_{1/4}} \geq F_{LE} \qquad (6.1)$$

where M_0 is the gross mass (kg); S is the wing area (m^2) and $\Lambda_{1/4}$ is the sweep of the 25% chord line.

Leading edge high lift devices may be used on combat aircraft to confer both reduced field length requirements and improve combat manoeuvre. Values of F_{LE} are suggested in Chapter 5, Eq (5.2).

The performance of high lift devices depends upon both the type used and the detail geometry of the layout employed. However, for initial work typical average values may be assumed based on the expectation that trim effects fall within the tolerances of the predicted coefficients.

6.2.4.2 Maximum lift coefficient

For this category of aircraft the maximum lift coefficient may be approximately represented by:

$$C_{LMAX} = (1.5 + \Delta_{LEL} + \Delta_{TEL})\cos\Lambda_{1/4} \tag{6.2}$$

where Δ_{LEL} is the increment due to the deployment of a leading edge high lift device in the landing setting. An average value may be taken as 0.65, but, of course, Δ_{LEL} is zero when leading edge devices are absent.

Δ_{TEL} is the increment due to the deployment of the trailing edge flaps to the landing setting. Typical practical three-dimensional values are given in Table 6.1.

The choice of 1.5 as the maximum value of the basic aerofoil lift coefficient is typical. However, higher values, up to around 2.0, are possible when a substantial camber is used. Such aerofoils have application to simple aircraft designs where it is desirable to avoid the use of variable geometry high lift devices, but their more general application is limited.

6.2.4.3 Landing approach conditions

For many aircraft the approach is made at a speed which is 1.3 times the stalling speed with the high lift devices in the landing configuration.

Thus the approach lift coefficient is:

$$C_{La} = 0.6(1.5 + \Delta_{LEL} + \Delta_{TEL})\cos\Lambda_{1/4} \tag{6.3}$$

Somewhat higher approach lift coefficients are possible if a lower approach speed can be accepted. For example the 0.6 factor in Eq (6.3) increases to nearly 0.7 when the approach is made at 1.2 times the stalling speed. This may be acceptable in some circumstances. When propeller slipstream effects are present over the wing, the value given by Eq (6.3) may be increased somewhat, say by 10 to 15%.

Table 6.1 Moderate to high aspect ratio wings

Typical lift increments from trailing edge flaps; Δ_{TEL} and Δ_{LEL}

FLAP TYPE	LIFT COEFFICIENT INCREMENTS*	
	Take-off (unstick) Δ_{TET}	Landing (approach) Δ_{TEL}
Plain	0.3	0.6
Single slotted	0.5	1.0
Double slotted and Fowler	0.7	1.35
Triple slotted	0.8	1.55

* Allowance is made for typical part span effects

6.2.4.4 Take-off condition - unstick lift coefficient

The unstick lift coefficient may be dependent upon the angle of nose-up rotation allowed by the geometry of the rear fuselage, but assuming that this is not a limitation and that the unstick speed is 1.1 times the relevant stalling speed, the unstick lift coefficient is:

$$C_{LUS} = 0.8(1.5 + \Delta_{LET} + \Delta_{TET})cos\Lambda_{1/4} \tag{6.4}$$

where Δ_{LET} can be taken as 0.4 when leading edge devices are fitted and set at their take-off position, but zero when they are not used.

Δ_{TET} is given in Table 6.1 for the trailing edge flaps at the take-off setting. Propeller slipstream effect may increase the value given by Eq (6.4) by some 15 to 20%.

6.2.4.5 Combat aircraft manoeuvre condition

Assuming that the high lift devices are not deployed in manoeuvres, the maximum lift coefficient with a margin over the stall for an instantaneous manoeuvre case may be taken as:

$$C_{LMAN} = 1.35cos\Lambda_{1/4} \tag{6.5}$$

6.2.4.6 Buffet limited cruise condition

The lift coefficient which may be assumed for cruise is limited by buffet considerations. Allowing a typical margin and assuming high lift devices are not extended the maximum cruise lift coefficient may be taken as:

$$C_{LC} = 0.65\cos\Lambda_{1/4} \tag{6.6}$$

Somewhat higher values may be possible with highly cambered aerofoils.

6.2.4.7 Climb ceiling condition

The climb ceiling may be limited by the usable lift coefficient. It is suggested that for initial calculations the value taken should not exceed 125% of that given by Eq (6.6).

6.2.5 Low aspect ratio wing configurations (A < 4.5)

6.2.5.1 General comments

Unlike the case of the moderate to high aspect ratio wing the lift characteristics of low aspect ratio wings are more critically dependent upon the aircraft configuration and other wing geometry parameters. The following particular conditions may be identified:

a) Low aspect ratio swept wing or delta wing in isolation.

b) Leading edge extension (LEX) on swept or delta wing.

c) Delta wing with cambered or variable geometry leading edge.

d) Low aspect ratio swept wing or delta wing with a close coupled canard.

e) Low aspect ratio or swept wing with cambered leading edge and a close coupled canard.

6.2.5.2 Maximum lift coefficient - combat manoeuvre

The maximum lift coefficient of a low aspect ratio wing is achieved at a high angle of attack. This might typically be about 30° for a wing having an aspect ratio of two.

For high speed flight where the maximum lift coefficient is important for manoeuvres the following relationship may be used for aspect ratio, $A < 4.5$:

$$C_{LH} = \left[\left(\frac{A}{2}\right)^{1/2}\cos\Lambda_{1/4} + \Delta_H\right](1 - 0.25M_N) \tag{6.7}$$

where Δ_H is an increment in lift coefficient appropriate to a given configuration as shown in Table 6.2.

6.2.5.3 Maximum lift coefficient - low speed flight

Usable lift coefficient at low speed is likely to be limited by restrictions of aircraft geometry and pilot vision.

For low speed flight during take-off and landing the usable angle of attack is likely to be around 12 to 14°, and this enables a datum maximum low speed lift coefficient to be established based on an angle of 20° when allowance is made for speed margins. This low speed datum lift coefficient, which is equivalent to a low speed maximum value, is approximately, for $1.5 \leq A \leq 4.5$:

$$C_{LL} = \left[\left(\frac{A}{2}\right)^{1/2} \cos\Lambda_{1/4} + \Delta_L\right] \tag{6.8}$$

where Δ_L is the configuration increment given in Table 6.2.

6.2.5.4 Take-off condition - unstick lift coefficient

Assuming lift off at 1.15 times the limiting condition for this configuration:

$$C_{LUS} = 0.75C_{LL} \tag{6.9}$$

6.2.5.5 Landing approach condition

For an approach at 1.2 times the limiting speed condition:

$$C_{La} = 0.7C_{LL} \tag{6.10}$$

Table 6.2 Low aspect ratio wings

Lift increments due to configuration Δ_L and Δ_H

CONFIGURATION BASIC WING PLUS (see paragraph 6.2.5.1)	LIFT COEFFICIENT INCREMENTS	
	Low speed Δ_L	High speed Δ_H
Leading edge extension (LEX)(b)	0.3	0.6
Variable geometry leading edge (c)	0.4	0.8
Close coupled canard (d)	0.5	0.8
Variable geometry leading edge and close coupled canard (e)	0.6	1.0

6.2.5.6 Cruise

In those cases where the buffet limited lift coefficient in cruise flight is of significance, it is suggested that performance analysis should be based on a limiting lift coefficient of:

$$C_{LC} = 0.4C_{LH} \tag{6.11}$$

6.3 Drag

6.3.1 Introduction

As is the case with initial lift evaluation it is helpful to reduce to a minimum the number of parameters needed to predict drag. Performance evaluation requires a knowledge of the drag characteristics of the aircraft in its cruise configuration over much of the speed range. In addition the aircraft drag with the high lift system at the take-off setting is needed to evaluate engine failed climb out performance, where this is relevant.

It is convenient to deal with drag prediction by distinguishing between the same two basic aircraft configurations as those defined in paragraph 6.2.2.

Total drag is made up of four components, which expressed in coefficient form are:

a) *Drag due to the shape and surface friction of the aircraft in incompressible flow conditions,* which is conveniently combined with:

b) *Compressibility wave drag due to the volume of the aircraft* to give a total zero lift drag, C_{DZ}.

c) *Vortex drag, or lift induced drag, in incompressible flow* which is conveniently combined with:

d) *Wave drag due to lift,* to give the total drag due to lift, C_{DI}.

Thus:

$$C_D = C_{DZ} + C_{DI} \tag{6.12a}$$

In practice it is found that C_{DI} is approximately a function of C_L^2 and conventionally the total drag coefficient is written as:

$$C_D = C_{DZ} + K_V\, C_L^2 \tag{6.12b}$$

where K_V is called the induced drag factor, and is the term in square brackets in Eqs (6.14) and (6.18).

K_V is often assumed to be a constant in a given flight condition but this can give erroneous drag values at both very high and very low values of lift coefficient. When the representation given by Eq (6.12b) is used it should be noted that C_{DZ} and K_V must both be based on the drag coefficient at moderate values of C_L. C_{DZ} derived in this way is often less than the true minimum value of the drag coefficient. If C_{DZ} and K_V are deduced from experimental results a C_D vs C_L^2 analysis should always be used.

The components of drag for a given speed and lift are dependent upon some, or all, of the following design parameters:

i) Wing reference area, S, which is a fundamental unknown in the synthesis process. It is therefore avoided as far is as possible in evaluation of the drag, although a relatively insensitive function of S, namely $S^{-0.1}$ is used. Total aircraft wetted area is related to S by an assumed factor, R_W, see Table 6.3.

ii) Aspect ratio, A.

iii) Thickness/chord ratio, t/c.

iv) Sweep, $\Lambda_{1/4}$.

v) Degree of wing laminar flow, defined by c_l (as a fraction of chord). Natural laminar flow is more easily achieved on unswept lifting surfaces operating at relatively lower Reynolds number.

vi) Overall effective length, of aircraft, l.

vii) Effective fuselage cross-section diameter, d.

Additionally certain factors are used for which values are suggested in given cases.

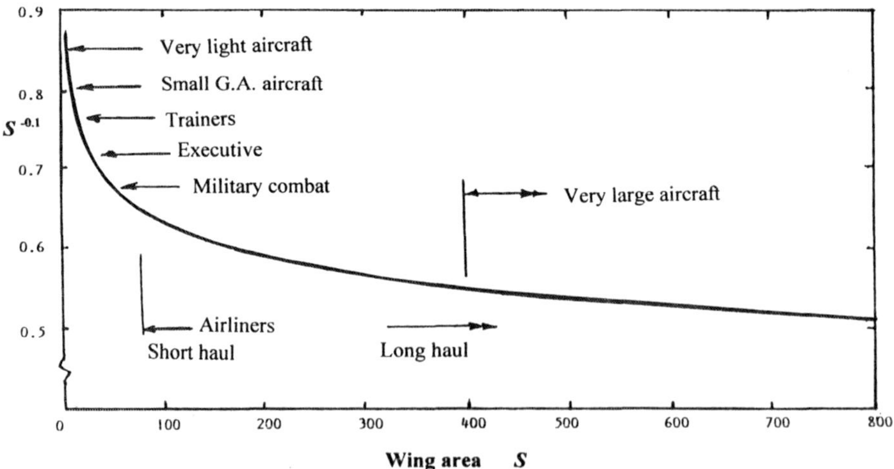

Figure 6.1 Typical values of $S^{-0.1}$ for aircraft type and wing area

6.3.2 Subsonic and transonic aircraft with moderate to high aspect ratio wings (A > 5)

6.3.2.1 Cruise configuration (landing gear and high lift devices retracted)

a) Zero lift drag coefficient:

$$C_{DZ} = 0.005\left(1 - \frac{2c_l}{R_W}\right)\bar{\tau}\left[1 - 0.2M_N + 0.12\left\{\frac{M_N(\cos\Lambda_{1/4})^{1/2}}{(A_f - t/c)}\right\}^{20}\right] R_W\, T_f\, S^{-0.1} \qquad (6.13a)$$

where A_f is an aerofoil factor which depends upon the aerofoil design. It may be as high as 0.93 for a specially designed advanced aerofoil, but can be as low as 0.75 for earlier aerofoils intended primarily for use in incompressible flow conditions; see Chapter 5, paragraph 5.2.2.4.

M_N is the flight Mach number

c_l is the fraction of chord of wing over which the flow is laminar

S, $\Lambda_{1/4}$ and t/c are defined in paragraph 6.3.1

R_W is a factor which is effectively the ratio of overall wetted area to the reference area, S

T_f is a type factor, which effectively allows for departure of the shape from the streamlined ideal.

$\bar{\tau}$ is a correction factor for wing thickness:

$$\bar{\tau} = \left[\frac{(R_W - 2)}{R_W} + \frac{1.9}{R_W}\left\{1 + 0.526\left(\frac{t}{c}\Big/0.25\right)^3\right\}\right] \qquad (6.13b)$$

The value of this factor is usually close to unity

Typical values of the parameter ($S^{-0.1}$) are given in Figure 6.1. Values of R_W and T_f for different classes of aircraft are suggested in Tables 6.3 and 6.4 respectively. Equation (6.13) applies only up to the critical Mach number, see Chapter 5, paragraphs 5.2.2.4 and 5.3.4.2.

b) Drag due to lift:

$$C_{DI} = \left[\frac{(1 + 0.12M_N^6)}{\pi A}\left\{1 + \frac{(0.142 + f(\lambda)A(10t/c)^{0.33}}{(\cos\Lambda_{1/4})^2} + \frac{0.1(3N_e + 1)}{(4 + A)^{0.8}}\right\}\right] C_L^2 \qquad (6.14a)$$

where C_L is the lift coefficient appropriate to the given flight condition and N_e is the number of engines, if any, which are located over the top surface of the wing.

The taper ratio function is given by:

$$f(\lambda) = 0.005\{1 + 1.5(\lambda - 0.6)^2\} \qquad (6.14b)$$

but for most purposes may be taken as 0.0062.

Table 6.3 Wetted area ratio, R_W

TYPE OF AIRCRAFT	R_W
Sailplane	3.0
Tailless types (delta)	2.5-3.0
Single-engine propeller type:	
i) Normal	3.75-4.0
ii) Agricultural	4.0-4.5
Small twin-engine propeller type:	
i) Low wing loading	4.0
ii) High wing loading	5.0
Bombers, jets	4.25
Jet trainers	4.5
Interceptors, strike aircraft, clean*	4.0-5.0
Strike aircraft with external stores, up to	6.0
Airliners, executive jets, freighters	5.5
Turboprop airliners	5.5

*The higher value is when the weapons are carried internally, although an integrated wing-fuselage configuration tends to have a lower value.

6.3.2.2 Climb out configuration - (landing gear retracted and high lift devices at take-off setting)

In this case:

$$C_D = C_{DZ} + \Delta C_{DT} + C_{DI} = (C_D)_{CO} + C_{DI} \qquad (6.15a)$$

where C_{DZ} and C_{DI} are as given by Eqs (6.13) and (6.14) respectively and $(C_D)_{CO}$ is the effective zero lift drag coefficient in the climb-out configuration. ΔC_{DT} is the total increment in drag due to deployment of the high lift devices.

Approximately:

$$\Delta C_{DT} = (0.03F_F - 0.004)/A^{0.33} \tag{6.15b}$$

F_F is a flap drag factor having the following typical values:

F_F = 1.0 for single slotted trailing edge flaps
1.2 for double slotted or Fowler flaps
1.5 for triple slotted or slotted Fowler flaps
0.133 when flaps are absent

Although it is not necessarily correct the drag produced by leading edge devices is assumed to be negligible in the take-off configuration.

Table 6.4 Aircraft type factor, T_f

TYPICAL AIRCRAFT	T_f
Very streamlined aircraft with negligible slipstream effects	1.0
Jet airliners, executive jets	1.1-1.2
Turboprop airliners	1.4
Combat aircraft, jet trainers, clean	1.2
Strike aircraft, trainers with external stores up to	1.85
Large freighters	1.2-1.3
General aviation aircraft with retractable landing gear	1.5
General aviation aircraft with fixed landing gear	2.0-2.5
Agricultural types	2.5

* Higher value applies when there are large fuselage fairways for landing gear stowage.

6.3.2.3 Aborted landing configuration - (landing gear extended and high lift devices at landing setting)

A typical drag coefficient for an extended landing gear is 0.03 and that for a flap at landing setting:

$$\Delta C_{DL} = 0.15F_F / A^{0.33} \tag{6.16a}$$

Thus the total drag coefficient becomes:

$$C_D = C_{DZ} + \Delta C_{DL} + 0.03 + C_{DI}$$

$$= (C_D)_{G0} + C_{DI}$$

where

$$(C_D)_{G0} = C_{DZ} + \Delta C_{DL} + 0.03 \tag{6.16b}$$

6.3.3 Transonic and supersonic configurations with low aspect ratio wings (A < 4.5)

6.3.3.1 General remarks

These configurations tend towards slender layouts with low values of thickness to chord ratio.

6.3.3.2 Cruise configuration, subsonic flight ($M_N \leq 0.95$)

a) Zero lift drag coefficient, C_{DZ}, the value given by using Eqs (6.13) with appropriate input data may be used for this case.

b) Drag due to lift, the value given by Eq (6.14a) is appropriate.

6.3.3.3 Climb out configuration

This class of aircraft usually have a relatively high installed thrust to weight ratio and hence may not use significant high lift device deployment during take-off and landing. When high lift devices are used for take-off there can be considerable variation in their performance and hence drag. In the absence of better information it is suggested that Eq (6.15a) may be used with F_F given a value of unity when flaps are used.

6.3.3.4 Cruise configuration, supersonic flight ($M_N \geq 1.05$)

a) Zero lift drag:

$$C_{DZ} = 0.005(1 - 0.2M_N)\,\bar{\tau}\,R_W\,T_f\,S^{-0.1} + C_{DW} \qquad (6.17a)$$

where C_{DW} is the volume wave drag contribution.

C_{DW} is dependent upon both fuselage and wing volume distributions and the following equation represents an attempt to reduce a complex issue to a form simple enough for use in initial design synthesis:

$$C_{DW} = K_0 . \frac{1}{(l/d)^2}\left[\frac{9.4K_F}{(l/d)^2(S/l^2)} + 1.2\left(\frac{K_F}{A}\right)^{1/2}\left(\frac{S}{l^2}\right)^{1/2}\left(\frac{t/c}{0.05}\right)\right]\left[1 + 0.034(3 - M_N)^{3.5}\right]$$

$$= \overline{C_{DW}}\left[1 + 0.034(3 - M_N)^{3.5}\right]$$

(6.17b)

K_o = 1.0 for an ideal area distribution
= 1.1 for a well-designed tailless delta or variable sweep aircraft
= 1.25 for a reasonably good area distribution, e.g. a supersonic airliner
= 1.7 to 2.0 for a combat aircraft with conformal weapons and not unduly "bumpy" area distribution
= 3.0 for a combat aircraft with "bumpy" area distribution, including external weapons

K_F = 1.0 for the ideal area distribution
= 0.87 for a typical combat aircraft taken with K_o = 1.9 approx
= 1.7 for a fuselage having a parallel section over about half the length taken with K_o = 1.25 approx.
(K_F is about 3.0 for a typical subsonic airliner fuselage)

l is effective body length
d is the effective maximum diameter of the body, for example:

$$d = 1.13\,(B^* H - A_p)^{1/2} \qquad (6.17c)$$

where B^* is average body width, say at engine intake face for fuselage-located engines, but not greater than $0.2(AS)^{1/2}$, that is 0.2 times wing span.

H is fuselage height

A_p is area of engine intake face

(B^* for a circular cross-section is $0.785d = H$)

For initial design it is convenient to avoid the need to allocate a value to the wing area, S, by assuming a typical value for (S/l^2) as indicated in Table 6.5.

Table 6.5 Typical values of the parameter (S/l^2)
(Supersonic designs)

TYPE OF AIRCRAFT	ASPECT RATIO RANGE	(S/l^2)
Combat (stealth and canard)	2.0-2.5	0.2
Combat (conventional tail)	3.0-4.0	0.14
Combat (tailless delta)	2.0	0.15
Combat (swing wing)	2.5-3.5	0.10
Airliner	<2.0	<0.10

b) Induced drag coefficient

$$C_{DI} = \left[\frac{0.24}{A} + K_W(M_N^2 - 1)^{1/2}\right] C_L^2 \qquad (6.18)$$

where K_W may be taken as 0.20 for an uncambered slender wing or 0.175 when leading edge camber is used or obtained by leading edge flap deployment.

6.3.3.5 Cruise configuration, transonic flight ($0.95 < M_N < 1.2$)

The drag at this speed may be important in the evaluation of transonic acceleration. Prediction of a reliable figure presents some difficulty.

a) In the case of zero lift drag it is suggested that the value used is:

i) Eq (6.13) with $M_N = 0.95$

ii) Eq (6.17) with $M_N = 1.00$ and above

iii) An arbitary smooth curve between these two values for $0.95 < M_N < 1.0$

b) Induced drag coefficient may be evaluated by using Eq (6.14) up to $M_N = 1.2$ and Eq (6.18) for $M_N = 1.25$ and above. It may be necessary to blend between the two.

6.4 Mass

6.4.1 Introduction

While the mass model must be simple it must also include all the major design parameters. A further discussion of mass prediction methods together with more detailed mass models may be found in Addendum 4.

The representation of the total mass of a given configuration consists of two parts:

a) An absolute, that is a defined, mass which comprises the fuselage structure and the payload including items directly related to the latter. For a given configuration it is fixed in terms of the total aircraft mass but may vary during a flight if payload is discharged, for example if bombs are dropped.

b) A variable mass, which for a given configuration is primarily dependent upon the take-off wing loading, the basic thrust to weight ratio and the fuel load required to meet the specified performance requirements. This latter variable also depends on the wing loading and thrust to weight ratio.

The total mass is thus:

$$M_0 = M_{FIXED} + M_{VARIABLE} \tag{6.19}$$

where $M_{FIXED} = M_{FUS} + M_{PAY} + M_{OP}$

$M_{VARIABLE} = M_{LIFTSUR} + M_{POWERPT} + M_{SYS} + M_{FUEL}$

M_{FUS}	is mass of fuselage structure
M_{PAY}	is mass of the payload; this is as specified but can include directly related items if necessary, such as provisions for passengers
M_{OP}	is mass of operational items (the difference between the basic empty and operating empty masses); this item is not to be confused with operators specified items which cover a greater mass contribution and usually include furnishings, etc.
$M_{LIFTSUR}$	is the total mass of the wing and the horizontal and vertical stabiliser/control surfaces
$M_{POWERPT}$	is the total mass of the installed powerplants
M_{SYS}	is the mass of the airframe systems, equipment, landing gear, etc., not included in M_{PAY}, but see Table 6.9
M_{FUEL}	is the fuel mass required to meet the design specification

6.4.2 Absolute mass contributions (fixed values in terms of total mass)

6.4.2.1 Payload

It is assumed that this is specified, either implicitly or explicitly, in the specification or requirements as outlined in Chapter 1.

6.4.2.2 Fuselage mass

This is based on the assumption that a fuselage layout has been prepared in accordance with Chapter 4 and therefore overall dimensions are known.

Table 6.6 Fuselage mass coefficient, C_2

CATEGORY OF FUSELAGE	C_2
Pressurised fuselages [Eq (6.20a)]	
Airliners, executive and feeder line aircraft of four or more abreast seating with wing-mounted landing gear	0.79
Airliners and related types with fuselage-mounted landing gear	0.81
Antisubmarine and regional airliners of less than four abreast seating	0.83
Freighter aircraft with fuselage-mounted landing gear and rear ramp door	0.87
Increment when engines are located on rear fuselage	0.01
Other fuselages [Eq (6.20b)]	
Land based combat aircraft with fuselage-mounted engines	0.04-0.036*
Naval combat aircraft with fuselage-mounted engines	0.043-0.039*
Bomber aircraft with wing-mounted engines	0.027
Single-engine light aircraft	0.06-0.04*
Twin-engine general aviation aircraft, with or without limited pressurisation	0.034

N.B. The values are all for metal construction except that those marked * make allowance for the possible benefits of using reinforced plastic materials where appropriate.

Two classes of fuselage configuration are identified.

a) Pressurised transport, executive and related types:

$$M_{FUS} = C_2\, p(9.75 + 5.84B)\left(\frac{2L}{(B + H)} - 1.5\right)(B + H)^2 \quad kg \qquad (6.20a)$$

where
- p is the cabin maximum working differential pressure, bar
- L is the overall fuselage length, m
- B is the maximum width of the fuselage, m
- H is the maximum height of the fuselage, m
- C_2 is a coefficient which depends upon the actual type of pressurised fuselage, see Table 6.6

b) Other aircraft:

$$M_{FUS} = C_2\left[L(B + H)V_D^{0.5}\right]^{1.5} \quad kg \qquad (6.20b)$$

where L is the overall length of the basic fuselage (aft of the engine bulkhead when a nose propeller engine is used), m
- V_D is the design maximum (diving) speed, m/s (EAS)
- C_2 is the coefficient given in Table 6.6

The term $(B + H)$ relates to half of the periphery of the maximum cross-section of the fuselage. For the datum case of a circular or elliptical cross-section the periphery is $\pi(B + H)/2$, the $\pi/2$ being allowed for in the coefficient C_2. For other cross-sections it is necessary to select an appropriate value for $(B + H)$. Usually H may be defined as the maximum depth excluding local excressences such as cockpit enclosures and other fairings. B now becomes the effective width of the cross-section. Two special cases may be identified:

i) Rectangular cross-section with corner radii; factor the actual width by up to 1.25, depending on size of corners, to get B

ii) Trapezoidal cross-section with sharp corners; factor the average width by 1.28 to get B

Discretion is necessary when fuselage air intakes are present. In most cases B should be taken as across the width of side intakes unless they are very short. In this latter case B should be taken as the mean between the width across the intakes and that of the basic fuselage. Likewise H should not necessarily include the full depth of dorsal or ventral air intakes.

For more precise information see Addendum 4, paragraph AD4.2.2.2

6.4.2.3 Mass of operational items

The items added to the basic empty mass to bring the aircraft to its empty, operating, condition include:

Crew and associated personal items
Safety equipment, such as emergency oxygen and life rafts
Freight equipment
Water and food, especially on transport types
Possibly residual fuel, but here this item is assumed to be included in the powerplant mass.

The following typical allowances are appropriate.

a) Passenger aircraft

$$85n_C + F_{OP}P \quad kg \tag{6.21a}$$

where P is the number of passengers
n_C is the number of crew
F_{OP} is the operating items factor and is of the order of:

i)	Feeder line aircraft, very short haul	7
ii)	Medium range	12
iii)	Very long range and executive	16

b) Freight aircraft

$$600 + 0.03(\text{Payload}) \tag{6.21b}$$

c) Other types

Crew provision of from 77 kg per person in light aircraft to a nominal 100 kg for combat types.

6.4.3 Variable mass contributions

6.4.3.1 Lifting surfaces

The total mass of the lifting surfaces is based on that of the wing by use of a coefficient, C_1, which includes a factor, C_5, to allow for the contribution from the other vertical and horizontal surfaces. This latter factor is dependent upon the type of aircraft and the configuration adopted, typical values being given in Table 6.10.

Table 6.7 Lifting surface mass coefficient, C_l

[Eq (6.2b)] $C_l = A^1 - B^1 M_0 \times 10^{-3}$

TYPE	$A^1 \times 10^3$	$B^1 \times 10^6$	Typical C_l
Subsonic transport aircraft; $M_0 > 5700$ kg			
Long range (nom. range > 5000 km)	0.72	0.5	Eq (6.23a)
Short/med. range; $M_0 > 46{,}000$ kg)	0.90	0	0.00090
Short/med. range; jet powered; $M_0 \leq 46{,}000$ kg	1.67	16.1	Eq (6.23b)
Short/med. range; propeller driven, $M_0 \leq 46{,}000$ kg	1.49	16.1	Eq (6.23c)
Executive jets (all ranges)	1.76	16.9	0.0016
General aviation types; $M_0 \leq 5700$ kg			
Propeller driven, cantilever wing;			
Single engine	2.0	100	0.00183
Twin engine	2.0	100	0.00164
Propeller driven, braced wing	1.74	112	0.0015
*Supersonic delta wings, all types			
$M_0 \leq 15{,}000$ kg	0.72 or 0.81	0	0.00072 or 0.00081
$M_0 > 30{,}000$ kg	0.48 or 0.55	0.64 or 0.72	0.00042 or 0.00047
(No evidence for $15{,}000 > M_0 > 30{,}000$ kg)			
Military jet strike/interceptors; $M_0 > 10{,}000$ kg			
Typically	0.62 to 0.74	0	0.00062 to 0.00074
Variable sweep, $M_0 < 40{,}000$ kg	1.24	14.9	0.00089
Naval aircraft with inboard wing fold	0.87	0	0.00087
Naval aircraft with outboard wing fold	0.75	0	0.00075
Military jet strike/interceptors; $M_0 \leq 10{,}000$ kg	1.18	50	0.00076
Military trainers and related types			
Jet powered; $M_0 < 10{,}000$ kg	1.73	105	0.0012
Propeller driven; $M_0 > 3100$ kg	1.49	0	0.00149
Propeller driven; $M_0 \leq 3100$ kg	4.0	800	0.002 to 0.003
Subsonic bombers			
Long range (nom. range > 10,000 km)	0.5	0	0.0005
Medium range	0.93	0	0.00093
Military freight aircraft			
Long range jet	0.72	0.5	Eq (6.23d)
Turboprop	0.77	0.53	Eq (6.23e)

N.B.*Delta wings are defined as $A \leq 2.5$ and $\lambda \leq 0.15$. The first, lower, value for delta wings is for tailless aircraft configurations. The values given are all for metal construction. A reduction of 15% is suggested when full use is made of fibre reinforced plastic materials in the construction of all lifting surfaces.

$$M_{LIFTSUR} = C_1 \left[A^{0.5} S^{1.5} sec\Lambda_E \left(\frac{1 + 2\lambda}{3 + 3\lambda} \right) \frac{M_0}{S} \bar{N}^{0.3} \left(\frac{V_D}{t/c} \right)^{0.5} \right]^{0.9} \quad kg \tag{6.22a}$$

where
M_0 is the total aircraft mass
A is the wing aspect ratio
S is the wing area, m^2
Λ_E is the effective sweep, usually 0.25 chord sweep but the mean of $\Lambda_{1/4}$ and Λ_{STRUCT} if the structural sweep is significantly different from basic aerodynamic sweep
λ is the ratio of the tip to centreline chords of the wing
$\bar{N}$ is 1.65 times the limit maximum manoeuvre acceleration factor, as given in the requirements, unless known to be overridden by gust considerations
V_D is the design maximum (diving) speed, m/s (EAS)
t/c is the thickness to chord ratio at wing centreline
C_1 is a coefficient depending on the type of aircraft

$$C_1 = A^1 - B^1 M_0 \times 10^{-3} \tag{6.22b}$$

Values of A^1 and B^1 are given in Table 6.7.

In many cases the dependancy of C_1 on B^1 is not great and for initial design work a typical, mean, value may be used. Typical values for some classes of aircraft are also given in Table 6.7. In the case of subsonic transport and freight aircraft it is possible to relate the take-off mass, and hence C_1, to the range, s, and payload requirements.

a) Long range, $s > 5000$ km

$$C_1 = 0.00072 - 0.0005(270 + 0.05s)P \times 10^{-6} \tag{6.23a}$$

b) Short/medium range; $M_0 \le 46{,}000$ kg

$$C_1 = 0.00167 - 0.016(370 + 0.03s)P \times 10^{-6} \tag{6.23b}$$

c) Turboprop; $M_0 \le 46{,}000$ kg

$$C_1 = 0.00149 - 5.8P \times 10^{-6} \tag{6.23c}$$

d) Long range jet freight aircraft

$$C_1 = 0.00072 - 0.0005(2.08 + 0.00038s)\, PAY \times 10^{-6} \qquad (6.23d)$$

e) Turboprop freight aircraft

$$C_1 = 0.00077 - 0.00053(2.08 + 0.00038s)\, PAY \times 10^{-6} \qquad (6.23e)$$

where P is the number of passengers to be carried over the design range, s, kilometers, and PAY is the corresponding payload, in kg, for freight aircraft.

In practice for one given configuration under study the parameters A, Λ_E, λ, $\bar{N}$, V_D and t/c will not be variable, hence:

$$M_{LIFTSUR} = \bar{C}_1 \left[S^{1.5} \frac{M_0}{S} \right]^{0.9} \quad kg \qquad (6.24a)$$

where:

$$\bar{C}_1 = C_1 \left[A^{0.5} sec\Lambda_E \left(\frac{1 + 2\lambda}{3 + 3\lambda} \right) \bar{N}^{0.3} \left(\frac{V_D}{t/c} \right)^{0.5} \right]^{0.9} \qquad (6.24b)$$

When it is necessary the individual contribution of the wing and tail surfaces may be isolated by the use of the coefficient, C_5, given in Table 6.10.

6.4.3.2 Powerplant

The powerplant mass consists of the basic powerplant factored to cover the additional mass of its mounting, exhaust system and where appropriate nacelles, pods, cowlings and propeller. The fuel system mass is also included here for convenience.

Table 6.8 Powerplant installation factors, C_3

TYPE OF AIRCRAFT	C_3
Executive jets and jet transports	1.56
Supersonic aircraft with variable geometry intakes	2.0
Turbopropeller transports	2.25
Propeller turbine trainers	2.0
General aviation, twin piston-engined types	1.80
All other types	1.40

The total installed powerplant mass is:

$$M_{POWERPLANT} = C_3 M_{ENG} \tag{6.25}$$

C_3 is a coefficient dependent upon the type of aircraft and engine, see Table 6.8

The mass of the basic engine, M_{ENG}, is best derived from actual engine data, if necessary scaled appropriately. In the absence of more accurate information the following expressions may be used. These are conveniently expressed in terms of thrust, or power, to weight ratio.

a) Military combat engines (turbojets or low bypass ratio turbofans)

$$\left.\begin{array}{lll} \text{Basic dry thrust rating:} & \dfrac{T_0}{(Mg)_{ENG}} & = 4.5 \textit{ to } 6.5 \\ \text{With typical afterburning:} & \dfrac{T_0}{(Mg)_{ENG}} & = 7 \textit{ to } 9 \\ \text{With provision for vectoring nozzles, etc.:} & \dfrac{T_0}{(Mg)_{ENG}} & = 4 \textit{ to } 6 \end{array}\right\} \tag{6.26a}$$

The higher values apply to more recent designs. T_0 is the equivalent sea level static thrust.

b) Civil transport engines (usually high bypass ratio turbofans)

$$\text{Sea level static rating:} \quad \frac{T_0}{(Mg)_{ENG}} = 5.0 \textit{ to } 6.5 \tag{6.26b}$$

The higher value applies to large, new technology, engines.

c) Advanced turbopropeller engines (including gearbox)

$$\frac{Power}{(Mg)_{ENG}} = 0.34 \textit{ to } 0.42 \; kW/N \tag{6.26c}$$

Where the higher figure applies to new, large engines, and power is the sea level static value.

d) Turboshaft engines (excluding reduction gearbox)

$$\frac{Power}{(Mg)_{ENG}} = 0.5 \text{ to } 0.8 \ kW/N \qquad (6.26d)$$

(0.34 to 0.42 kW/N with gearbox)

e) Piston engines

Small unsupercharged; 0 < Power < 150 kW:

$$\frac{Power}{(Mg)_{ENG}} = 0.057(1 + 0.006 \ kW) \ kW/N \qquad (6.26e)$$

Unsupercharged; Power > 150 kW:

$$\frac{Power}{(Mg)_{ENG}} = 0.12 \ kW/N \qquad (6.26f)$$

Supercharged; Power > 150 kW:

$$\frac{Power}{(Mg)_{ENG}} = 0.1 \ kW/N \qquad (6.26g)$$

f) Small rotary engines

$$\frac{Power}{(Mg)_{ENG}} = 0.135 \ kW/N \qquad (6.26h)$$

6.4.3.3 Systems, equipment, landing gear, etc.

In some cases it may be possible that special equipment is carried as part of a payload. However, this is not usually the case, the equipment carried being mainly determined by the operational aspects of the design. For convenience the equipment mass is considered to be aggregated with that of the aircraft systems and the landing gear, but fuel system mass is included in the powerplant item.

$$M_{SYS} = C_4 M_0 \qquad (6.27)$$

where C_4 is a coefficient depending upon the type of aircraft and is given in Table 6.9, M_0 being total aircraft mass. Where relevant passenger furnishings are included.

Table 6.9 Systems factor, C_4

Includes equipment, landing gear and passenger furnishing

TYPE OF AIRCRAFT	C_4
General aviation light single-engined types	0.12
Subsonic bombers	0.12
Subsonic freighters	0.12
Long range supersonic aircraft	0.12
Long range jet transports	0.14
Small regional transports	up to 0.22
General aviation, larger single- and twin-engine types	0.16
Executive aircraft	0.21 to 0.3*
Propeller turbine trainers	0.32
All other types	0.19

*Higher figure includes luxury furnishing.

6.4.3.4 Fuel

The fuel mass is a function of performance and is determined from the performance analysis in terms of aircraft total mass. The equations necessary to undertake this are dealt with in Chapter 7.

6.4.3.5 Secondary lifting surfaces

The mass of the secondary lifting surfaces is included with that of the wing in paragraph 6.4.3.1 being related to it by the coefficient C_5 given in Table 6.10.

Table 6.10 Lifting surface factor, C_5

TYPE OF AIRCRAFT	C_5
Tailless delta	1.10
Long haul jets transport	1.16
Short/medium haul jet transports	1.20
Executive jet aircraft	1.30
All other types	1.24

C_5 = mass of all lifting surfaces/mass of wing

6.4.4 Total mass

The individual contributions to the total mass derived from Eqs (6.20), (6.21), (6.24), 6.25) and (6.27) together with the masses of the payload and fuel may be substituted into Eq (6.19) to obtain the total mass:

$$M_0 = \left[C_2 \, f\left(L,\ B,\ H,\ p\ or\ V_D\right) + M_{OP} + M_{PAY}\right] + \bar{C}_1\left[S^{1.5}\left(\frac{M_0}{S}\right)\right]^{0.9} + C_3 M_{ENG} + C_4 M_0 + M_{FUEL} \qquad (6.28)$$

The $S^{1.5}$ term may be written in terms of the total mass and wing loading:

$$S^{1.5} = \left(\frac{S}{M_0}\right)^{1.5} M_0^{1.5} \qquad (6.29a)$$

so that the term in the square brackets relating to $\bar{C}_1$ becomes:

$$\frac{M_0^{1.5}(M_0/S)}{(M_0/S)^{1.5}} = \frac{(M_0)^{1.5}}{(M_0/S)^{0.5}} \qquad (6.29b)$$

The M_{ENG} term may be written in terms of the static thrust to weight ratio of the engine and that for the aircraft as a whole:

$$\begin{aligned} M_{ENG} &= \frac{M_{ENG}}{T_0}\left(\frac{T_0}{M_0 g}\right) M_0 g \\ &= M_0\left(\frac{T_0}{M_0 g}\right) \Big/ \left\{\frac{T_0}{(Mg)_{ENG}}\right\} \end{aligned} \qquad (6.30)$$

$\{T_0 / (Mg)_{ENG}\}$ may be derived from known engine characteristics or from Eqs (6.26). In the case of propeller-engined aircraft it is convenient to convert the static power to equivalent static thrust, see for example, Chapter 3, paragraphs 3.6.2.3 and 3.6.2.4.

Substitution of Eqs (6.29b) and (6.30) into Eq (6.28) gives:

$$M_0 = \left[C_2 f\,(L,\ B,\ H,\ p\ or\ V_D) + M_{OP} + M_{PAY}\right] + \bar{C}_1\left[(M_0)^{1.35} \Big/ \left(\frac{M_0}{S}\right)^{0.45}\right]$$

$$+ \bar{C}_3\, M_0 \left(\frac{T_0}{M_0\, g}\right) + C_4\, M_0 + M_{FUEL} \tag{6.31}$$

where $$\bar{C}_3 = C_3 \Big/ \left\{\frac{T_0}{(Mg)_{ENG}}\right\}$$

For a given class of aircraft, type of engine and wing geometry, $\bar{C}_1$, C_2, $\bar{C}_3$ and C_4 are defined. Reference to Chapter 4 enables the fuselage parameters to be established. Hence the mass of the aircraft, M_0, is a function only of the payload, fuel, wing loading and thrust to weight ratio:

$$M_0\left[1 - \bar{C}_3\left(\frac{T_0}{M_0 g}\right) - C_4 - \frac{M_{FUEL}}{M_0}\right] - \bar{C}_1\left[(M_0)^{1.35} \Big/ \left(\frac{M_0}{S}\right)^{0.45}\right] \tag{6.32}$$

$$= C_2 f\,(L,\ B,\ H,\ p\ or\ V_D) + M_{OP} + M_{PAY}$$

The values of wing loading and thrust to weight ratio are derived from an analysis of the performance requirements, as is the fuel mass ratio, see Chapters 7 and 8.

Chapter 7

Performance estimation

7.1 Introduction

7.1.1 General

The size and mass of an aircraft are critically dependent upon the performance requirements that are stated in the specification. In order to estimate the performance and relate it to that required it is necessary to be in possession of the following data:

i) Aircraft mass in the various phases of the flight
ii) Lift and drag characteristics
iii) Powerplant characteristics

At the initial synthesis stage of a project some aspects of these data are not known. In particular such items as overall aircraft mass and installed power must be regarded as outputs of the synthesis process. For this reason the first phase of performance estimation is intended to establish the various combinations of the two fundamental parameters:

a) (T/Mg): the thrust to weight ratio

b) (Mg/S): the wing loading, which is the ratio of the weight to the wing reference area,

which enable the requirements to be met. The acceptable combinations of these two parameters will in general be different for each of the various performance requirements as well as being dependent upon other aircraft characteristics such as the powerplant and wing configuration.

In the absence of specific powerplant data the powerplant models suggested in Chapter 3 may be used. The lift, drag and, subsequently, the mass information needed may be derived from the appropriate equations given in Chapter 6.

The performance estimation equations presented here differ in some respects from those which may be found in other references as they have been specifically derived and presented in such a way as to facilitate the determination of the two fundamental parameters referred to above.

7.1.2 Performance phases

The categories, or phases, of the performance which it is necessary to examine include some or all in the following list.

7.1.2.1 Take-off and initial climb

Take-off may be considered as consisting of:

i) The ground run during which the aircraft accelerates from rest to a speed at which it can lift off.
ii) The initial, first segment, climb from lift off to clear a specified obstacle height.
iii) A second segment climb subsequent to the retraction of the landing gear.

The first two of these items give the take-off distance.

Take-off performance evaluation is complicated in the case of multi-engined aircraft since it is necessary to consider the effect of a failure of a critical engine. It is necessary to establish a decision speed above which the take-off and climb out must be continued. The second segment climb performance requirement is based on this engine failed case.

The take-off field length is typically taken as 1.15 times the distance for an all engines operating take-off to the obstacle height or 1.0 times the distance needed to accelerate up to the decision speed and then bring the aircraft to a safe stop, see paragraph 7.2.3.2.

7.1.2.2 Climb to operating altitude and ceilings

Performance requirements are specified for a third segment climb which takes the aircraft to an initial operating altitude, but these are not frequently critical. The more significant aspects of climb performance are:

i) The fuel used during the climb.
ii) The rate of climb, both initially and what is available when the operating altitude is reached, usually at least 1 m/s.
iii) The ceiling, or altitude, at which the rate of climb has fallen to an unacceptable level. The service ceiling is sometimes set where climb rate has fallen to 0.5 m/s and the combat ceiling where it has fallen to 2.5 m/s.
iv) The time to operating height, which may be important for military interceptor aircraft.
v) The ground distance covered during the climb in that it may reduce the range required to be achieved during the cruise phase or radius of operation.

The climb may be limited either by the condition where the thrust is equal to the drag or when the implied lift coefficient reaches its maximum usable value.

7.1.2.3 Cruise/normal operating flight

For many aircraft the most important flight performance is that of cruising flight. The major exceptions to this are those aircraft whose role demands a significant manoeuvre capability, but even in these cases the fuel required to reach the operational radius is of importance. The aim in cruising flight is to achieve the lowest possible rate of fuel usage in a prescribed condition.

Cruise performance is usually specified in terms of distance for a given payload, but endurance rather than the distance covered is more important for aircraft intended to operate in a search role. The cruising height may be determined by operational considerations or more likely by the need to minimise fuel consumption. When the latter is the case, and where it is relevant, the design cabin pressure differential should be established to enable the aircraft to operate up to the altitude required for the most efficient flight at the end of the cruise phase.

7.1.2.4 Normal operating/maximum speed

The normal operating and maximum speed are usually closely related, but not identical. In the case of transport and related types the normal operating speed is established by operational requirements but may be compromised by the penalties implied in making it higher than is essential. For example there may be negligible gain in increasing the cruise speed of a short haul transport aircraft beyond a certain value as the total flight time tends to be greatly influenced by other parts of the flight.

Military combat aircraft frequently have a requirement for maximum speeds at different operating altitudes, such as a high subsonic speed at low level and supersonic performance at altitude.

7.1.2.5 Manoeuvre

Manoeuvre requirements are a major factor in the design of combat and aerobatic aircraft. Manoeuvre capability may be defined in terms of acceleration normal to the flight path, turn rate, or specific excess power, see paragraph 7.5.3. These are all related through the forward speed and flight altitude. Frequently the specification of turn rate in a given flight condition is most appropriate for combat aircraft, while normal acceleration is more relevant for aerobatic designs.

There are two main categories of flight manoeuvre, which are:

i) Maximum instantaneous normal acceleration or turn rate. This is determined either by the structural design limits or by the maximum usable lift coefficient in the given flight condition.

ii) Sustained normal acceleration or turn rate, which is determined by the thrust available to match the drag in the manoeuvre.

7.1.2.6 Descent

The performance during descent is not usually of great importance. Fuel consumption is relatively low since the engines operate under near idling conditions. Speed brakes may be required to increase the drag and consequently the rate of descent, especially for aircraft such as trainers and combat types where reduced sortie time is a consideration.

7.1.2.7 Approach and landing

The approach speed requirement frequently has a dominant effect upon the wing size. While a low approach speed is associated with a short landing distance there are frequently operational circumstances where the design approach speed must be set above a certain value. For example Air Traffic Control requires that all aircraft landing on a given runway have similar approach speeds, typically in the range of 60 to 72 m/s for civil transports. While a few special cases may be tolerated these are the exception rather than the rule. The stalling characteristics of the aircraft in the landing configuration are established such that the required approach speed is usually 1.3 times the stalling speed, but see paragraph 7.2.4.1

The angle of the descent on the approach is determined by Air Traffic Control/navigational aids and is typically 3°. This ensures that the rate of descent near to the ground is acceptably low.

The arbitrary landing distance is defined as:

i) The ground distance covered during the descent from a height of 15.3 m at the approach angle. Often the speed is reduced somewhat during the phase.
ii) The distance required for the aircraft to "flare" from descending to level flight, ideally contacting the ground at the end of the flare with zero vertical velocity.
iii) The distance covered before the retarding devices are operated.
iv) The stopping distance, usually assuming that the runway is hard, smooth and dry. In addition to brakes the use of other devices such as thrust reversers and parachutes are allowed providing their reliable integrity can be demonstrated.

For most aircraft the required landing runway length is 1.67 times the arbitrary landing distance, see paragraph 7.2.4.2.

7.1.2.8 Baulked landing/missed approach

There is an important performance requirement for the aircraft to be able to climb away should the approach be terminated as late as the point at which it arrives at 15.3 m height above the ground. In this condition both flaps and landing gear will be in the landing condition and the engines are assumed to be at idle. The aircraft has to be able to climb at an angle of about 2° at a speed no greater than the approach speed nor less than the minimum control speed, see paragraph 7.2.4.3.

7.2 Definition of aircraft speeds and associated conditions

7.2.1 Introduction

Numerous speed definitions are used in specifying aircraft performance, handling and structural requirements. The summary here covers those which are important for the purposes of initial design.

7.2.2 General speeds

V_{TAS} True air speed, which is important in that it defines the distance covered and is directly related to Mach number.

V_{EAS} Equivalent air speed, which is the speed which has the same value of dynamic pressure at sea level density. $V_{EAS} = V_{TAS}\sigma^{0.5}$ where σ is the relative air density.

V_V The true vertical speed, as in a climb.

V_S The stalling speed, that is the minimum speed at which the aircraft can maintain controlled level flight. While there are several definitions of V_S, for initial design purposes it may be considered as being directly related to the maximum lift coefficient in a given flight condition. V_S will have different values depending upon such matters as the configuration of high lift devices, powerplant slipstream and Mach number. Because stalling speed is mostly associated with low speed flight performance adjacent to the ground it is usually defined as an equivalent airspeed .

V_{MD} The minimum drag speed. This is the speed at which the value of the lift coefficient, C_{LMD}, is such that the zero lift and induced drag effects have the same magnitude, see Chapter 6, paragraph 6.3. It usually implies a relatively high value of lift coefficient which may be in a buffet region. For this reason prolonged flight at, or below, the minimum drag speed is not usual.

V_A The manoeuvre speed, that is the lowest speed at which the aircraft can achieve its specified normal acceleration factor, n_l. It is related to the stalling speed ($V_A = (n_l)^{0.5}V_S$ assuming that the maximum lift coefficient is constant).

M_{NCRIT} The value of V_{TAS} divided by the local speed of sound, a, at which the aircraft experiences unacceptably large drag increase or lift/drag divergence. See Chapter 5, paragraph 5.2.2.4.

V_C/M_C The normal maximum operating speed/Mach number as used for structure load definition.

V_{MO}/M_{MO} The maximum operating speed/Mach number as used for flight performance and handling. It is approximately the same as M_{NCRIT} and is related to (V_C/M_C).

V_{NE}/M_{NE} The never exceed speed/Mach number used to define an operational limitation for structural load purposes.

V_D/M_D The maximum design (dive) speed/Mach number used as the highest speed or Mach number considered for structural load purposes. In terms of speed it may be up to 25% higher than V_C and at least $\Delta M_N = 0.05$ greater than M_C.

(Note that speeds used for structural design purposes are normally quoted as equivalent airspeeds.)

7.2.3 Take-off (Refer to Figure 7.1)

7.2.3.1 Speeds

The definition of speeds during the take-off phase is associated with the situation of the failure of an engine during take-off.

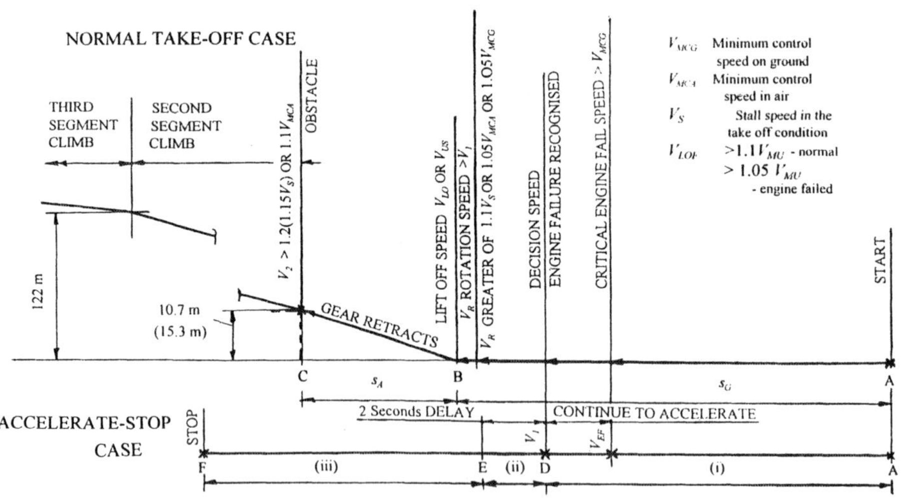

Figure 7.1 Take-off definitions

V_{MCG} — This is the lowest speed at which the aircraft can be controlled while it is in contact with the ground (excluding taxying conditions).

V_{MCA} — This is the lowest speed at which the aircraft can be controlled once it is airborne.

V_{EF} — The most critical speed at which engine failure may occur. It must not be less than V_{MCG}.

V_I — The speed reached at some time after V_{EF} before the pilot is assumed to recognise the engine failure. It is the decision speed. At and above this speed the take-off must be continued.

V_R — The speed at which the aircraft is normally rotated for take-off. It must not be less than V_I but otherwise is the greater of:

1.05 V_{MCA} or 1.05 V_{MCG}
or 1.1 V_S in the take-off configuration.

V_{LOF} or V_{US} — The speed at which the aircraft can lift off the ground or unsticks. In practice it is marginally greater than V_R.

V_{MU} — The minimum speed at which the aircraft can safely lift off. Military requirements specify that V_{LOF} is a margin above this speed.

V_2 — The speed reached at the point where the aircraft has climbed to 10.7 m height. It is known as the take-off safety speed and must not be less than 1.2 V_S or 1.1 V_{MCA}.
(Not less than 1.15V_S for four-engined civil aircraft).
(Small military aircraft have to achieve 15.3 m height at V_2).

7.2.3.2 Ground distances

s_G — The ground distance from the start point A, to the point where V_{LOF} is reached with all engines operating, B.

s_A — The air distance from lift-off, point B, to the point where V_2 is achieved at the required height, point C.

ToL — The total take-off length. This is the sum of s_G and s_A, that is the distance from point A to point C, factored by 1.15 when all engines are operating, but by only 1.0 when an engine has failed.

ASL The accelerate-stop length. It is the sum of the distances:

i) Point A to point D where the decision speed, V_1, is reached.

ii) Point D to Point E, the distance covered in 2 s to allow for pilot reaction time.

iii) Point E to Point F, the distance needed to bring the aircraft safely to a stop after Point E.

No factor is required on *ASL*.

7.2.3.3 Second segment climb - engine failed

The second segment climb is the climb from the point where V_2 is reached at the required height up to, normally, 122 m height. The flight is assumed to continue at speed V_2 with the landing gear retracted and the high lift devices in the take-off position. With the critical engine failed the aircraft must be able to achieve the following climb gradients:

2 engines 2.4% (climb angle 1.38° approx.)
3 engines 2.7% (climb angle 1.55° approx.)
4 engines 3.0% (climb angle 1.72° approx.)

7.2.3.4 Third segment climb - engine failed

Above 122 m height the required climb gradients are reduced to about half of those stipulated in the second segment climb.

7.2.4 Approach and landing (Refer to Figure 7.2)

7.2.4.1 Speeds

V_a The approach speed. It is typically 1.3 times the stalling speed with the high lift devices in the landing condition. However, it may be less, such as $1.2V_S$ for a land based combat type and possibly $1.15V_S$ for aircraft carrier operation. It must be maintained down to a point 15.3 m above the ground.

V_{TD} The speed at touch down, which is at the conclusion of a "flare" intended to reduce the vertical velocity to nominally zero. It may be as low as $1.1V_S$.

7.2.4.2 Ground distance

s_A In this case s_A is the air distance covered from the point A, where the aircraft passes through 15.3 m height, to the point B, where the glide slope, typically 3°, ends and the flare commences.

s_F The distance covered during the flare, to ground contact at point C.

s_G The ground distance required to apply the brakes and any retarding device at point D and then bring the aircraft to a safe stop at point E. Brake forces are limited to avoid undue tyre wear. This usually implies a maximum braking coefficient of just under 0.4 on a smooth, dry surface.

LL The total landing distance, is the sum of s_A, s_F and s_G, that is the distance between points A and E, factored typically by 1.67. (A factor of 1.43 is used for single and light twin propeller aircraft). Lower factors are used when the stopping distance is based on low braking coefficients such as may occur in wet or icy conditions.

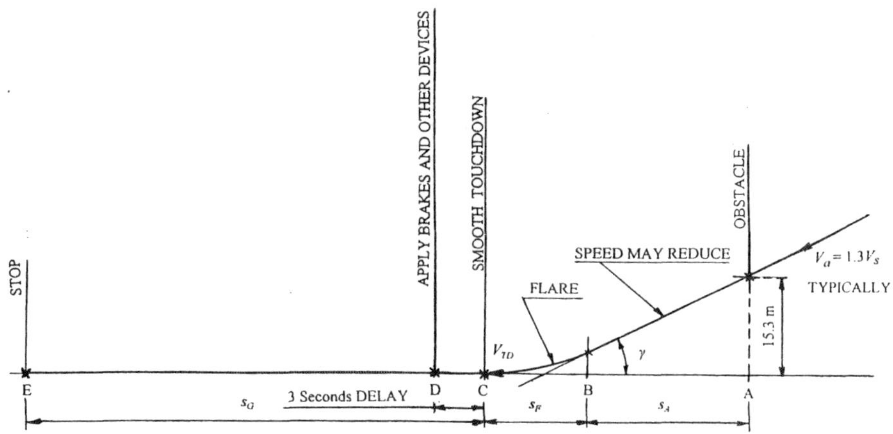

Figure 7.2 Landing definitions

7.2.4.3 Baulked landing/missed approach

A missed approach is assumed to commence when the aircraft is at 15.3 m height on a normal approach. With the high lift devices in the landing configuration and the landing gear extended the aircraft has to climb away at an angle of about 2°. Initially the engines are assumed to be at idle setting, but this may be increased to the level achievable 8 s after the decision to abort the landing. The climb out speed must be between the minimum control speed in this configuration and V_a.

In the event of a missed approach being undertaken when one engine is inoperative the required climb gradients are:

2 engines 2.1% (climb angle 1.20° approx.)
3 engines 2.4% (climb angle 1.38° approx.)
4 engines 2.7% (climb angle 1.55° approx.)

7.3 Take-off and second segment climb

7.3.1 Normal take-off - all engines operating - landplane

The field length required is approximately:

$$ToL = \frac{k_e}{C_{LUS}}\left(\frac{T}{Mg}\right)_0^{-1.35}\left(\frac{Mg}{S}\right)_0 + 6\left(\frac{Mg_0}{SC_{LUS}}\right)^{1/2} + 120\left[1 - \left(\frac{T}{Mg}\right)_0\right] \quad m \qquad (7.1a)$$

where the suffix 'o' refers to conditions at the start of the take-off.

k_e is 0.1 for jet, fan and piston propeller engines
0.12 for turbopropeller engines with no thrust coefficient limitation (see Chapter 3, paragraph 3.6.2.5)
0.09 for turbopropeller engines where thrust is limited and approximately constant during take-off

C_{LUS} is the unstick lift coefficient, see Chapter 6, paragraph 6.2.4.4

The ToL includes the required 1.15 factor. The first term represents the value of s_G, the ground run to the lift-off speed, the second term the distance covered during rotation and the third term the distance to climb to the 10.7 m prescribed obstacle height. For cases where 15.3 m obstacle height is required the 120 should be replaced by 170. For an aircraft designed specifically for short take-off the factor 6 in the rotation term should be reduced to 2.3. It should be noted that both $(Mg/S)_0$ and $(T/Mg)_0$ appear in two terms and hence when the ToL is known and the dependence of these two parameters is required it is necessary to invert the equation and use an iterative solution.

$$\left(\frac{T}{Mg}\right)_0^{-1.35} = \frac{C_{LUS}}{k_e(Mg/S)_0}\left[ToL - 120\left\{1 - \left(\frac{T}{Mg}\right)_0\right\}\right] - \frac{6}{k_e}\left(\frac{C_{LUS}}{(Mg/S)_0}\right)^{1/2} \qquad (7.1b)$$

Reference to Table 3.2 of Chapter 3 enables an initial value of $(T/Mg)_0$ to be selected for insertion into the right-hand side of the equation.

In the case of propeller driven aircraft T_0 is the static value as given in Chapter 3, paragraph 3.6.2.4, and when this is substituted into Eq (7.1b) the following relationships are derived:

a) Turboprop aircraft:

$$N_E \frac{P_0}{(Mg)_0} = \frac{1.15 \times 10^4}{(nD_p)^{2.983}} \left(\frac{Mg}{S}\right)_0^{0.105} \left(S^{-0.1}\right)^{-1.05} \left(\frac{T_0}{z^{0.15} N_E (Mg)_0}\right)^{1.105} N_E \qquad (7.1c)$$

where P_0 is rated power of each engine
N_E is number of engines
z is number of propeller blades
(nD_p) is propeller speed times propeller diameter in m/s

Typical values of ($S^{-0.1}$) are given in Chapter 6, Figure 6.1.

b) Piston engined aircraft:

$$N_E \frac{P_0}{(Mg)_0} = 0.0123 \left(\frac{Mg}{S}\right)^{-1.76} \left(S^{-0.1}\right)^{-1.76} \left(\frac{T_0}{N_E (Mg)_0}\right)^{1.176} N_E \qquad (7.1d)$$

7.3.2 Normal take-off - all engines operating - seaplane

In this case, making a typical allowance for the total drag, approximately:

$$ToL = \frac{0.072}{C_{LUS}} \left(\frac{T}{Mg}\right)_0^{-2} \left(\frac{Mg}{S}\right)_0 + 6\left(\frac{Mg_0}{SC_{LUS}}\right)^{1/2} + 120\left[1 - \left(\frac{T}{Mg}\right)_0\right] \quad m \qquad (7.2a)$$

The comments made in paragraph 7.3.1 apply here also, and:

$$\left(\frac{T}{Mg}\right)_0^{-2} = \frac{13.9 C_{LUS}}{(Mg/S)_0} \left[ToL - 120\left\{1 - \left(\frac{T}{Mg}\right)_0\right\}\right] - 83.4\left(\frac{C_{LUS}}{(Mg/S)_0}\right)^{1/2} \qquad (7.2b)$$

7.3.3 Engine failed take-off balanced field length - landplane

It is possible to derive an expression to relate the decision speed, V_1, to the unstick speed, V_{US} to meet the requirements and give a balanced condition. Typical practical values are found to be:

$$0.91 V_{US} \le V_1 \le 0.95\, V_{US}$$

with the higher values for aircraft with only two engines.

In practice the installed thrust is more likely to be determined by the second segment climb requirement, see paragraph 7.3.4. In view of this if it assumed that $V_I = 0.94V_{US}$, the accelerate-stop length is approximately:

$$ASL = \frac{0.82k_e}{C_{LUS}}\left(\frac{Mg}{S}\right)_0\left\{\left(\frac{T}{Mg}\right)_0^{-1.35} + 2.2\right\} \quad m \tag{7.3a}$$

where the first term represent the ground distance to the decision point and the second term the braking distance assuming a mean braking coefficient of 0.4.

Again in the case of propeller driven aircraft T_0 is the static value, as given in Chapter 3, paragraph 3.2.6.3. Equation (7.3a) may be rearranged as:

$$\left(\frac{T}{Mg}\right)_0^{-1.35} = \frac{1.22C_{LUS}}{k_e(Mg/S)_0}ASL - 2.2 \tag{7.3b}$$

7.3.4 Second segment climb with one engine failed

This may well be a critical case in the determination of the installed thrust to power ratio of twin engined transport aircraft. The design condition is when the engine has failed after the decision speed in hot and high environmental conditions. While requirements vary somewhat it is suggested that the "hot and high" conditions may be represented by taking a value for relative density of $\sigma = 0.8$. For most aircraft the climb lift coefficient is about 0.84 C_{LUS}. The landing gear is assumed to have been retracted and the drag of the aircraft with the flaps at the take-off setting is given in Chapter 6, paragraphs 6.3.1 and 6.3.2.2 as:

$$C_D = (C_D)_{CO} + (K_V)_0 C_L^2$$

In this case let:

$$(\bar{\beta})_{CO} = (C_D)_{CO}\,(K_V)_0$$

where $(C_D)_{CO}$ is the zero lift drag coefficient and $(K_V)_0$ is the induced drag factor for the appropriate flight case.

The climb gradient to be achieved is dependent upon the number of engines, as given in paragraph 7.2.3.3.

The thrust, T_{CO} required to meet this condition is:

$$\frac{T_{CO}}{(Mg)_0} = \alpha\left\{\left[\frac{(C_D)_{CO}}{C_{LUS}}\right] + \frac{0.71(\bar{\beta})_{CO}}{\left[(C_D)_{CO}/C_{LUS}\right]} + \gamma\right\} \tag{7.4a}$$

For Eq (7.4a) only:

2-engined aircraft	$\alpha = 2.74$	$\gamma = 0.020$
3-engined aircraft	$\alpha = 2.06$	$\gamma = 0.023$
4-engined aircraft	$\alpha = 1.83$	$\gamma = 0.025$

a) Low bypass ratio and fan engines:-

$$T_{CO} = T_0 \tau_{CO} \tag{7.4b}$$

where τ_{CO} is the value of τ at the climb out Mach Number ($0.00324V_{US}$) and at standard atmosphere conditions ($\sigma = 1$), see Chapter 3, paragraph 3.6.2.2.

b) Propeller driven aircraft:

i) Piston engines

The total climb out thrust is related to the static value by:

$$\left(\frac{T}{Mg}\right)_0 = 0.72N_E V_{US}^{0.51}\left(\frac{Mg}{S}\right)_0^{-0.15}\left(S^{-0.1}\right)^{1.5}\left(\frac{T_{CO}}{N_E M g_0}\right)^{0.85} \tag{7.4c}$$

ii) Turbopropeller engines:

Assuming that the power output is "flat rated" to about 35°C:

$$\left(\frac{T}{Mg}\right)_0 = 4.8N_E\, z^{0.15}(nD_p)^{3.062}V_{US}^{0.55}\left(\frac{Mg}{S}\right)_0^{-0.095}\left(S^{-0.1}\right)^{0.95}\left(\frac{T_{CO}}{N_E\, M g_0}\right)^{0.905} \times 10^{-7} \tag{7.4d}$$

When the powerplant is thrust coefficient limited thoughout the take-off and climb out it is reasonable to assume that the static thrust value and the second segment climb condition are of the same magnitude.

The equivalent power to Eqs (7.4c) and (7.4d) may be derived from Eqs (7.1d) and (7.1c) respectively.

7.3.5 Deflected thrust take-off - low bypass ratio and fan engines

The effect of deflecting the thrust during take-off is two-fold:

i) It reduces the force available for forward motion by a factor *cos* θ, where θ is the angle of thrust deflection relative to the horizontal.

ii) It increases the vertical force for lift-off, i.e. there is a supplement to the aerodynamic lift.

These effects can be allowed for in the ground roll distance by modifying Eq (7.1a) to:

$$s_G = 0.087\left(\frac{Mg}{S}\right)_0 \left(\frac{(Mg)_0}{T_0\ cos\theta}\right)^{1.35} \frac{\left(1 - T_0\ sin\theta/(Mg)_0\right)}{C_{LUS}} \quad m \tag{7.5a}$$

The 1.15 factor is not included in this ground roll distance and C_{LUS} is only the aerodynamic effect.

This is more conveniently expressed as:

$$s_G = \frac{0.087}{C_{LUS}}\left(\frac{Mg}{S}\right)_0 \left(\frac{T}{Mg}\right)_0^{-1.35} \frac{\left(1 - T_0\ sin\theta/(Mg)_0\right)}{(cos\theta)^{1.35}} \quad m \tag{7.5b}$$

which enables a direct comparison to be made with the usual case.

Equation (7.5b) may be rearranged as:

$$\left(\frac{Mg}{S}\right)_0 = 11.5 C_{LUS}\ s_G \left(\frac{T}{Mg}\right)_0^{1.35} \Bigg/ \left[\frac{1 - T_0 \sin\theta/(Mg)_0}{(cos\theta)^{1.35}}\right] \tag{7.5c}$$

The evaluation of θ to give a minimum value of s_G is somewhat complex but is approximately:

$$\theta = sin^{-1}\left[(T/Mg)_0^{1.5}\right] \quad for\ 0.4 \le (T/Mg)_0 \le 0.8 \tag{7.5d}$$

Thrust deflection at lower thrust to weight ratio is only marginally beneficial while above a value of 0.8 the configuration is tending towards vertical take-off. It must be noted that when the thrust is also used for control the required thrust to weight ratio should be increased by about 8% for short take-off and 15% for vertical take-off.

7.3.6 Fuel used during take-off

This may vary considerably according to the type of powerplant and operating conditions. It is suggested that a typical allowance is 1% of the normal take-off weight, that is the third segment climb commences with a weight of 0.99 $(Mg)_0$.

7.4 Approach and landing

7.4.1 General

In general the evaluation of approach and landing performance is less complex than that of take-off since, with the exception of a missed approach condition, the powerplant performance is only of secondary importance.

One difficulty is the determination of the ratio of the design landing mass to the take-off mass. This is not the mass remaining after the climb, cruise and descent fuel has been used, and where appropriate, load disposed of. The need for flexibility in operation, such as large payloads over short distances, implies that the design landing mass will, in general, be higher than the mass at the end of the "design" sortie. On the other hand the lower it is the less severe will be the landing design conditions. In an attempt to simply this issue, which is essentially an operational as well as a design decision, an analysis of existing aircraft has been undertaken and the results are summarised in Table 7.1. In the absence of more precise information this information may be used to establish a first value of the ratio (M_L/M_0), suffix L indicating the landing condition.

Chapter 6, paragraph 6.2.4.3 gives expressions to enable an initial estimate to be made of the lift coefficient relevant to the approach to land.

Table 7.1 Ratio of landing to take-off masses

AIRCRAFT TYPE	Range limit	M_L/M_0 (landing/take-off mass)
Short haul transports Short range tactical military transports	$s \geq 1000$ $s < 1000$	$0.98-2(s-1000) \times 10^{-5}$ 0.98
Medium/long haul transports Military long range aircraft		$1-2s \times 10^{-5}$
Executive aircraft: a) Intended for general operations b) Intended for short ranges c) Intended solely for long ranges	 $s \geq 3500$ $s < 3500$ $s > 4500$	 $0.98-2.7(s-3500) \times 10^{-5}$ 0.98 $1-2.7s \times 10^{-5}$
Military combat aircraft - depends on actual sortie role but typically		0.65 to 0.8
Trainers: a) Basic b) Advanced and light combat Combat developments		 1.0 0.85 to 0.95 0.8

(s is design range in km)

7.4.2 Ground distance covered during approach and landing

The usual requirement is for a factor of 1.67 to be applied to the ground distance covered between the aircraft passing through 15.3 m height on the approach and coming to a stop. Assuming that:

i) The normal acceleration in the flare is $0.15g$

ii) The speed at the end of the flare, touch down, is $0.9V_a$

iii) There is a 3 s delay before the brakes and the other retarding devices are applied.

iv) The stopping distance is reduced by 6% to allow for aerodynamic drag.

The factored landing length is:

$$LL = \frac{25.55}{\tan \gamma} + 4.5V_a + 0.0255\ L_L\ V_a^2 \qquad m \tag{7.6a}$$

where

$$L_L = \left[\frac{1}{\left(\dfrac{\mu_G}{0.38}\right)} + \frac{5.59}{\left\{\left(\dfrac{\mu_G}{0.38}\right) + 1.2\left(\dfrac{T}{Mg}\right)_0 \left(\dfrac{M_0}{M_L}\right)\right\}} + 20.6 \tan \gamma \right] \qquad m$$

and γ is the descent angle on the approach at the speed V_a

μ_G is the braking coefficient

(M_L/M_0) is the ratio of landing to take-off masses

$(T/Mg)_0$ is the static installed thrust to mass ratio, and allows for the use of reverse thrust. It is taken as zero when reverse thrust is not used.

Equation (7.6a) may be arranged as:

$$V_a = -\left(\frac{88.2}{L_L}\right) + \left[\left(\frac{88.2}{L_L}\right)^2 - 2\left\{\frac{19.6}{L_L}\left(\frac{25.55}{\tan \gamma} - LL\right)\right\}\right]^{1/2} \tag{7.6b}$$

and also:

$$V_a = 1.278\left[\frac{1}{C_{La}}\left(\frac{Mg}{S}\right)_0 \left(\frac{M_0}{M_L}\right)\right]^{1/2} \qquad m/s \tag{7.6c}$$

$$or \qquad \left(\frac{Mg}{S}\right)_0 = 0.612\ V_a^2\ C_{La}\left(\frac{M_0}{M_L}\right) \qquad (7.6d)$$

C_{La} is the approach lift coefficient and $(Mg/S)_0$ the take-off wing loading; and for the common civil transport case when $\gamma = 3°$:

$$LL = 488 + 4.5V_a$$

$$+ 0.0255\ V_a^2\left[\frac{1}{\left(\frac{\mu_G}{0.38}\right)} + \frac{5.59}{\left\{\left(\frac{\mu_G}{0.38}\right) + 1.2\left(\frac{T}{Mg}\right)_0\left(\frac{M_0}{M_L}\right)\right\}} + 1.08\right]\ m \qquad (7.6e)$$

and for a typical value of $\mu_G = 0.38$ and no reverse thrust, $L_L = 7.67$ and:

$$LL = 488 + 4.5V_a + 0.196V_a^{\,2} \quad m \qquad (7.6f)$$

for which:

$$V_a = -11.5 + [132.2 - 5.11(488 - LL)]^{1/2} \quad m/s \qquad (7.6g)$$

Equation (7.6) may be solved to give a value of V_a to meet a required landing fielding length, LL, with appropriate values of μ_G, etc. This value of V_a may then be used to derive a value of take-off wing loading commensurate with the landing conditions. This case may well place a maximum limit on the design wing loading.

For a steep approach where $\gamma = 6°$ the equations become:

$$LL = 242 + 4.5V_a + 0.23V_a^{\,2} \quad m \qquad (7.6h)$$

for which:

$$V_a = -9.7 + [94.9 - 4.32(242 - LL)]^{1/2} \quad m/s \qquad (7.6i)$$

7.4.3 Missed approach

The missed approach requirement discussed in paragraphs 7.1.2.8 and 7.2.4.3 may be marginally critical in some cases. However, it is not straightforward to analyse at the preliminary design stage for the following reasons:

i) It requires a knowledge of the engine response to sudden throttle movement, which is difficult to generalise.

ii) It requires a knowledge of the landing gear drag, which may have considerable variation.

iii) Small changes in the landing flap setting may have considerable effect upon the drag, without a comparable major alteration to the lift characteristics. This effect requires a more detailed analysis than presented here, and may provide a way of overcoming a problem should one exist.

It is therefore suggested that this requirement is not considered at the very preliminary stage of a conceptual design. However, it is essential that it be investigated as soon as sufficient design information has become available for a realistic analysis to be undertaken.

7.4.4 Fuel used during descent, approach and landing

It is suggested that for conventional aircraft a fuel mass of $0.01M_0$ be allowed for the descent, approach, landing and final reserve. In the case of a vertical landing it is appropriate to assume that the powerplants are operated for 1 min at full power in addition to the $0.01M_0$.

7.5 Generalised flight performance representation

7.5.1 General

It is convenient to express the performance achieved by the aeroplane at a given time by the so-called "energy height", and the performance potential in terms of the "specific excess power". With the exception of those performance requirements related to take-off and landing the energy height and specific excess power representations may be adapted to apply to point performance requirements.

7.5.2 Energy height

7.5.2.1 General

The energy height, h_e, is defined as the sum of the actual geopotential height of the aircraft at a given time and the height equivalent of the kinetic energy of forward velocity:

$$h_e = h + \frac{V^2}{2g} \tag{7.7}$$

where h is geopotential height and V is the forward velocity.

Figure 7.3 shows lines of constant energy height as a function of actual height and Mach number. From this it can be seen that at low subsonic speeds the majority of the energy possessed by an aircraft is geopotential, while at speeds of the order of Mach number 2 and above the greatest contribution is from the kinetic term. In terms of performance potential the criterion is the rate of change of energy height with time:

$$\frac{dh_e}{dt} = \frac{V}{g}\frac{dV}{dt} + \frac{dh}{dt} \tag{7.8}$$

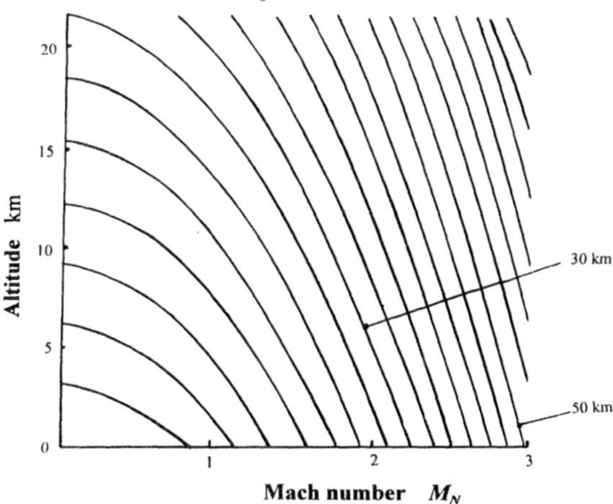

Figure 7.3 Lines of constant energy height

Figure 7.4 illustrates an aircraft in straight climbing flight at an angle γ to the horizontal. For simplicity it is assumed that the lift acts through the centre of gravity and pitching moments have been trimmed out.

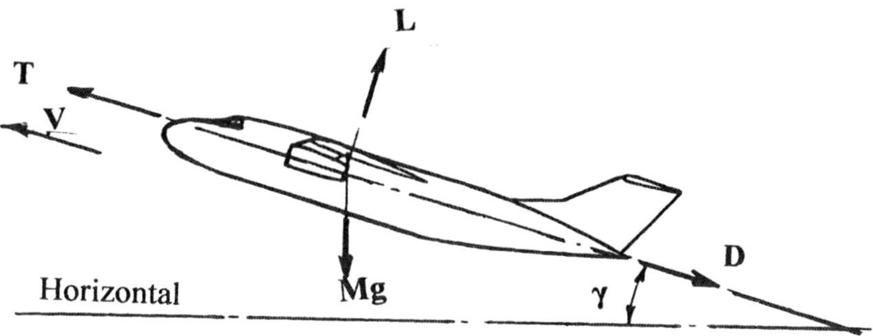

Figure 7.4 Aircraft in climb

Resolution of forces along and perpendicular to the flight path gives:

$$T - D - Mg\ sin\gamma - \frac{MdV}{dt} = 0 \tag{7.9a}$$

$$L = Mg\ cos\gamma \tag{7.9b}$$

where T is the thrust
L is the lift
M is the mass

The rate of change of geopotential height with time is:

$$\frac{dh}{dt} = V\ sin\ \gamma \tag{7.10}$$

and substituting for $sin\gamma$ into Eq (7.9a) yields:

$$(T - D) = \frac{Mg}{V}\frac{dh}{dt} + \frac{MdV}{dt}$$

$$(T - D)\frac{V}{Mg} = \frac{dh}{dt} + \frac{V}{g}\frac{dV}{dt} = dh_e \tag{7.11a}$$

and using Eq (7.8), Eq (7.11a) may be written as:

$$(T - D)\frac{V}{Mg} = \frac{dh}{dt}\left(1 + \frac{V}{g}\frac{dV}{dh}\right) \tag{7.11b}$$

whence:

$$\frac{dh}{dt} = \left((T-D)\frac{V}{Mg}\right) \Big/ \left(1 + \frac{V}{g}\frac{dV}{dh}\right) \tag{7.11c}$$

The term in the denominator is effectively a kinetic energy correction factor, KE. When the climb is undertaken at a constant true airspeed, (dV/dh), and hence KE, are zero thus:

$$\text{rate of climb} = \frac{dh}{dt} = \frac{V}{Mg}(T - D) \tag{7.11d}$$

For constant Mach number climb at altitudes of 11 km $\leq h \leq$ 20 km, the speed of sound is constant and Eq (7.11d) applies. At altitudes below 11 km the speed of sound decreases with height increase and hence a constant Mach number climb implies a reducing true airspeed, ($d\,V/d\,h$) is negative, and the actual rate of climb is greater than that given by Eq (7.11d). However, a constant equivalent airspeed climb implies an increasing true velocity with altitude increase, so that the actual rate of climb is less than that given by Eq (7.11d).

7.5.2.2 Constant airspeed climb case

At altitudes up to about 14 km the variation of relative air density with altitude is approximately given by:

$$\sigma = e^{-h/9000}$$

where h is in metres, and hence:

$$d\sigma/dh = -\sigma/9000$$

and since $V_{TAS} = V_{EAS}\,\sigma^{-1/2}$, approximately:

$$\frac{d\,V_{TAS}}{dh}\left(V_{EAS}\,\sigma^{-1/2}\right) = V_{EAS}\,d\left(\sigma^{-1/2}\right)\frac{d\sigma}{dh}$$

$$= \sigma^{-1/2}\,V_{EAS}\Big/18{,}000$$

$$and\ \frac{V}{g}\frac{dV}{dh} = 5.66\times 10^{-6}\,V^2_{EAS}/\sigma$$

Therefore for constant equivalent airspeed climb the kinetic energy correction factor is approximately given by:

$$KE = \left(1 + \frac{V}{g}\frac{dV}{dh}\right) = \left(1 + 5.66\times 10^{-6}\,V^2_{EAS}/\sigma\right) \qquad (7.11e)$$

7.5.2.3 Constant Mach number climb condition

Likewise up to 11 km altitude the speed of sound is approximately equal to 340 $\sigma^{0.117}$ m/s so that approximately:

$$V_{TAS} = 340 M_N\,\sigma^{0.117}$$

where M_N is the constant Mach number climb condition.
This relationship may be used to give:

$$\frac{dV_{TAS}}{dh} = \frac{dV_{TAS}}{d\sigma}\frac{d\sigma}{dh}$$

$$= 0.117 \times 340\ M_N \times \sigma^{-0.883}\left(\frac{-\sigma}{9000}\right)$$

$$= -\ 4.42 \times 10^{-3}\ M_N\ \sigma^{0.117}$$

$$and\ \frac{V}{g}\frac{dV}{dh} = -\ 0.153 M_N^2\ \sigma^{0.234}$$

The corresponding kinetic energy correction factor is thus approximately:

$$KE = \left(1\ -\ 0.153\ M_N^2\ \sigma^{0.234}\right) \tag{7.11f}$$

and an average value between sea level and 11 km altitude is $\left(1\ -\ 0.134 M_N^2\right)$.

Between 11 and 20 km altitude the speed of sound is constant and there is no need to apply a correction factor in a constant Mach number climb.

Above 20 km altitude the factor becomes approximately:

$$KE = (1+0.016\ M_N^2) \tag{7.11g}$$

for a constant Mach number climb.

7.5.3 Specific excess power

The specific excess power, SEP or p_S, is used to derive the climb, level acceleration or normal manoeuvre potential of the aeroplane. However, it is expressed identically to the rate of change of energy height, that is from Eq (7.11b):

$$p_S = \frac{V}{Mg}(T\ -\ D)$$

This may be expressed in alternative forms. For example:

$$p_S = V\left[\frac{T}{Mg} - \frac{D}{Mg}\right]$$

$$p_S = V\left[\frac{T}{Mg} - \frac{n}{L/D}\right] \tag{7.12a}$$

if the aircraft has a normal manoeuvre factor of n.

Alternatively using lift and drag coefficient representation:

$$p_S = \frac{V}{Mg}\left[T - \frac{qSnC_D}{cos\theta}\right] = \frac{V}{Mg}\left[T - qSC_D\right] \tag{7.12b}$$

where q is the dynamic pressure ($\rho SV^2/2$, ρ being air density)
S is the wing area
C_D is the drag coefficient
$n = cos\theta$ if flight path is straight

and using the conventional representation for drag, see Chapter 6, paragraph 6.3.1:

$$p_S = \frac{V}{Mg}\left[T - qS\left(C_{DZ} + K_V\ C_L^2\right)\right] \tag{7.12c}$$

where C_{DZ} is the zero lift drag coefficient (includes the C_{DW} term for supersonic flight)
C_L is the lift coefficient
K_V is the induced drag factor

The lift coefficient, C_L, may be expressed as:

$$C_L = \frac{nMg}{qS} \tag{7.13a}$$

and defining:

$$\bar{\beta} = K_V\ C_{DZ} \tag{7.13b}$$

$$p_S = V\left[\frac{T}{Mg} - \left\{\frac{qSC_{DZ}}{Mg} + \bar{\beta}\frac{n^2}{qC_{DZ}}\left(\frac{Mg}{S}\right)\right\}\right] \tag{7.13c}$$

Alternatively:

$$p_S = V\left|\frac{T}{Mg} - \left(\bar{D} + \frac{\bar{\beta}n^2}{\bar{D}}\right)\right| \tag{7.13d}$$

$$and \qquad \bar{D} = qC_{DZ}/(Mg/S) \tag{7.13e}$$

where $\bar{D}$ is effectively a zero lift drag to weight ratio.

Equation (7.13d) is found to be a useful form of the expression for specific excess power when evaluating climb, level speed and manoeuvre performance, and hence:

$$\left(\frac{T}{Mg}\right)_0 = \left[\frac{p_S}{V} + \left(\bar{D} + \frac{\bar{\beta}n^2}{\bar{D}}\right)\right]\left(\frac{k_m}{\tau_m}\right) \tag{7.13f}$$

where τ_m is the value of τ appropriate to the given flight Mach number and altitude, as defined in Chapter 3

k_m is the ratio of the mass at the specified condition (often combat mass) to that at take-off

The maximum lift to drag ratio occurs when:

$$C_{DZ} = K_V\, C_L^2 \tag{7.14a}$$

therefore:

$$\left(\frac{L}{D}\right)_{MAX} = \frac{C_{LMD}}{2C_{DZ}} = \frac{1}{2(\bar{\beta})^{1/2}} \tag{7.14b}$$

where C_{LMD} is the lift coefficient to give the minimum drag condition.

The corresponding minimum drag speed is:-

$$V_{MD} = \left[\frac{2}{\rho}\,\frac{Mg}{S}\,\frac{(\bar{\beta})^{1/2}}{C_{DZ}}\right]^{1/2} \tag{7.14c}$$

7.6 Flight performance - climb and ceilings

7.6.1 General

The accurate prediction of climb performance is complex due to the fact that many flight parameters change during the climb. Thrust, drag and weight all vary with altitude, forward speed and as fuel is used. While analytical expression of these variations is possible, analytical integration over the height of climb is fraught with difficulty. The problem may be overcome to some extent for the purposes of overall aircraft synthesis by:

a) Limiting the climb characteristics to one of three practical, basic, forms:

- i) Constant equivalent airspeed at subsonic speed, usually starting from nominal sea level conditions.
- ii) Constant Mach number.
- iii) Linear variation of Mach number with altitude at supersonic speed.

b) Utilising a form of Eq (7.13d) which is based on mean values of climb performance, derived from numerical analysis of typical aircraft and powerplant characteristics. In some cases empirically derived correction factors are required. A special, but important, climb condition which is not covered by the above assumption is the gradient achieved immediately after take-off with a failed engine. This is dealt with in paragraph 7.4 above.

c) Assuming that the climb angle does not exceed about 30 ° to the horizontal.

d) Assuming that the variation of thrust with altitude is proportional to the relative density of the atmosphere, for all other conditions constant, see Chapter 3, paragraph 3.6.2.2.

If the definition is made that thrust is proportional to (σ^s) then typically:

0-11 km altitude	s is 0.85 to 0.95 for low bypass ratio engines 0.6 to 0.7 for high bypass ratio, fan engines
Above 11 km altitude	s is unity, but for simplicity assume that approximately:-
0-13 km altitude	s is about 0.7 as a mean value for fan engines
0- 4 km altitude	s is about 1.1 for piston engines based on power.

In the case of turbopropeller engines there is an interaction with forward speed, see Chapter 3, paragraph 3.6.2.4.

Often it is not necessary to predict climb performance in detail at the preliminary phase of a design, but an estimate of fuel consumed during the climb is required. Table 7.2 may be used to obtain an approximate indication of the reduction in total aircraft mass during a typical climb. The reduction in mass is given as a function of the change in relative density during the climb and is dependent upon the type of powerplant and climb characteristic.

In all of the analysis the terminology used is:

i) Suffix 0 indicates sea level static, dry, engine conditions and all up mass, corresponding wing loading and incompressible flow drag characteristics.

ii) Suffix 1 indicates the starting climb conditions at a height H_1 defined by relative density σ_1 at appropriate Mach number, M_{N1} as well as thrust and mass, etc. It is convenient to define H in km.

iii) Suffix 2 indicates the corresponding conditions at the end of that particular phase of the climb.

Table 7.2 Typical climb mass ratio factors

(Ratio of mass at end of climb, point 2, to that at beginning of climb phase, point 1)

TYPE OF CLIMB	POWERPLANT	MASS RATIO
Constant equivalent airspeed	Propeller - piston Propeller - turbine High bypass ratio, fan Low bypass ratio, dry	$\sigma_2^{0.03}$ $\sigma_2^{0.02}$ $\sigma_2^{0.016}$ $\sigma_2^{0.018}$
Constant Mach number (subsonic: ($M_N \leq M_{CRIT}$)	High bypass ratio, fan Low bypass ratio, dry Low bypass ratio, wet	$(\sigma_2/\sigma_1)^{0.02}$ $(\sigma_2/\sigma_1)^{0.02}$ $(\sigma_2/\sigma_1)^{0.024}$
Variable Mach number (supersonic: $1.2 \geq M_N \geq M_{CRUISE}$)	Low bypass ratio, dry Low bypass ratio, wet	$(\sigma_2/\sigma_1)^{0.035}$ $(\sigma_2/\sigma_1)^{0.045}$

σ is the relative density.

7.6.2 Constant equivalent airspeed climb

7.6.2.1 General

The basic assumptions in this case are:

a) The climb commences at sea level and is terminated at, or below, 11 km altitude.

b) The Mach number during the climb does not exceed the critical value, as defined in Chapter 5, paragraphs 5.2.2.4 and 5.3.4.2. Should this occur the climb should be continued at constant Mach number.

In the case of civil aircraft operation Air Traffic Control requirements may limit the initial climb to 127 m/s EAS (250 knots) up to 3.05 km height (10,000 ft).

The mean rate of climb is given by:

$$(V_V)_{MEAN} = \frac{0.73X_I ZQ_V}{\left[1 + 0.12\left(\frac{ZQ_V}{100}\right)^2\right]}\left[f(Thrust) - f(Drag)\right] \quad m/s \qquad (7.15a)$$

The forward speed in the climb is V_{EAS} ($M_{NI} = V_{EAS}/340$ at sea level).

$$Q_V = \left(\left(\frac{Mg}{S}\right)_I \frac{\bar{\beta}_0^{1/2}}{(C_{DZ})_0}\right)^{1/2} \quad m/s \qquad (7.15b)$$

where

$$Z = V_{EAS}/1.458Q_V \text{ and } \bar{\beta}_0 = (C_{DZ})_0 (K_V)_0 \qquad (7.15c)$$

(Z is unity when the climb is at 1.14 times the minimum drag speed, V_{MD}). X_I is a correction factor for propellor engines, see Eq. (7.18a). $f(Thrust)$ depends upon the type of powerplant, see below.

$$f(Drag) = 1.16\bar{\beta}_0^{1/2}\left[Z^2\left(1 + 0.9\sigma_2^{-0.42}\right) + \frac{0.66n^2}{Z^2}\left(1 + \sigma_2^{-0.38}\right)\right] \qquad (7.15d)$$

where n is the normal acceleration factor in the climb, usually approximately unity.

7.6.2.2 Low bypass ratio and fan engines ($X_I = 1.0$)

a) Thrust factor:

$$f(Thrust) = \frac{\tau_0}{(Mg)_I}\tau_{MNI}\left(1 + Y\right) \qquad (7.16a)$$

i) Low bypass ratio engines, without reheat in use:

$$Y = \frac{(0.992 - 0.134R)}{(0.92 - 0.031R)}\sigma_2^{(S - 0.42)} \qquad (7.16b)$$

ii) Low bypass ratio engines, with reheat:

$$Y = \frac{(1.383 + 0.025R)}{(1.27 - 0.052R)} \sigma_2^{(S - 0.42)} \tag{7.16c}$$

iii) High bypass ratio, fan engines:

$$Y = \frac{(0.64 - 0.016R)}{(0.76 - 0.016R)} \sigma_2^{(S - 0.42)} \tag{7.16d}$$

where R is the bypass ratio and τ_{MNI} is the thrust factor at the start of climb conditions and is derived from Chapter 3, Eqs (3.7b), (3.7e) or (3.8b) as appropriate, and for simplicity it may assumed that the climb is in the range $0.4 \le M_N \le 0.9$, even though the Air Traffic Control restriction referred to in paragraph 7.6.2.1 implies an initial climb Mach number of 0.37.

Equation (7.16a) may be substituted into Eq (7.15a) to obtain the mean rate of climb, and the result may be rearranged to give:

$$\frac{T_0}{(Mg)_I} = \left[\frac{1.37(V_V)_{MEAN}}{ZQ_V} \left\{ 1 + 0.12\left(\frac{ZQ_V}{100} \right)^2 \right\} + f\,(Drag) \right] \Bigg/ \left[\tau_{MNI}(1 + Y) \right] \tag{7.16e}$$

where T_0 is the total installed sea level, dry, static thrust with all engines operating.

b) Ceiling. It is unlikely that this class of aircraft will have a ceiling requirement in constant equivalent airspeed conditions. See paragraph 7.6.3.2.

c) The fuel consumed during the climb is:

$$\frac{W_F}{(Mg)_0} = \frac{0.1ac_0H_2}{(V_V)_{MEAN}} \frac{T_0}{(Mg)_0} \tau_{MNI} \left[1 + \gamma\sigma_2^{(S + 0.08)} + \beta\left\{ 1 + \gamma\sigma_2^{(S - 0.537)} \right\} \left(\frac{V_{EAS}}{100} \right) \right] \tag{7.17}$$

(H_2 in km)

For the purposes of this equation only:

i) Low bypass ratio engine, dry:
$\alpha = 1.45$ $\beta = 0.82$ $\gamma = 1.0$
$c_0 = c'\ (1 - 0.15R^{0.65})$, as defined in Chapter 3, paragraph 3.6.3.

ii) Low bypass ratio engines, with afterburning:
$\alpha = 1.10$ $\beta = 0.059$ $\gamma = 1.1$
$c_0 = (1.47 + 0.34R) / (1.05 - 0.15R)$

iii) High bypass ratio engine
$\alpha = 1.45$ $\beta = 0.22$ $\gamma = 0.82$
$c_0 = c'\ (1 - 0.15R^{0.65})$

7.6.2.3 Propeller engines

The climb may be assumed to be at $1.14V_{MD}$ ($Z = 1$), however, the following expressions are general.

The thrust correction factor, X_I for propeller engines is:

$$X_I = \left[\frac{(T/Mg)_I - 2.07\bar{\beta}_0^{1/2}}{(T/Mg)_I Y - 2.02\bar{\beta}^{1/2}\sigma_2^{-0.38}}\right]^{-0.12} \quad (7.18a)$$

a) Thrust factor:

$$f\,(Thrust) = \left(\frac{T}{Mg}\right)_I (1 + Y) \quad (7.18b)$$

where: i) Piston engines $Y = \sigma_2$ (7.18c)

ii) Turbine engines $Y = \dfrac{\left(\sigma_2^{1.08} + 0.0032Q_V\sigma_2^{0.313}\right)}{\left(1 + 0.0032Q_V\right)}$ (7.18d)

Equation (7.18b) may be used with Eq (7.15a) to give the thrust to weight ratio needed for a given mean rate of climb:

$$\left(\frac{T}{Mg}\right)_I = \left[\frac{1.37(V_V)_{MEAN}}{ZX_I Q_V}\left\{1 + 0.12\left(\frac{ZQ_V}{100}\right)^2\right\} + f\,(Drag)\right] \Big/ (1 + Y) \quad (7.18e)$$

where X_1 may be assumed to be unity in the first instance, and the resulting value of thrust to weight ratio used to calculate a value of X_1 from Eq (7.18a) to be used in Eq (7.18e) until convergence is achieved. Equation (7.18e) may be used to define the equivalent static thrust to weight ratio:

1) Piston engines:

$$\left(\frac{T}{Mg}\right)_0 = 0.71 N_E V_{EAS}^{0.51}\left(\frac{Mg_I}{Mg_0}\right)^{0.85}\left(\frac{Mg}{S}\right)^{-0.15}\left(S^{-0.1}\right)^{1.5}\left(\frac{T_I}{N_E Mg_I}\right)^{0.85} \tag{7.18f}$$

from Chapter 3, Eq (3.11e) where it has been assumed that the propeller advance ratio is less than unity, and N_E is the number of engines. Equation (7.1d) may be used to determine the power required.

2) Turboprop engines, using Chapter 3, Eq (3.10):

$$\left(\frac{T}{Mg}\right)_0 = \frac{1.27 N_E \left(nD_p\right)^{2.7} z^{0.15}}{\left[1 + 0.0032 Q_V\right]^{0.862}} Q_V^{0.862}\left(\frac{Mg_I}{Mg_0}\right)^{0.862}\left(\frac{Mg}{S}\right)_0^{-0.138}\left(S^{-0.1}\right)^{1.38}\left(\frac{T_I}{N_E Mg_I}\right)^{0.862} \times 10^{-6} \tag{7.18g}$$

when the advance ratio is one or more and:

$$\left(\frac{T}{Mg}\right)_0 = \frac{6.36 N_E \left(nD_p\right)^{3.062} z^{0.15}}{\left[1 + 0.0032 Q_V\right]^{0.905}} Q_V^{0.543}\left(\frac{Mg_I}{Mg_0}\right)^{0.905}\left(\frac{Mg}{S}\right)_0^{-0.095}\left(S^{-0.1}\right)^{0.95}\left(\frac{T_I}{N_E Mg_I}\right)^{0.905} \times 10^{-7} \tag{7.18h}$$

when the advance ratio is less than one.

Equation (7.1c) may be used to convert these thrust to weight ratios to power to weights ratios.

b) Ceiling. The vertical velocity at the end of the climb is:

$$\left(V_V\right)_2 = \frac{1.46 Z Q_V}{\left[1 + 0.12\left(\frac{Z Q_V}{100}\right)^2\right]}\left[\left(\frac{T}{Mg}\right)_I Y - 1.15\bar{\beta}_0^{\,1/2}\left(Z^2 \sigma_2^{-0.62} + \frac{0.73}{Z^2}\sigma_2^{-0.38}\right)\right] \ m/s \tag{7.19a}$$

(The correction term X_I is not required here.)

Equation (7.19a) may be rearranged to give the total thrust to weight ratio at the start of the climb required to give a specific rate of climb, or ceiling value $(V_V)_C$ at the end of the climb:

$$\left(\frac{T}{Mg}\right)_I = \left[\frac{0.685(V_V)_C}{ZQ_V}\left\{1 + 0.12\left(\frac{ZQ_V}{100}\right)^2\right\} + 1.15\bar{\beta}_0^{1/2}\left(Z^2\sigma_2^{-0.62} + \frac{0.73}{Z^2}\sigma_2^{-0.38}\right)\right] \Big/ Y$$

(7.19b)

This equation may be used to define the equivalent static thrust to weight ratio, and hence power to give the required final rate of climb $(V_V)_C$:

1) Piston engines:
   Use Eq (7.19b) to substitute into Eq (7.18f) to obtain the static thrust to weight ratio, and then use Eq (7.1d) to get the power to weight ratio.
2) Turbopropeller engines:
   Use Eq (7.19b) with either of Eqs (7.18g) or (7.18h), dependent upon the advance ratio, to obtain the static thrust to weight ratio and then use Eq (7.1c) to get the power to weight ratio.

c) Fuel consumed during the climb:

1) Piston engines - climb at 85% maximum rated power:

$$\frac{W_F}{(Mg)_0} = \frac{0.585 N_E H_2}{(V_V)_{MEAN}} \frac{P_0}{(Mg)_0}\left(1 + 0.27\sigma_2^{1.1}\right)$$

(7.20a)

$$(H_2 \text{ in km})$$

2) Turbopropeller engines:

$$\frac{W_F}{Mg)_0} = \frac{0.375 N_E H_2}{(V_V)_{MEAN}} \frac{(P_0)}{(Mg)_0}\left[1 - 0.025(P_0)_{EQ} \times 10^{-3}\right]\left[1 + \sigma_2 + 0.321\sigma_2^{0.88}\left(\frac{V_{EAS}}{100}\right)\right]$$

$$(H_2 \text{ in km})$$

(7.20b)

7.6.2.4 Ground distance covered during climb

In all cases of constant equivalent airspeed climbs the ground distance covered during the climb is approximately:

$$s_G = \frac{0.5 H_2 V_{EAS}}{\left(V_V\right)_{MEAN}} \left(1 + \sigma_2^{-1/2}\right) \cos\left[\sin^{-1}\left\{\frac{2\left(V_V\right)_{MEAN}}{V_{EAS}\left(1 + \sigma_2^{-1/2}\right)}\right\}\right] \quad km \qquad (7.21)$$

The cosine term is approximately unity unless the climb is unusually steep, and H_2 is in km.

7.6.3 Constant Mach number climb

7.6.3.1 General

It is assumed that a constant Mach number climb is only appropriate to aircraft powered by low bypass ratio and fan engines as it is unlikely that propeller powered types will have good climb performance at higher forward speeds.

It is feasible for an aircraft to commence a constant Mach number climb at sea level but it is more likely that the initial climb will be at a constant equivalent airspeed as covered in the previous section. The constant Mach number climb may be assumed to commence at an altitude where the true speed is approaching the cruise Mach number.

It is possible that climb at higher altitude may imply an unacceptably high value of the lift coefficient, C_L. This has two effects:

i) The estimation of $(V_V)_{MEAN}$ may be in error, especially if the C_L implied is more than $1.5C_{LMD}$ (C_L at minimum drag speed). This may be overcome by dividing the climb into two sections, firstly up to an altitude implied by C_L equal to C_{LMD}, and secondly above that altitude.

ii) The required C_L may be beyond a usable value. It is suggested that as a first approximation the usable value of C_L should be assumed to be 1.25 times the cruise buffet limited values quoted in Chapter 6, paragraph 6.2.4.6.

7.6.3.2 Low bypass ratio and fan engines - climb up to 11 km altitude

a) Mean rate of climb up to 11 km altitude.

$$(V_V)_{MEAN} = \frac{170 M_N}{\left(1 - 0.153 M_N^2 \sigma_1^{0.23}\right)} \left[\frac{T_0}{(Mg)_1} \frac{\tau_1}{\sigma_1^{(S - 0.02)}} \left\{\sigma_1^{(S + 0.1)} + \sigma_2^{(S + 0.12)}\right\} - Q_M\left(\sigma_1^{1.38} + \sigma_2^{1.35}\right) - \frac{\bar{\beta}}{Q_M}\left(\sigma_1^{-1.1} + \sigma_2^{-1.08}\right)\right] \quad m/s \qquad (7.22a)$$

where T_0 is the sea level, dry static thrust
τ_I is the value at Mach number M_N and at start of climb altitude defined by σ_I, with allowance for installation losses, see Chapter 3, paragraph 3.6.4
$\bar{\beta} = (C_{DZ} K_V)$ where C_{DZ} and K_V are the values appropriate to M_N

$$Q_M = \frac{70910 M_N^2 C_{DZ}}{(Mg/S)_I} \tag{7.22b}$$

Equation (7.22a) may be inverted to give the initial thrust to weight ratio required to give a specified mean rate of climb:

$$\frac{T_0}{(Mg)_I} = \left\{\frac{\sigma_I^{(S-0.02)}}{\sigma_I^{(S+0.1)} + \sigma_2^{(S+0.12)}}\right\}\left[\frac{0.0059(1 - 0.153 M_N^2 \sigma_I^{0.23})}{M_N}(V_V)_{MEAN} + Q_M(\sigma_I^{1.38} + \sigma_2^{1.35}) + \frac{\bar{\beta}}{Q_M}(\sigma_I^{-1.1} + \sigma_2^{-1.08})\right] \Big/ \tau_1 \tag{7.22c}$$

b) Ceiling up to 11 km altitude. The vertical velocity at the end of climb is:

$$(V_V)_2 = \frac{340 M_N}{(1 - 0.153 M_N^2 \sigma_2^{0.23})}\left[\frac{T_0}{(Mg)_I}\tau_I \frac{\sigma_2^{(S+0.1)}}{\sigma_I^S} - Q_M \sigma_2^{1.33} - \frac{\bar{\beta}}{Q_M}\sigma_2^{-1.1}\right] \quad m/s \tag{7.23a}$$

If $(V_V)_C$ is the required vertical velocity at the ceiling then the value of the sea level, dry, static thrust required to achieve it is:

$$\frac{T_0}{(Mg)_1} = \frac{\sigma_I^S}{\tau_I}\left[\frac{0.00294(1 - 0.153 M_N^2 \sigma_2^{0.23})\sigma_2^{-(S+0.1)}(V_V)_C}{M_N} + Q_M \sigma_2^{(1.23-S)} + \frac{\bar{\beta}}{Q_M}\sigma_2^{-(1.2+S)}\right] \tag{7.23b}$$

c) Fuel consumed in climb up to 11 km altitude:

$$\frac{W_F}{(Mg)_0} = \frac{0.139(H_2 - H_1)\,\tau_1 c_1}{(V_V)_{MEAN}} \frac{T_0}{(Mg)_0}\left\{1 + \left(\frac{\sigma_2}{\sigma_1}\right)^{(S+0.08)}\right\} \tag{7.24}$$

$$(H_2 \text{ and } H_1 \text{ in km})$$

where c_1 is the specific fuel consumption appropriate to Mach number M_N and at start of climb altitude defined by σ_1.

d) Ground distance covered up to 11 km altitude:

$$s_G = \frac{170M_N}{(V_V)_{MEAN}}(H_2 - H_1)\left(\sigma_1^{0.117} + \sigma_2^{0.117}\right)\cos\left[\sin^{-1}\left\{\frac{(V_V)_{MEAN}}{170M_N\left(\sigma_1^{0.117} + \sigma_2^{0.117}\right)}\right\}\right] \text{ km}$$

$$(H_2 \text{ and } H_1 \text{ in km}) \tag{7.25}$$

7.6.3.3 Low bypass ratio engines - climb from 11 to 20 km altitude

a) Mean rate of climb from 11 km altitude:

$$(V_v)_{MEAN} = 148X_2M_N\left[\frac{T_0\tau_1}{(Mg)_1}\left\{1 + \frac{\sigma_2^{(S-0.02)}}{0.297^{(S-0.02)}}\right\} - 0.223Q_M\left(1 + 3.29\sigma_2^{0.98}\right) - 4.48\frac{\bar{\beta}}{Q_M}\left(1 + 0.304\sigma_2^{-0.98}\right)\right] \tag{7.26a}$$

where $(V_V)_{MEAN}$ is in m/s. X_2 is a correction factor to allow for low relative rate of climb as the ceiling is approached:

$$X_2 = \left[\frac{\left\{\frac{T_0\tau_1}{(Mg)_1} - 0.223Q_M - 4.48\frac{\bar{\beta}}{Q_M}\right\}}{\left\{3.36\sigma_2^{0.98}\left(\frac{T_0\tau_1}{(Mg)_1} - 0.22Q_M\right) - \frac{1.36\bar{\beta}}{Q_M\sigma_2^{0.98}}\right\}}\right]^{-0.11} \tag{7.26b}$$

For the usual case where s is equal to unity, Eq (7.26a) may be inverted to express the thrust to weight ratio required to give a specified mean rate of climb:

$$\frac{T_0}{(Mg)_1} = \left\{1 + \frac{3.36\sigma_2^{0.98}}{\sigma_2^{0.98}}\right\}\left[\frac{0.0068}{X_2 M_N}(V_V)_{MEAN} + 0.223 Q_M\left(1 + 3.29\sigma_2^{0.98}\right) + 4.48\frac{\bar{\beta}}{Q_M}\left(1 + 0.304\sigma_2^{-0.98}\right)\right] \Big/ \tau_1 \qquad (7.26c)$$

Since X_2 implicitly involves the thrust to weight ratio, Eq (7.26c) is best solved by initially assuming X_2 to be unity, and then repeating the process with the derived thrust to weight ratio, using Eq (7.26b), until convergence is achieved.

b) Ceiling above 11 km altitude, but less than 20 km.

The vertical velocity at the end of climb for $s = 1.0$ is:

$$(V_V)_2 = 993 M_N\left[\frac{T_0\tau_1}{(Mg)_1}\sigma_2^{0.98} - 0.223 Q_M\sigma_2^{0.98} - 0.404\frac{\bar{\beta}}{Q_M}\sigma_2^{-0.98}\right] \ m/s \qquad (7.27a)$$

and the value of T_0 to a give ceiling vertical velocity of $(V_V)_C$ is:

$$\frac{T_0}{Mg_1} = \left[\frac{0.001\sigma_2^{-0.98}}{M_N}(V_V)_C + 0.223 Q_M + 0.404\frac{\bar{\beta}}{Q_M}\sigma_2^{-1.96}\right] \Big/ \tau_1 \qquad (7.27b)$$

c) Fuel consumed for climb from 11 km up to 20 km altitude is:

$$\frac{W_F}{(Mg)_0} = \frac{0.139(H_2 - 11)}{(V_V)_{MEAN}}\tau_1 c_1\frac{T_0}{(Mg)_0}\{1 + 3.37\sigma_2\} \qquad (7.28)$$

$(H_2$ *in km)*

where c_1 refers to conditions at M_N and 11 km altitude.

d) Ground distance covered from 11 to 20 km altitude is:

$$s_G = \frac{295M_N}{(V_V)_{MEAN}}(H_2 - 11)\cos\left[\sin^{-1}\left\{\frac{(V_V)_{MEAN}}{295M_N}\right\}\right] \quad km \tag{7.29}$$

$(H_2 \ in \ km)$

7.6.3.4 Low bypass ratio engines - climb above 20 km altitude

a) Mean rate of climb from 20 km altitude; s = 1.0

$$(V_V)_{MEAN} = \frac{148X_3M_N}{(1 + 0.016M_N^2)}\left[\frac{T_0}{(Mg)_1}\tau_I(1 + 12.73\sigma_2^{0.967}) - 0.0542Q_M(1 + 11.9\sigma_2^{0.94}) - 18.4\frac{\bar{\beta}}{Q_M}(1 + 0.786\sigma_2^{-0.967})\right] \tag{7.30a}$$

where $(V_V)_{MEAN}$ is in m/s and τ_I refers to conditions at M_N and 20 km altitude.

For this case the correction factor X_3 is given by:

$$X_3 = \left[\left\{\frac{T_0}{Mg_I}\tau_I - 0.0542Q_M - 18.4\frac{\bar{\beta}}{Q_M}\right\} \Big/ \left\{\frac{12.73T_0\tau_I}{(Mg)_I}\sigma_2^{0.967} - 0.643Q_M\sigma_2^{0.94} - 1.44\sigma_2^{-0.967}\right\}\right]^{-0.11} \tag{7.30b}$$

Equation (7.30a) may be inverted to give the initial thrust to weight ratio needed to yield a given mean rate of climb, it being necessary to assume that X_3 is unity in the first instance:

$$\frac{T_0}{(Mg)_I} = \frac{(1 + 12.73\sigma_2^{0.967})}{\tau_I}\left[\frac{0.0068}{X_3M_N}(1 + 0.016M_N^2)(V_V)_{MEAN} + 0.0542Q_M(1 + 11.9\sigma_2^{0.94}) + 18.4\frac{\bar{\beta}}{Q_M}(1 + 0.786\sigma_2^{-0.967})\right] \tag{7.30c}$$

The first value of X_3 derived may be used for subsequent correction.

b) Ceiling above 20 km altitude. The vertical velocity at the end of the climb is:

$$\left(V_V\right)_2 = \frac{3762M_N}{\left(1 + 0.016M_N^2\right)}\left[\frac{T_0}{(Mg)_I}\tau_I\sigma_2^{-0.97} - 0.0504Q_M\sigma_2^{0.94} - 0.114\frac{\bar{\beta}}{Q_M}\sigma_2^{-1.94}\right] \quad m/s \tag{7.31a}$$

and hence the value of thrust to weight ratio needed to achieve a ceiling vertical velocity, $(V_V)_C$ is :

$$\frac{T_0}{(Mg)_I} = \left[\frac{0.000266\left(1 + 0.016M_N^2\right)}{M_N}\sigma_2^{-0.97}\left(V_V\right)_C + 0.0504Q_M\sigma_2^{0.03} + 0.114\frac{\bar{\beta}}{Q_M}\sigma_2^{-1.94}\right] \tag{7.31b}$$

c) Fuel consumed in climb above 20 km.

$$\frac{W_F}{(Mg)_0} = \frac{0.139\left(H_2 - 20\right)\tau_I c_I}{\left(V_V\right)_{MEAN}}\frac{T_0}{(Mg)_0}\left(1 + 13.9\sigma_2\right) \tag{7.32}$$

$$(H_2 \ in \ km)$$

where c_I refers to conditions at M_N and 20 km altitude.

d) Ground distance covered above 20 km altitude.

$$s_G = \frac{296\left(H_2 - 20\right)}{\left(V_V\right)_{MEAN}}\cos\left[\sin^{-1}\left\{\frac{\left(V_V\right)_{MEAN}}{296M_N}\right\}\right] \quad km \tag{7.33}$$

$$\left(H_2 \ in \ km\right)$$

7.6.4 Linearly varying Mach number climb

A linear variation of Mach number with altitude has been selected as typical of supersonic climb profiles although there are many other possibilities. The equations may be used to cover the case of nominally constant altitude supersonic acceleration, but see the next paragraph.

It is assumed that the aircraft will accelerate through the transonic region in level flight, see paragraph 7.7.4. The climb commences at a Mach number M_{N1} , which is typically 1.2, and increases so that the required supersonic operating Mach number M_{N2} is reached at a specified operating altitude. Such a climb profile is reasonably appropriate for aircraft intended to have a relatively long range supersonic cruise. The analysis only applies to engines of low bypass ratio. Initial and final heights are H_1 and H_2 km defined by σ_1 and σ_2 respectively. The formulation of the equations is such that it is necessary to introduce a small, nominal, altitude increment even when the flight is at constant height.

a) Mean rate of climb:

$$(V_V)_{MEAN} = \frac{1.42M_{N1}}{\left(1 + M_{N1}M'_N\right)}\left[\frac{T_0}{(Mg)_1}\tau_{MN1}\bar{A} - 0.752Q_{VM}\sigma_1\bar{B} - \frac{1.33\bar{\beta}_1\bar{C}}{Q_{VM}\sigma_1}\right] \quad m/s \tag{7.34a}$$

where $$Q_{VM} = \frac{70910(C_{DZ})_1 M_{N1}^2}{(Mg/S)_1} \quad and \quad \bar{\beta}_1 = (C_{DZ})_1(K_V)_1 \tag{7.34b}$$

(compare with. Eq (7.22))

$$\bar{A} = 1 + \bar{\tau}\,\bar{M}_N\left(\sigma_2/\sigma_1\right)^{0.96} \tag{7.34c}$$

$$\bar{B} = 1 + \Delta\bar{M}_N\left(M_{N2}/M_{N1}\right)^2\left(\sigma_2/\sigma_1\right)^{0.96} \tag{7.34d}$$

$$\bar{C} = 1 + \bar{K}_V\,\bar{M}_N\left(M_{N1}/M_{N2}\right)^2\left(\sigma_1/\sigma_2\right)^{0.96} \tag{7.34e}$$

($\bar{A}$ $\bar{B}$ and $\bar{C}$ respectively are the contributions of thrust, zero lift and induced drag)

$$M_N' = 8.87(M_{N2} - M_{N1})/(H_2-H_1) \tag{7.34f}$$

$$\bar{\tau} = (1 + 0.91M_{N2})/(1 + 0.91M_{N1}) \tag{7.34g}$$

$$\bar{M}_N = \frac{M_{N2}(1 + M_N'M_{N1})}{M_{N1}(1 + M_N'M_{N2})} \tag{7.34h}$$

$$\bar{C}_{DW} = C_{DW}/(1 + 0.034(3 - M_N)^{3.5}) \tag{7.34i}$$

$$\Delta = \frac{(C_{DZ})_2}{(C_{DZ})_1} = \frac{(C_{DZ})_0(1 - 0.2M_{N2}) + \bar{C}_{DW}\{1 + 0.034(3 - M_{N2})^{3.5}\}}{(C_{DZ})_0(1 - 0.2M_{N1}) + \bar{C}_{DW}\{1 + 0.034(3 - M_{N1})^{3.5}\}} \tag{7.34j}$$

$$\bar{K}_V = \{1 + 1.2(M_{N2}^2 - 1)^{1/2}\}/\{1 + 1.2(M_{N1}^2 - 1)^{1/2}\} \tag{7.34k}$$

where $(K_V)_1$, C_{DW} and $(C_{DZ})_0$ are defined in Chapter 6, paragraph 6.3

T_0 is the sea level static, dry, installed thrust

τ_{MN1} is the value at M_{N1} and height H_1 with allowance for installation losses

$(Mg/S)_1$ is the wing loading at the start of this phase of the climb.

The thrust to weight ratio required to give a specified mean rate of climb may be obtained by rearrangement of Eq (7.34a):

$$\frac{T_0}{(Mg)_1} = \left[\frac{0.007(V_V)_{MEAN}}{M_{N1}}(1 + M_{N1}M_N') + 0.752Q_{VM}\sigma_1\bar{B} + \frac{1.33\bar{\beta}_1\bar{C}}{Q_{VM}\sigma_1}\right] \Big/ (\tau_{MN1}\,\bar{A}) \tag{7.35}$$

b) Ceiling. The vertical velocity at the end of the climb is:

$$(V_V)_2 = \frac{284M_{N1}}{(1 + M_{N1}M_N')}\left[\frac{T_0}{(Mg)_1}\tau_{MN1}(\bar{A} - 1) - 0.752Q_{VM}\,\sigma_1\bar{B} - \frac{1.33\bar{\beta}_1\bar{C}}{Q_{VM}\sigma_1}\right] \quad m/s \tag{7.36a}$$

The installed dry static thrust to give a ceiling vertical velocity of $(V_V)_C$ is:

$$\frac{T_0}{(Mg)_1} = \left[\frac{0.0035(V_V)_C\left(1 + M_{N1}M_N'\right)}{M_{N1}} + 0.752Q_{VM}\sigma_1(\bar{B} - 1) + \frac{1.33\bar{\beta}_1}{Q_{VM}\sigma_1}(\bar{C} - 1)\right] \Big/ \tau_{MN1}(\bar{A} - 1) \tag{7.36b}$$

c) Fuel consumed during the climb

$$\frac{W_F}{(Mg)_0} = \frac{0.139(H_2 - H_1)}{(V_V)_{MEAN}} \tau_{MN1} \frac{T_0}{(Mg)_0} c_{MN1}\left\{1 + \bar{\tau}\,\bar{c}\left(\frac{\sigma_2}{\sigma_1}\right)\right\} \tag{7.37a}$$

$$(H_2 \text{ and } H_1 \text{ in km})$$

where c_{MN1} is the specific fuel consumption at the start of this phase of the climb, and for dry engine conditions:

$$\bar{c} = (1 + 0.285M_{N2})/(1 + 0.285M_{N1}) \tag{7.37b}$$

or when reheat is used

$$\bar{c} = (1 + 0.2M_{N2})/(1 + 0.2M_{N1}) \tag{7.37c}$$

d) Ground distance covered in the climb is approximately given by:

$$s_G = \frac{60(2M_{N1} + 3M_{N2})(H_2 - H_1)}{(V_V)_{MEAN}} \cos\left[\sin^{-1}\left\{\frac{(V_V)_{MEAN}}{148(M_{N1} + M_{N2})}\right\}\right] \text{ km} \tag{7.38}$$

$$(H_2 \text{ in km})$$

In many cases the cosine term is close to unity.

7.7 Maximum level speed, manoeuvre and transonic acceleration

7.7.1 Turn rate and manoeuvre acceleration factor

The turn rate:

$$\psi = \frac{g(n^2-1)^{1/2}}{V} \qquad rad/\text{sec} \tag{7.39a}$$

$$where \quad n = \left[\left(\frac{\psi V}{g}\right)^2 + 1\right]^{1/2} \tag{7.39b}$$

n is the total normal acceleration factor. Thus given a turn rate requirement n may be evaluated for a given flight condition.

The radius of the turn is the velocity divided by the rate of turn and follows from Eq (7.39a).

7.7.2 Maximum level speed and sustained manoeuvre

7.7.2.1 General

Both maximum level speed and sustained manoeuvre imply a situation where the specific excess power, p_S, is zero. The difference is that the maximum level speed case is in straight, level flight so that the manoeuvre factor, n, is unity while this is clearly not the case in a sustained manoeuvre.

Equation (7.13d), for $p_S = 0$ gives:

$$\frac{T}{Mg} = \bar{D} + \frac{\beta n^2}{\bar{D}} \tag{7.40a}$$

which directly enables (T/Mg) to be found for a given set of conditions defined by n and $\bar{D}$.

Rearranging: $$\bar{D}^2 - \left(\frac{T}{Mg}\right)\bar{D} + \beta n^2 = 0 \tag{7.40b}$$

These equations may be used in various ways:

a) To find the minimum thrust to weight ratio and corresponding wing loading to meet a given condition. Using Eq (7.40a):

$$(T/Mg)_{MIN} \quad occurs\ when\ \bar{D} = \frac{qC_{DZ}}{(Mg/S)} = n\bar{\beta}^{\,1/2} \tag{7.40c}$$

$$and \quad (T/Mg)_{MIN} = 2n\bar{\beta}^{\,1/2} = 2n(C_{DZ}K_V)^{1/2} \tag{7.40d}$$

This case is met for a given true velocity, V, when:

$$\left(\frac{Mg}{S}\right) = \frac{\rho V^2}{2n}\left(\frac{C_{DZ}}{K_V}\right)^{1/2} \tag{7.40e}$$

b) Generally Eq (7.40b) may be solved to give:

$$\bar{D} = \left[\left(\frac{T}{Mg}\right) \pm \left\{\left(\frac{T}{Mg}\right)^2 - 4\bar{\beta}n^2\right\}^{1/2}\right] \bigg/ 2 = \frac{qC_{DZ}}{(Mg/S)} \tag{7.41a}$$

and the corresponding wing loading for a velocity, V, is:

$$\left(\frac{Mg}{S}\right) = \rho V^2 C_{DZ} \bigg/ \left[\left(\frac{T}{Mg}\right) + \left\{\left(\frac{T}{Mg}\right)^2 - 4\bar{\beta}n^2\right\}^{1/2}\right] \tag{7.41b}$$

c) In level, straight flight n = 1, therefore, using Eqs (7.38e) and (7.38d):

$$V = 1.41\left[\left(\frac{Mg}{S}\right)\left(\frac{K_V}{C_{DZ}}\right)^{1/2}\right]^{1/2} \tag{7.42a}$$

for which

$$\left(\frac{T}{Mg}\right)_{MIN} = 2(C_{DZ}K_V)^{1/2} \tag{7.42b}$$

and generally, the maximum speed for a given wing loading and thrust to weight ratio may be derived from Eq (7.41b):

$$V_{MAX} = \left[\frac{(Mg/S)}{\rho C_{DZ}}\left\{\left(\frac{T}{Mg}\right) + \left\{\left(\frac{T}{Mg}\right)^2 - 4\bar{\beta}\right\}^{1/2}\right\}\right]^{1/2} \quad m/s \tag{7.42c}$$

(True air speed)

7.7.2.2 Relationship of thrust required to sea level static value

a) Low bypass ratio and fan engine propelled aircraft. The values of thrust to weight ratio derived from Eq (7.40) may be related to the sea level take-off conditions by simply multiplying by the factor (k_m/τ_m), defined at Eq (7.13f), for the appropriate conditions.

b) Propeller propelled aircraft. The relationship is more complex in this case due to the dependence of thrust on power and propeller characteristics. The equations given under cruise conditions in paragraph 7.8.3.6 may be used with the appropriate conditions.

7.7.3 Instantaneous manoeuvre

As mentioned in paragraph 7.1.2.5, the intensity of an instantaneous manoeuvre is limited either by the maximum usable lift coefficient in that flight condition or by the structural design normal acceleration factor, n_1.

Hence either $\qquad (n)_{INST} = n_1$

or $\qquad (n)_{INST} = \frac{\rho}{2}V^2 C_{LUSE} / (Mg/S) \qquad$ (7.43)

whichever is least, C_{LUSE} being the usable lift coefficient.

Equation (7.43a) implies that:

$$(Mg/S)_0 = \frac{\rho}{2} V^2 C_{LUSE} / n_{INST} k_m$$

Thus given a value of $(n)_{INST}$, either directly, or through turn rate using Eq (7.39), the maximum (Mg/S) which enables it to be achieved may be found.

7.7.4 Transonic acceleration

7.7.4.1 General

The transonic acceleration characteristics are important for a supersonic design in that:

i) It is necessary to ensure that the aircraft can reach supersonic flight.

ii) The time taken for transonic acceleration must not be unduly high.

The estimation of transonic drag is complex because of the changes in flow regime around $M_N = 1$. An analysis of some typical supersonic designs of slender and not so slender configuration suggests that in the range $0.9 \le M_N \le 1.2$ it is satisfactory to assume:

a) For $0.9 \le M_N \le 1.0$: C_{DZ} should be based on a subsonic representation.

b) For $1.0 \le M_N < 1.2$: C_{DZ} should be based on a supersonic representation.

c) The induced drag factor, K_V, is best based on a subsonic representation up to a Mach number of about 1.2 above which a supersonic representation is more satisfactory. However, in some circumstances this may underestimate the induced drag at $M_N = 1.2$.

These assumptions assume that the drag expressions are those of Chapter 6, paragraph 6.3.

7.7.4.2 Mean acceleration factor

The above drag representation will be assumed together with a definition of the transonic range as being $0.9 \le M_N \le 1.2$.

For an aircraft in level flight, Eq (7.13d) may be written as:

$$\frac{p_S}{V} = \left[\frac{T}{Mg} - \left(\bar{D} + \frac{\bar{\beta}}{\bar{D}}\right)\right]$$

where(p_S/V) is the acceleration factor in forward flight, that is the forward acceleration is $(9.81p_S/V)$ m/s.

Using the previously quoted representations of thrust and drag, together with the assumptions outlined above, the mean acceleration factor between $M_N = 0.9$ and $M_N = 1.0$ is:

$$m_a = \bar{T}_a \frac{T_0}{(Mg)_9} - \bar{D}_a \tag{7.44a}$$

$$\text{where } \bar{T}_a = 0.5\left(\tau_{MN9} + \tau_{MN10}\right)$$

$$and\ \bar{D}_a = \frac{51920(C_{DZ})_o}{(Mg/S)_9}\left[1 + 0.0081\left\{\frac{(\cos\Lambda_{1/4})^2}{A_f - t/c}\right\}^{20} + 0.942\frac{\bar{C}_{DW}}{(C_{DZ})_o}\right]\sigma^{1.235}$$

$$+ 1.727(K_V)_o(Mg/S)_9\ \sigma^{-1.235}\ x10^{-5}$$

The mean acceleration factor between $M_N = 1.0$ and $M_N = 1.2$:

$$m_b = \bar{T}_b\frac{T_0}{(Mg)_9} - \bar{D}_b$$

$$where\ \bar{T}_b = 0.505(\tau_{MN10} + \tau_{MN12}) \qquad (7.44b)$$

$$and\ \bar{D}_b = \frac{67844(C_{DZ})_o}{(Mg/S)_9}\left[1 + 1.69\frac{\bar{C}_{DW}}{(C_{DZ})_o}\right]\sigma^{1.235}$$

$$+ 1.44\ (K_V)_o\ (Mg/S)_9\ \sigma^{-1.235}\ x10^{-5}$$

where suffix 9 indicates conditions at M_N=0.9, 10 at M_N=1 and 12 at M_N=1.2.

τ_{MN} values here at appropriate M_N and altitude. These may be related to the static conditions by using the factor (k_m/τ_m) defined at Eq (7.13f).

$(C_{DZ})_0$ and $(K_V)_0$ are the incompressible values. See Chapter 5, paragraphs 5.2.2.4 and 5.2.6.2 for the definition of $\Lambda_{1/4}$, A_f and t/c.

Allowance has been made for fuel used during the acceleration, typically $0.01M_0$.

The above equations are applicable up to 13 km altitude, although 11 km is a typical condition. Further it is likely that reheat will be necessary for transonic acceleration.

For a typical low bypass ratio engine (R = 0.8) Eqs (7.44) may be simplified to give: at 11 km, R = 0.8 and s = 0.8, using Chapter 3, Eqs (3.7) and (3.8):

$$and\ \bar{D}_a = \frac{11600(C_{DZ})_o}{(Mg/S)_9}\left[1 + 0.0081\left\{\frac{(\cos\Lambda_{1/4})^2}{A_f - t/c}\right\}^{20} + 0.942\frac{\bar{C}_{DW}}{(C_{DZ})_o}\right] \qquad (7.44c)$$

$$+ 7.727(K_V)_o(Mg/S)_9\ x10^{-5}$$

$$\bar{T}_b = 0.6241F_\tau$$

and
$$\bar{D}_b = \frac{15160(C_{DZ})_0}{(Mg/S)_9}\left[1 + 1.69\frac{\bar{C}_{DW}}{(C_{DZ})_0}\right] + 6.456(K_V)_0(Mg/S)_9 \times 10^{-5} \qquad (7.44d)$$

7.7.4.3 Time of transonic acceleration

The time taken to accelerate between $M_N = 0.9$ and $M_N = 1.2$ is:

$$t = 0.0102a\left[\frac{1}{m_a} + \frac{2}{m_b}\right] \quad s \qquad (7.45a)$$

where a is the local speed of sound and at 11 km altitude the time taken is:

$$t = 3\left[\frac{1}{m_a} + \frac{2}{m_b}\right] \quad s \qquad (7.45b)$$

7.7.4.4 Thrust required

Substitution from Eq (7.44) for m_a and m_b enables the required thrust to weight ratio to be determined for a given time for the transonic acceleration:

$$\frac{T_0}{(Mg)_9} = \bar{E} + \left(\bar{E}^2 - 2\bar{F}\right)^{1/2} \qquad (7.45c)$$

where
$$\bar{E} = \frac{0.0051a}{\bar{T}_a\bar{T}_b t}\left[\frac{98t}{a}\left(\bar{T}_a\bar{D}_b + \bar{T}_b\bar{D}_a\right) + 2\bar{T}_a + T_b\right]$$

(7.45d)

$$\bar{F} = \frac{0.0051a}{\bar{T}_a\bar{T}_b t}\left[\frac{98t}{a}\bar{D}_a\bar{D}_b + 2\bar{D}_a + \bar{D}_b\right]$$

And for the case at 11 km altitude given by Eqs (7.44c) and (7.44d):

$$\bar{E} = \left[0.8\left(\bar{D}_b + 1.11\bar{D}_a\right) + 7.5/t\right] \Big/ F_\tau$$

$$\bar{F} = \left[1.428\bar{D}_a\bar{D}_b + 4.3\left(2\bar{D}_a + \bar{D}_b\right)/t\right] \Big/ F_\tau^2 \qquad (7.45e)$$

7.7.4.5 Fuel used during acceleration

As mentioned previously, for an aircraft with reasonable transonic performance, the fuel used during the acceleration from a Mach number of 0.9 to 1.2 is likely to be equivalent to about 1% of the weight at the commencement of the acceleration.

It is found that a more accurate evaluation is given by taking mean values of the thrust and specific fuel consumption, that is values equivalent to a Mach number of 1.05.

Fuel used :

$$\frac{W_F}{(Mg)_0} = 2.78\frac{T_0}{(Mg)_0}t\left(\tau_{MEAN}\ c_{MEAN}\right) \times 10^{-4} \qquad (7.46a)$$

where τ_{MEAN} and c_{MEAN} refer to conditions at a Mach number of 1.05 and appropriate altitude and t is total time of the acceleration.

For the case of transonic acceleration at 11 km altitude, using reheat with engines of bypass ratio 0.8:

$$\frac{W_F}{(Mg)_0} = 2.74\frac{T_0}{(Mg)_0}t \times 10^{-4} \qquad (7.46b)$$

where use has been made of the engine models given in Chapter 3.

7.8 Cruise - range and performance

7.8.1 General

It is usual to assume that the aircraft is in steady level flight during the cruise phase. While this is not always strictly true it is, for all practical purposes, a sufficient assumption. The range, that is the distance flown during the cruise, is dependent to some extent on the flight technique employed. For example in some instances a sacrifice of potential range may be made in order to achieve a higher speed. The overall efficiency in cruising flight is simply the ratio of the work done in propelling the aircraft to the energy content of the fuel used to achieve it, thus:

$$\eta_o = -\frac{T\Delta s}{\bar{H}\Delta M} \tag{7.47a}$$

where T is the thrust, Δs the incremental distance travelled, ΔM is the rate of fuel used, expressed in terms of the decrement of aircraft mass, M, and $\bar{H}$ is the energy content of unit mass of fuel.

In the limit:

$$ds = -\eta_0 \bar{H}(dM/T) \tag{7.47b}$$

In steady level flight the thrust and drag are equal and the lift is equal to the weight so that

$$ds = -\eta_o \bar{H} \frac{L}{D} \frac{dM}{Mg} \tag{7.47c}$$

From Eq (7.47a), since $ds = V\,dt$:

$$\eta_0 \bar{H} = -\frac{TV}{dM/dt} \tag{7.47d}$$

where $(\eta_o \bar{H})$ may be regarded as a propulsion efficiency (L/D) is the lift to drag ratio which is a measure of the aerodynamic efficiency

(dM/M) is effectively an airframe efficiency since it is a measure of the ratio of fuel used to total mass. For a given all up mass and payload the lower the zero fuel mass the greater will be the fuel mass available for cruise

Now $(dM/dt)/T$ is a measure of the specific fuel consumption since it is the rate of fuel mass usage in terms of thrust. In terms of the rate of fuel weight used rather than mass, the specific fuel consumption is:

$$c = (dMg/dt)/T$$

substituting this into Eq (7.47d) gives:

$$\eta_o \bar{H} = gV$$

and hence from Eq (7.47c):

$$ds = \frac{V}{c}\frac{L}{D}\frac{dM}{M} \tag{7.48}$$

This is the basic form of the well known Breguet range equation. (Note here c is N/N/s units).

It is clear that for given powerplant and airframe characteristics the greatest distance flown occurs when (VL/D) is maximised.

Thus, as a general conclusion, it may be stated that the maximum range occurs when the velocity is as high as possible consistent with maintaining a maximum value of the lift to drag ratio. This usually means flying at, or close to the critical Mach number. However, there are often operational restrictions which influence the flight conditions.

When (VL/D) and c are constant Eq (7.48) may be integrated to give the distance travelled in cruise, s:

$$s = 3600\frac{V}{c}\frac{L}{D}\log_e\frac{M_1}{M_2}$$

where c is now in the more usual units of N/N/h for a jet engine, and M_1 and M_2 are the masses at the beginning and end of the cruise respectively, or:

$$s = 8.29\frac{V}{c}\frac{L}{D}\log_{10}\frac{M_1}{M_2} \qquad km \tag{7.49}$$

Clearly the 8.29 constant depends upon the units used to define the specific fuel consumption. For example if c is in mass units, kg/N/h, the constant becomes 0.845 while if it is quoted as mg/N/s the constant is 234.7. In the case of propeller driven aircraft where c is quoted in terms of power rather than thrust the equation takes a different form, see paragraph 7.8.3.

While range is frequently calculated by using Eq (7.49) there are circumstances where the assumption of constant values is invalid.

7.8.2 Jet and fan engine propelled aircraft

7.8.2.1 Flight limitations and relationship to sea level static conditions

Cruise flight conditions may be limited in a number of ways:

a) *Mach number limit* which is effectively the critical Mach number as determined by drag rise and buffet.

b) *Powerplant thrust available* which may be inadequate to enable the critical Mach number to be achieved at a particular altitude.

c) *Flight altitude* which may be determined by Air Traffic Control instructions and, in the limit, by the design pressure differential in a cabin.

The thrust in cruise is equal to the drag and hence it is directly defined by the lift to drag ratio appropriate to a given flight condition. The thrust to weight ratio in cruise may be related back to the sea level static value by multiplying by the factor (k_m/τ_m), defined at Eq (7.13f), with the appropriate conditions.

7.8.2.2 Mach number limited flight

This is a common case. At a given altitude the velocity of Eq (7.49) is uniquely defined in terms of Mach number as $(M_N\, a)$ where a is the speed of sound at that altitude and so Eq (7.49) may be written as:

$$s = 8.29a\frac{M_N}{c}\frac{L}{D}\log_{10}\frac{M_1}{M_2} \quad km \qquad (7.50a)$$

and between 11 and 20 km altitude where a is constant:

$$s = 2445\frac{M_N}{c}\frac{L}{D}\log_{10}\frac{M_1}{M_2} \quad km \qquad (7.50b)$$

In this condition the maximum range occurs when the altitude of flight is chosen such that the lift coefficient is the value corresponding to the minimum drag condition, thereby giving the maximum value of lift to drag ratio. As fuel is used the lift required to balance the weight decreases so in order to maintain the maximum lift to drag ratio it is necessary to reduce the dynamic pressure. Since the assumption is that the Mach number is constant this can only be done by allowing the aircraft to climb. The implied reduction in drag, but not drag coefficient, is more or less balanced by the reduction in thrust. This so-called "climbing cruise" is the most efficient flight pattern associated with constant Mach number. Air Traffic Control usually demands that aircraft fly at a constant flight level for a given period and so a stepped climb is often the nearest approximation to a true climbing cruise that can be achieved. There are exceptions, for example supersonic and some high performance executive aircraft which can operate at altitudes above the usual Air Traffic Control levels which are typically up to 43,000 ft (13.11 km).

Equation (7.50b) may be developed to give the wing loading appropriate to maximum lift to drag ratio at a given Mach number and altitude.

Let
$$\bar{\beta}_{Cr} = \left(C_{DZ}\right)_{Cr}\left(K_V\right)_{Cr} \tag{7.50c}$$

where suffix Cr indicates cruise conditions.

Then, from Eq (7.14b) the maximum lift to drag ratio is $\left(0.5/\left(\bar{\beta}_{Cr}\right)^{1/2}\right)$ and from Eq (7.14c):

$$\left(\frac{Mg}{S}\right)_{Cr} = 0.613\sigma a^2 M_N^2 \left(C_{DZ}\right)_{Cr} / \left(\bar{\beta}_{Cr}\right)^{1/2} \tag{7.50d}$$

Between 11 and 20 km altitude the speed of sound, a, is constant and in this case:

$$\left(\frac{Mg}{S}\right)_{Cr} = 53300\sigma M_N^2 \left(C_{DZ}\right)_{Cr} / \left(\bar{\beta}_{Cr}\right)^{1/2} \tag{7.50e}$$

When a stepped climb is necessary, over a given constant altitude part of the flight the lift coefficient decreases as fuel is used and hence the lift to drag ratio cannot be constant at the maximum value. Flight below the minimum drag condition may aggravate buffet effects so it is generally best to start the stage as near to the maximum lift to drag condition as possible. There is only a slow reduction in lift to drag ratio as the fuel is used but when there has been sufficient reduction in mass to enable flight at the next level to be achieved, either 2000 or 4000 ft higher, a step can be contemplated.

7.8.2.3 Thrust limited - increasing Mach number flight

In some circumstances the available thrust at a given altitude may be insufficient for the aircraft to achieve the critical Mach number. When this is so the maximum speed attainable is:

$$V = \left(\frac{2T}{\rho S C_D}\right)^{1/2} \tag{7.51a}$$

where T is the limited thrust corresponding to velocity, V, and the altitude defined by the air density, ρ. C_D is the total drag coefficient at that velocity.

The corresponding lift coefficient is:

$$C_L = \frac{MgC_D}{T}$$

so that the attainable lift to drag ratio is (Mg/T) which is clearly not constant. The speed also is not constant in that C_D will reduce as Mg reduces due to the reduction in lift coefficient.

Using Eq (7.48):

$$ds = \frac{1}{c}\left(\frac{2T}{\rho S C_D}\right)^{1/2} \frac{Mg}{T}\frac{dM}{M} = \frac{g}{cT^{1/2}}\left(\frac{2}{\rho S C_D}\right)^{1/2} dM \qquad (7.51b)$$

Analytical development of Eq (7.51b) is difficult because of the variation of C_D with C_L which implies an increase in speed and thus a variation of thrust. Over small increments of the flight where C_D and T may be assigned mean values:

$$\Delta s = \frac{35.3}{cT^{1/2}}\left(\frac{2}{\rho s C_D}\right)^{1/2}(M_1 - M_2) \quad km$$

where c is now N/N/hour and M_1 and M_2 are the masses at the beginning and end of that short segment of the flight.

In practice as the speed of the aircraft increases as the fuel is used it may well be that at some stage the critical Mach number is reached so that Eq (7.50) applies. An analysis similar to that suggested for propeller-powered aircraft in paragraph 7.8.3.3. may be used.

7.8.2.4 Diversion and standby

The range formula may be used to evaluate diversion and standby fuel, using the appropriate speed and altitude conditions.

7.8.2.5 Endurance

If it is assumed that the specific fuel consumption is a constant in cruise then it follows that the maximum endurance corresponds with minimum thrust, that is minimum drag at a given altitude. Hence the maximum endurance coincides with the maximum range. In practice the specific fuel consumption is not constant, tending to increase with speed. The true maximum endurance may occur at a speed somewhat lower than that which gives maximum range.

7.8.3 Propeller-powered aircraft

7.8.3.1 Specific fuel consumption and propeller efficiency

In the case of propeller-driven aircraft it is common for the specific fuel consumption to be stated in terms of the power of the basic engines and for propeller efficiency to be used to define the thrust in a given flight condition.

The engine power in a given flight condition is:

$$P = TV/\eta$$

where η is now the propeller efficiency, see Chapter 3, paragraph 3.6.2.3.

Equation (7.48) now takes the form:

$$ds = \frac{\eta}{c}\frac{L}{D}\frac{dM}{M} \tag{7.52}$$

where here c is defined in terms of N/W/sec.

Note that although speed does not appear explicitly in Eq (7.52) it is implied by the need to develop a definitive value of lift , L, at a given altitude.

7.8.3.2 Range in constant conditions

For constant conditions, should they apply, Eq (7.52) may be integrated to give:

$$s = 8290\frac{\eta}{c}\frac{L}{D}\log_{10}\left(\frac{M_1}{M_2}\right) \quad km \tag{7.53a}$$

where now c is in N/kW/h.

The constant becomes 845 when c is given in units of kg/kW/h and 234 when it is in µg/J.

This equation is often the most convenient form to use for evaluating the range of propeller-driven aircraft. The appropriate values of propeller efficiency, η, as derived from Chapter 3, paragraph 3.6.2.5 give:

a) For advance ratio, J, less than unity (usual case for piston-engined aircraft); using Chapter 3, Eq (3.11h), Eq (7.53a) becomes:

$$s = \frac{3316}{c}V_{Cr}^{0.16}\frac{L}{D}\log_{10}\left(\frac{M_1}{M_2}\right) \quad km \tag{7.53b}$$

b) For advance ratio, J, greater than unity (usual case for turbine-engined aircraft), using Chapter 3, Eq (3.9e), Eq (7.53a) becomes:

$$s = \frac{4890}{c} P_0^{0.05} \frac{L}{D} \log_{10}\left(\frac{M_1}{M_2}\right) \quad km \tag{7.53c}$$

where P_0 is the rated static power of one engine.
Equation (7.53a) may be developed to give the optimum wing loading for a given engine power setting and altitude.

$$D = T = \eta P/V$$

also

$$D = \rho V^2 S\,(C_{DZ})_{Cr}$$

for the maximum lift drag ratio case. Hence:

$$V = \left[\frac{\eta P}{\rho S (C_{DZ})_{Cr}}\right]^{1/3} \quad m/s \tag{7.53d}$$

and using Eq (7.14c) for the same condition

$$\left(\frac{Mg}{S}\right)_{Cr} = 0.613\sigma \frac{(C_{DZ})_{Cr}}{(\bar{\beta}_{Cr})^{1/2}} \left[\frac{0.816\eta P}{\sigma S (C_{DZ})_{Cr}}\right]^{1/2} \tag{7.53e}$$

7.8.3.3 Range in practical conditions

The assumption of constant conditions used to obtain Eq (7.53) is unlikely to apply in practice. As mass reduces the aircraft must be allowed to climb to maintain constant velocity and lift to drag ratio or the power setting must be changed. In fact allowing the aircraft to climb also changes the power conditions. The range may be evaluated in short steps, as suggested in the case of thrust limited jet engine flight, paragraph 7.8.2.3. A somewhat better approach is to use numerical, or graphical, integration of specific cruise conditions analysed to give range increment per unit fuel mass used.

The numerical integration process requires some initial assumptions to be made:

a) A given engine setting, say maximum continuous cruise power.

b) For ease of flying the aircraft the flight pattern used is often based on a constant equivalent airspeed. However, constant true airspeed or constant Mach number conditions are possible, as is a constant altitude cruise.

The procedure is to select a series of conditions, for example given values of aircraft mass and true speed. For each set of values the drag may be evaluated as a function of altitude, propeller efficiency evaluated and drag matched against the thrust to:

i) determine the altitude corresponding to that condition for the given power setting.

ii) calculate the rate of fuel usage in that condition.

iii) by using Eq (7.52), calculate the distance flown per unit mass of fuel.

Figure 7.5 illustrates a typical set of results from such a procedure. For completeness a buffet boundary determined by usable lift coefficient is shown. Three possible flight paths are shown:

A) Fast climbing cruise at constant equivalent airspeed.

B) Long range climbing cruise at constant true airspeed which may be difficult to fly but is found to give greater range.

C) Constant altitude cruise which implies a speed change, but may be required by Air Traffic Control.

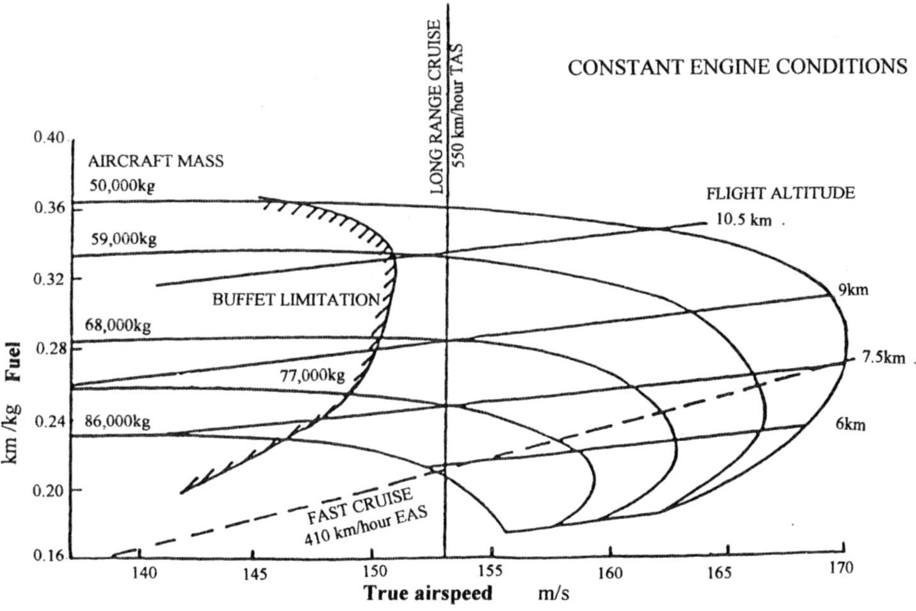

Figure 7.5 Turboprop cruise performance

Figure 7.6 shows the ranges achieved by integrating conditions along the selected cruise lines.

This type of procedure is not really applicable to the preliminary, initial, synthesis phase of design since insufficient aircraft data are available. However, Figure 7.5 does show that the rate of fuel usage does not vary a great deal when the cruise is at constant altitude. Thus for initial design Eq (7.53) may be used with mean values in as far as they are available.

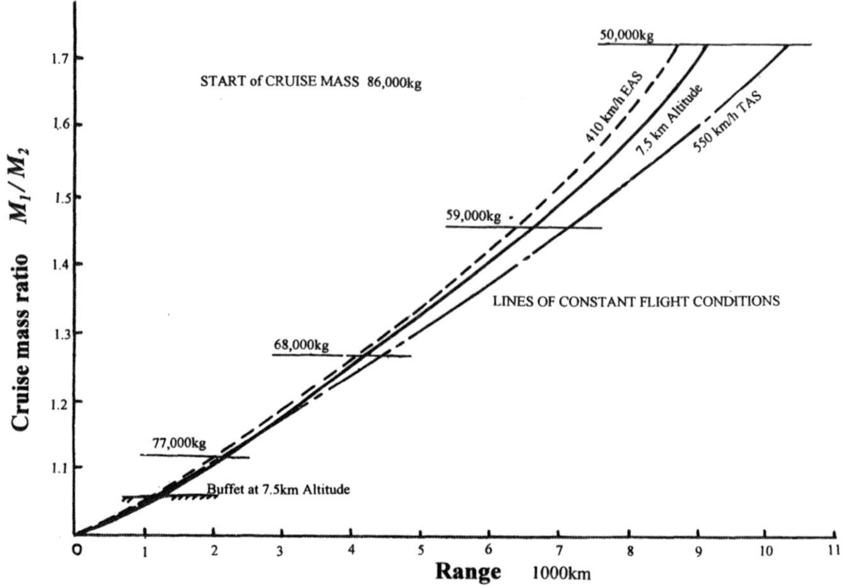

Figure 7.6 Turboprop range

7.8.3.4 Diversion and standby

The fuel used during diversion and standby is evaluated using the cruise formulae with the appropriate altitude and speed conditions. Reserves for these purposes vary considerably according to specific requirements.

7.8.3.5 Endurance - minimum power conditions

It can be shown that the speed to give the minimum flight power is $0.759 V_{MD}$ where V_{MD} is the minimum drag speed. However, this does not necessarily represent the condition for minimum engine power since it is likely that at such a low flight speed the propeller efficiency will be much reduced. Furthermore, flight at this theoretical minimum power speed implies a high lift coefficient which is likely to be above the usable value in buffet terms. Low engine power settings imply higher specific fuel consumption so that the practical speed to obtain maximum endurance may be close to the maximum lift to drag ratio condition, that is the condition which gives the maximum range.

7.8.3.6 Equivalent static thrust and power in cruise

The thrust in cruising flight is equal to the drag at any given time. The power required follows directly from a knowledge of the velocity and propellor efficiency. However, it is convenient to relate cruise thrust to the equivalent static thrust and thence to power.

1) Piston engines

In the majority of cases it is likely that the cruise speed will be sufficiently low for the advance ratio, *J*, to be less than unity. When this is the case the equivalent sea level static thrust to weight ratio is:

$$\frac{T_0}{(Mg)_0} = \frac{0.74 N_E V_{Cr}^{0.51}}{\sigma_{Cr}^{0.935}} \left(\frac{Mg_{Cr}}{Mg_0}\right)^{0.85} \left(\frac{Mg}{S}\right)_0^{-0.15} \left(S^{-0.1}\right)^{1.5} \left(\frac{T_{Cr}}{N_E Mg_{Cr}}\right)^{0.85} \qquad (7.54a)$$

V_{Cr} is the true air speed in cruise at an altitude defined by the relative density σ_{Cr}
Mg_{Cr} is the appropriate aircraft weight.

Should the advance ratio be greater than unity, say $V_{Cr} > 90$ m/s:

$$\frac{T_0}{(Mg)_0} = \frac{0.3 N_E \; V_{Cr}^{0.714}}{\sigma_{Cr}^{0.935}} \left(\frac{Mg_{Cr}}{Mg_0}\right)^{0.85} \left(\frac{Mg}{S}\right)_0^{-0.15} \left(S^{-0.1}\right)^{1.5} \left(\frac{T_{Cr}}{N_E Mg_{Cr}}\right)^{0.85} \qquad (7.54b)$$

2) Turboprop engine

In this case the advance ratio is most likely to be greater than unity and:

$$\frac{T_0}{(Mg)_0} = \frac{1.4 \times 10^{-4} \; z^{0.15} (nD_p)^{2.7} N_E}{\left[\sigma_{Cr}^{0.883}/M_N + 0.75\sigma_{Cr}^{0.733}\right]^{0.862}} \left(\frac{Mg_{Cr}}{Mg_0}\right)^{0.862} \left(\frac{Mg}{S}\right)_0^{-0.138}$$

$$\times \left(S^{-0.1}\right)^{1.38} \left(\frac{T_{Cr}}{N_E \; Mg_{Cr}}\right)^{0.862} \qquad (7.55a)$$

where $M_N = 0.00294 \; V_{Cr} / \sigma_{Cr}^{0.117}$

Should the advance ratio be less than unity:

$$\frac{T_0}{(Mg)_0} = \frac{1.23 \times 10^{-5}\ z^{0.15}(nD_p)^{3.062} N_E M_N^{0.362}}{\left[\sigma_{Cr}^{0.93}/M_N + 0.75\sigma_{Cr}^{0.78}\right]^{0.905}} \left(\frac{Mg_{Cr}}{Mg_0}\right)^{0.905}$$
$$\times \left(\frac{Mg}{S}\right)_0^{-0.095} \left(S^{-0.1}\right)^{0.95} \left(\frac{T_{Cr}}{N_E\ Mg_{Cr}}\right)^{0.905} \qquad (7.55b)$$

The power equivalent of the sea level static thrusts given by Eqs (7.54) and (7.55) may be derived from Eqs (7.1d) and (7.1c) respectively.

7.8.4 Engine failed cruise

When an engine fails during cruise it is usually necessary to fly at lower altitude and Mach number, but the same basic equations apply. There is an increase in zero lift drag due to the failed engine, say about 5%.

7.9 Descent

The fuel used during the descent is relatively low as the engines operate at near idle conditions. It is suggested that it is included in the landing and final reserve as covered in paragraph 7.4.4.

The distance covered during the descent can be of significance for longer range aircraft when the descent angle is not steep. It is suggested that in the absence of better information a typical descent angle of 4° be assumed which gives a typical rate of descent of about 14 m/s.

Then the ground distance covered during descent is approximately:

$$s_G = H_2 / \tan 4° = 14H_2 \quad km \qquad (H_2 \text{ in km}) \qquad (7.56)$$

Chapter 8

Parametric analysis and optimisation

8.1 Introduction

The aim of parametric analysis is to examine the influence of the various design requirements on the configuration of an aircraft and, ultimately, to derive an optimised design. It is quite feasible to undertake the whole operation as a seamless process and such an approach will be considered subsequently. However, there is merit in dividing the complete task into two separate phases. The first stage, which is introduced in paragraph 8.2.2, is to establish the combinations of installed thrust to weight ratio (T/Mg) and wing loading (Mg/S) which satisfy all the prescribed requirements. The second stage, see paragraphs 8.7 and 8.8, is to select the most appropriate combinations of these two dominant parameters and to determine which of them yields a design optimised to have a minimum mass, see Chapter 1, paragraph 1.4.

The first stage involves consideration of both the overall aircraft configuration and the detail geometry such as aspect ratio and sweep. For each set of parameters the wing loading appropriate to the lowest required thrust to weight ratio can be established. While this may not always correspond to the best design it frequently does and hence is a good basis for further study at the second phase of the process. The advantage of limiting the initial analysis to this procedure is that it gives a good indication of the more sensitive parameters and requirements and hence an understanding of the critical aspects of a particular concept. Furthermore, the output may be made essentially non-dimensional whereas it is inevitable that the subsequent second stage analysis yields absolute results.

8.2 Procedure for parametric analysis (first stage)

8.2.1 General

The nature of the calculations required for the parametric analysis is such that some form of computational assistance is highly desirable. While it is perfectly possible to write a self-contained program there is considerable merit in adopting an approach based on the use of a suitable spreadsheet. This gives a much more open view of the procedure and greatly facilitates interaction in the design process. The implication of changes to parametric values may immediately be seen.

The formulation of the spreadsheet is dependent to a great degree upon the powerplant and flight regime under consideration. The four basic powerplant models derived in Chapter 3 may be combined with the subsonic and supersonic flight regimes to give five possible categories:

A Subsonic flight with piston engines
B Subsonic flight with turbopropeller engines
C Subsonic flight with fan engines
D Subsonic flight with low bypass ratio jet engines
E Subsonic and supersonic flight, including transonic acceleration, with low bypass ratio jet engines. This category includes the use of reheat.

Although category D is a special case of category E the much greater complexity of the latter justifies the separation of the former.

The procedure for the parametric analysis is best illustrated by example. A relatively simple case study in category C follows in detail so that the various aspects of the process can be explained. The results obtained for designs which fall in the other categories are included in Addendum 5.

8.2.2 First stage parametric analysis

The procedure for the first stage of the parametric analysis is outlined below.

a) Select one, or more, overall aircraft concepts for investigation. Chapter 2 may be referred to for this.

b) Identify the most probable type of powerplant suited to the role of the aircraft. This is discussed in Chapter 3 and in paragraph 8.3 below.

c) Prepare a preliminary layout for the fuselage, and in particular establish the overall dimensions, using Chapter 4 as a guide. The fuselage layout may well have some effect upon the choice of the overall concept. In the case of supersonic aircraft the fuselage dimensions are needed for the first stage of the parametric analysis, but otherwise the information is not required until the second stage.

d) Select a number of wing geometric configurations for investigation, referring to Chapter 5 as a guide. The variable parameters in this selection will most likely include:

i) A range of aspect ratio.
ii) A range of thickness to chord ratio.
iii) A range of sweep when the aircraft is required to operate at high subsonic or supersonic speed. In many cases the sweep and thickness to chord ratio are directly related and may be treated as a single parameter. The type of aerofoil may need to be considered.
iv) Types of high lift devices likely to be needed.

In most cases it is adequate to assume initially a single, typical, value of taper ratio, see Chapter 5, paragraph 5.3.3.

e) The lift characteristics for the selected range of wing geometry and high lift devices are then evaluated, using Chapter 6 paragraph 6.2. The information is given in a form which enables approximate values to be derived as a function only of sweep for a given configuration. In the first instance it is convenient to estimate zero sweep values, which may subsequently be corrected as necessary.

f) The drag coefficients are given in Chapter 6, paragraph 6.3. They are dependent upon the flight speed and wing geometry so the initial evaluation is limited to establishing basic expressions.

g) Reference to Table 5.3 of Chapter 5 enables a range of likely wing loading to be selected for investigation. In evaluating certain performance requirements it is simplest to use the thrust to weight ratio as an input and, although this is the exception rather than the rule, Table 3.2 of Chapter 3 may be used to give a guide to a possible range of values.

h) The relevant performance requirements are interpreted from the equations given in Chapter 7. The most useful form of these equations is considered in paragraph 8.4 below. The data derived by the above procedure are applied individually to each set of parameters selected, in order to establish:

i) The thrust to weight ratio corresponding to a given wing loading, or vice versa, for each relevant performance requirement.
ii) Limited values of wing loading which are independent of thrust to weight ratio, such as instantaneous manoeuvre or landing approach conditions.

i) For each set of parameters it is possible to produce a series of curves giving the variation of thrust to weight ratio with wing loading to meet each performance requirement. Overlaying of these curves enables a "design space" to be identified within which all of the requirements are met as illustrated in Figure 8.1. A design point can be selected from within this space using the following guidelines:

i) The highest value of wing loading consistent with the lowest value of thrust to weight ratio which meets all the requirements.

ii) In some cases there may not be a clear minimum point, for example the critical value of thrust to weight ratio increases with wing loading up to some limiting value of the latter parameter. When this is the case it is best to select the highest allowable wing loading condition on the basis that this could result in the lightest final solution for that set of parameters. However, this is not inevitably the case and hence the assumption must be checked in the subsequent analysis. This difficulty is automatically overcome when the whole analysis process is mathematically optimised, see paragraph 8.8.2.

j) The design points derived for each set of parameters are then used in graphical format, preferably of the carpet type, to summarise the results. Any overriding constraints, such as the "structural parameter" *SP* referred to in Chapter 5, may be overlaid on this graph. It must be noted that the points on the graph are not overall optimum values, each being simply a set of conditions which fulfils the requirements. They do form the basis for the second phase of the parametric analysis.

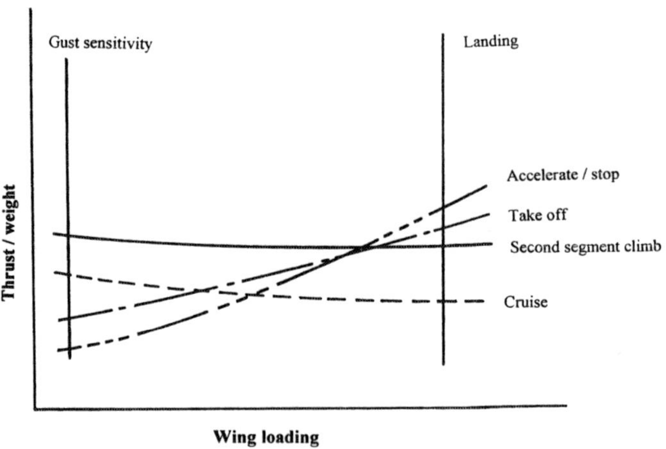

Figure 8.1 Thrust to weight ratio as a function of wing loading

It is helpful to consider the form of the input data needed to commence the analysis and this is covered in the next two paragraphs.

8.3 Powerplant representation

Chapter 3 gives methods for relating the variation of thrust characteristics in different flight conditions to the sea level static value. In the case of propeller-driven aircraft this may then be translated into a power. The form of the thrust characteristics depends upon the type of powerplant.

a) Low bypass ratio and fan engines. The performance of these engines is expressed in terms of the product of the sea level static thrust and a parameter, τ, which is a function of altitude and Mach number as well as engine operating conditions. To a large extent the influence of the latter has been covered by assuming typical operating conditions for given flight modes.

b) Propeller-propelled aircraft. When propellers are used the engine performance is measured in terms of power and the consequent thrust depends upon forward velocity and propeller efficiency. For the present purposes the engine thrust has been determined by assuming a typical propeller efficiency as function of advance ratio, J, and engine power as a function of altitude and speed. The thrust in any give flight condition is related back to sea level static conditions through the power relationship and so the evaluation of sea level static thrust in terms of power is of considerable significance.

i) Turbine engines: as there may be considerable variation in propeller design characteristics it is necessary to specify, or assume, both the number of propeller blades, z, and the propeller tip speed, (nD_p). Paragraph 3.6.2.5 of Chapter 3 gives typical values of these for different classes of aircraft. A typical variation of disc loading with power then enables the static thrust to be expressed by Eq (3.10a) of Chapter 3.

ii) Piston engines: there is usually much less variation of propeller operating conditions due to the relatively narrow power range of this class of powerplant. For initial design purposes it is possible to assume typical values of number of blades and a typical tip speed of 90 m/s. Using a disc loading representation the static thrust is given by Eq (3.11a) in Chapter 3.

8.4 Selection of performance equations

8.4.1 General

The performance requirements and the corresponding equations will mostly be considered in the order in which they appear in Chapter 7. This order is not intended to indicate their relative importance which will change according to the particular set of requirements. The notation and equation references of Chapter 7 are retained.

8.4.2 Take-off

8.4.2.1 General

The equations relevant to the take-off phase of the flight are to be found in paragraph 7.3 of Chapter 7. They cover:

i) Normal take-off, landplane - Eq (7.1). These equations include the power as well as the thrust requirements where this is relevant for propeller driven aircraft.
ii) Normal take-off, seaplane - Eq (7.2).
iii) Engine failed take-off, landplane - Eq (7.3).
iv) Second segment climb with one engine failed - Eq (7.4).
v) Deflected thrust take-off - Eq (7.5).

The fuel used during take-off and the initial climb out can be very variable, depending upon operational procedures. For simplicity it is suggested that an assumption of 1% of the take-off mass is adequate for the purposes of preliminary design, and possibly somewhat conservative.

8.4.2.2 Normal take-off - landplane

While Eq (7.1a) gives the take-off distance for a given set of aircraft characteristics, for initial design work Eq (7.1b), which gives the thrust to weight ratio needed to meet a given take off-length as a function of wing loading, is more useful. Even so there is a complication in that the thrust to weight ratio also appears on the right-hand side of the equation. While further rearrangement is possible the result is algebraically complex. A simpler approach is to use a typical value of thrust to weight ratio, as given in Table 3.2 of Chapter 3, for the right-hand side term only. The resulting approximate value of thrust to weight ratio may then be used for a second iteration. Unless the thrust to weight ratio has a very unusual value the second result is of sufficient accuracy.

In the case of propeller driven aircraft, Eqs (7.1c) and (7.1d) enable the thrust to weight ratios to be converted to power to weight ratios for turboprop and piston engines respectively.

8.4.2.3 Normal take-off - seaplane

Equation (7.2b) is comparable with Eq (7.1b) quoted above for landplane. The difference reflects the different drag characteristics of water-borne take-off. The same approach in dealing with the initial estimation of thrust to weight ratio is appropriate.

8.4.2.4 Engine failed take-off - landplane

Equation (7.3b) explicitly gives the thrust to weight ratio needed for a landplane to complete an "accelerate-stop" aborted take-off within a given distance. The result is directly comparable with Eq (7.1b) and Eqs (7.1c) or (7.1d) may be used to convert the value to power when it is appropriate.

8.4.2.5 Second segment climb

The second segment climb requirement with a failed engine is frequently a critical design case, especially for two-engined transport aircraft. Equation (7.4a) states the basic expression for climb out thrust to weight ratio. Equations (7.4b), (7.4c) and (7.4d) relate this to the take-off, that is sea level static, value for jet/fan, piston and turboprop engines, respectively. Equation (7.4a) makes an allowance for "hot and high" conditions, and the subsequent equations include this. One special case not covered by Eq (7.4d) is that of a turbopropeller engine which is thrust, rather than power, limited. In this case it is suggested that the static thrust value should be assumed to be the same as the second segment climb value. Again Eq (7.1c) or (7.1d) may be used to relate the equivalent static thrust to power.

8.4.2.6 Deflected thrust take-off

The ground distance to rotation when the thrust is deflected is given by Eq (7.5b) and is seen to include a somewhat complex function of thrust to weight ratio. Further, the optimum thrust deflection to give minimum take-off length is also a function of thrust to weight ratio, as seen from Eq (7.5d). For these reasons this case is an exception to the usual in that it is easiest to evaluate the wing loading corresponding to given, assumed, thrust to weight ratios, rather than the reverse. Thus Eq (7.5c) is a rearrangement of Eq (7.5b) for this purpose.

8.4.3 Approach and Landing

8.4.3.1 General

The approach and landing conditions are given in paragraph 7.4 of Chapter 7:

i) Approach speed, as a function of wing loading and lift coefficient - Eq (7.6c).

ii) Corresponding landing field length - Eq (7.6a) as a general expression and Eqs (7.6e) (7.6f), (7.6g) and (7.6h) for frequently met particular conditions.

In practice the approach speed may be specified, or can be assumed, and the landing wing loading then follows. Alternatively the landing field length may be specified and the approach speed required to achieve this may be estimated, the wing loading and approach lift coefficient being derived from it.

8.4.3.2 Approach speed

For design purposes the most satisfactory procedure is to determine the approach speed in the first instance. This is done by finding the lower value of that required to meet the given landing condition, and any specified maximum value. The value of the approach speed required to meet the given landing field length is derived from Eq (7.6c) or for the special case of a typical 3° approach angle, with no reverse thrust and a braking deceleration factor of 0.38, Eq (7.6g). The wing loading corresponding to a given value of the approach speed and approach lift coefficient is given by Eq (7.6d).

8.4.3.3 Landing field length

For a given set of conditions the landing field length comes from one of Eqs (7.6a), (7.6e), (7.6f) or (7.6g), whichever is appropriate. It will be noted that when there is no requirement for reverse thrust the landing conditions are independent of thrust to weight ratio. Since landing is most unlikely to determine the installed thrust it is initially convenient to base the landing conditions on those relevant to no reverse thrust. Subsequently the impact of reverse thrust may be evaluated by using the thrust to weight ratio determined from other performance cases.

8.4.3.4 Missed approach

For simplicity it is suggested that the missed approach requirement covered in Chapter 7, paragraph 7.4.3 is not considered for the preliminary design synthesis on the basis that its accurate evaluation is dependent upon a number of secondary issues. However, it is of importance and must be checked as soon as the configuration of the aircraft has been established and sufficient data are forthcoming.

8.4.4 Climb

8.4.4.1 General

That the evaluation of climb performance is complex is clear from paragraph 7.6 of Chapter 7. In many cases, however, it is reasonable to simplify the procedure for the first stage of the parametric analysis. Climb performance does have to be analysed fully when it is an important requirement, as may be the case for some combat aircraft, and also at the second stage of the parametric analysis when the ground distance covered in climb is significant.

In all cases it is necessary to specify the type of climb, final climb altitude and intermediate details such as transition from a constant equivalent airspeed to a constant Mach number climb. Where climb performance is critical these details demand investigation. The conditions should be chosen so that the end of climb velocity more or less coincides with the start of cruise condition.

8.4.4.2 Simplified approach to climb performance

When a simple approach is adequate the two values which must be ascertained are:

a) The fuel used, both up to the commencement of the climb and during it. Paragraph 7.3.6 of Chapter 7 suggests that the weight of the aircraft at the beginning of climb can be assumed to be $0.99(Mg)_0$, see also para 8.4.2.1 above. Table 7.2 of Chapter 7 gives approximate weight ratios for different forms of climb as a function of climb altitude defined in terms of relative density. Thus the ratio of the weight at the end of climb (start of cruise) to the take-off value can be simply estimated.

b) The end of climb thrust to weight ratio. This may be regarded as the service ceiling condition, typically where the rate of climb has fallen to 0.5 m/s or a somewhat larger value, say 1.5 m/s for civil transport to give operational flexibility. The relevant final rate of climb equations are given in Chapter 7.

i) Constant equivalent airspeed climb (constant EAS). It is assumed that this applies only to propeller driven aircraft, fan/turbojet powered aircraft having transferred to a constant Mach number condition at some point in the climb. The relevant equations are Eqs (7.19a) and (7.19b) used in conjunction with Eqs (7.18f), (7.18g) or (7.18h) as appropriate. Which equation of Eq (7.18) is relevant is determined by the type of powerplant and whether the advance ratio, J, is greater or less than unity. An advance ratio greater than unity is most likely to be appropriate for turboprop aircraft. As the design becomes defined it is necessary to check that the correct assumptions have been made.

ii) Constant mach number climb (constant M_N).This is assumed to be appropriate for low bypass ratio and fan engines. It is convenient to handle this case in three distinct altitude ranges, as determined by the variation of the speed of sound.

1) Up to 11 km altitude - Eqs (7.23a) and (7.23b).
2) Between 11 and 20 km altitude - Eqs (7.26a) and (7.26c).
3) Above 20 km altitude - Eqs (7.30a) and (7.30c).

Because the changes above 20 km altitude are small it may be adequate to extend the application of Eqs (7.26a) and (7.26c) to somewhat greater altitude than 20 km without significant loss of accuracy.

iii) Linearly varying Mach number climb supersonic (variable M_N)
Only low bypass ratio engines are covered by this condition. The vertical velocity at the end of the climb, which enables the ceiling to be established, is given by Eq (7.36a) and the corresponding dry static thrust at sea level is evaluated from Eq (7.36b).

When more accurate climb performance is necessary, the analysis referred to in the subsequent paragraphs 8.4.4.3. to 8.4.4.5. should be used.

8.4.4.3 Evaluation of climb characteristics: 1 - constant equivalent airspeed case

a) Rate of climb. When the rate of climb is not a primary requirement the achieved value may be estimated by using the thrust to weight ratio and wing loading determined from another more critical performance condition. The relevant expressions are to be found in paragraph 7.6.2 of Chapter 7, where it should be noted that the Q_V factor includes the wing loading. Equations (7.15) are the basic equations, supplemented by:

i) low bypass ratio and fan engines - Eqs (7.16a) to (7.16d)
ii) propeller engines - Eqs (7.18a) to (7.18d)

b) Thrust required to give a mean rate of climb, or a given climb time. The relevant rate of climb equations may be rearranged to give the thrust to weight ratio required to achieve a given climb time.

i) Low bypass ratio and fan engine - Eq (7.16e).
ii) Propeller engines - Eq (7.18e).

It will be seen that in the propeller-engined case the evaluation of the thrust to weight ratio is complicated by the presence of the correction term, X_1 [Eq (7.18a)]. This is a factor which assumes importance as the climb approaches the ceiling condition. In the first instance it is acceptable to assume that it has a value of unity, and then the consequently derived value of thrust to weight ratio may be used to obtain a more accurate estimate for subsequent recalculation.

The thrust to weight ratios given by Eqs (7.16) and (7.18) have to be converted to equivalent take-off values:

1) low bypass ratio and fan engines require a simple correction to allow for the ratio of start of cruise to take-off weights.
2) propeller engine conversions are given by Eqs (7.18f), (7.18g) and (7.18h) for piston and turbopropeller applications respectively.

c) Ceiling. The relevant equations for the evaluation of the ceiling conditions for propeller engined aircraft are referred to in paragraph 8.4.4.2, above.

d) Fuel used. The fuel used during constant equivalent airspeed climb is expressed in terms of the mean rate of climb and the height increment completed. For convenience it is given as a ratio of the take-off weight in the following equations:

i) low bypass ratio and fan engines - Eq (7.17).
ii) piston engines - Eq (7.20a).
iii) turbopropeller engines - Eq (7.20b).

e) Ground distance covered during climb. The approximate ground distance covered during the climb for all classes of powerplant and for constant equivalent airspeed conditions is given by Eq (7.21). The cosine term is close to unity when the mean rate of climb is an order of magnitude less than the forward speed.

8.4.4.4 Evaluation of climb characteristics: 2 - constant Mach number case

This case is covered by the equations given in paragraph 7.6.3 of Chapter 7. These equations apply only to low bypass ratio and fan engines.

When a given time to height is required it should be noted that the climb may well have to include an initial constant equivalent airspeed segment and, for a supersonic aircraft, transonic acceleration and supersonic climb segments. It may therefore be necessary to consider a number of different combinations in order to establish the best overall climb profile.

a) Rate of climb. This is given by Eqs (7.22a), (7.26a) and (7.31a) for the altitude ranges up to 11 km, and 11 to 20 km; above 20 km respectively.

b) Thrust to weight ratio for a given rate of climb. Equations (7.22c), (7.26c) and (7.31b) may be used to derive the thrust to weight ratio required to achieve a given rate of climb in the three separate altitude ranges. The thrust to weight ratio is in terms of sea level static thrust and the weight at the beginning of a given segment of the climb. The thrust factor, τ_l, is relevant to the climb Mach number and the altitude at the beginning of the particular climb segment.

c) Ceiling. The relevant equations are referred to in paragraph 8.4.4.2. above.

d) Fuel used in climb. This is given in terms of take-off weight by Eqs (7.24), (7.28) and (7.32) for the altitude ranges up to 11 km, and 11 to 20 km; above 20 km, respectively.

e) Ground distance covered in the climb. Equations (7.25), (7.29) and (7.33) are the relevant expressions for the three altitude ranges as appropriate.

8.4.4.5 Evaluation of climb characteristics: 3 - Linearly varying Mach number at supersonic speed

The relevant equations are to be found in paragraph 7.6.4 of Chapter 7. They apply only to aircraft powered by low bypass ratio engines.

a) Rate of climb. If the installed thrust to weight ratio is known from other cases, such as transonic acceleration, the mean rate of climb achieved during the assumed climb pattern is given by Eq (7.34a). The assumed climb pattern is based on a linear variation of Mach number with altitude commencing, for example, at $M_N = 1.2$ at 11 km.

b) Thrust to weight ratio for a given rate of climb. Equation (7.34a) is rearranged to the form of Eq (7.35) in order to estimate the thrust to weight ratio. The result is in terms of sea level static equivalent dry thrust and the weight at the start of the supersonic climb. The latter requires conversion to the take-off weight for purposes of comparison.

c) Ceiling. The relevant equation is referred to in paragraph 8.4.4.2. above.

d) Fuel used. Fuel used during the supersonic climb is given by Eq (7.37a).

e) Ground distance covered during climb. The ground distance covered during the supersonic climb can be obtained from Eq (7.38).

8.4.5 Manoeuvres and maximum level speed

8.4.5.1 General

The manoeuvre equations are given in paragraph 7.7 of Chapter 7, having been derived from the generalised flight performance to be found in paragraph 7.5 of Chapter 7.

8.4.5.2 Specific excess power

The specific excess power in a given flight condition is given by Eq (7.13d) in Chapter 7. The manoeuvre factor, n, is unity in level flight. Equation (7.13f) is derived from Eq (7.13d) and gives the take-off static thrust to weight ratio needed to achieve a given specific excess power in defined flight conditions.

8.4.5.3 Maximum level speed

The maximum level speed in a given flight condition coincides with zero specific excess power in that condition, and the acceleration factor, n, is unity. Chapter 7, Eq (7.42c), gives the wing loading and thrust to weight ratio relationship appropriate to a given maximum speed and altitude. This equation involves a quadratic expression of thrust to weight ratio and hence it is a case where it is simplest to evaluate the wing loading appropriate to a given thrust to weight ratio, using Eq (7.41b) with n equal to one. The assumed thrust to weight ratio may be related to the datum sea level static condition by use of the factor (k_m/τ_m) defined in Eq (7.13f).

8.4.5.4 Sustained manoeuvre

Manoeuvre requirements are often specified in terms of rate of turn, in which case Eq (7.39b) may be used to convert it to a normal acceleration factor, n. Sustained manoeuvre is another situation where the specific excess power is zero. Equations (7.40c) and (7.40d) are relevant but it is simplest to use Eq (7.13f) with n equal to one to relate the requirement to the initial take-off conditions.

8.4.5.5 Instantaneous manoeuvre

Instantaneous manoeuvre is independent of thrust. The two possible limiting conditions are given in paragraph 7.7.3 of Chapter 7. The available manoeuvre may be determined either by the structural design limit factor, n_l, or the maximum usable lift coefficient associated with the given wing loading.

8.4.6 Transonic acceleration

8.4.6.1 General

In some cases it may be sufficient to follow a procedure similar to that suggested for climb performance in paragraph 7.4.4.2. This involves making an assumption of the fuel used during transonic acceleration and undertaking a later check on the thrust needed to achieve the manoeuvre. Chapter 7, paragraph 7.7.4.5, suggests that the fuel used during transonic acceleration may be of the order of 1% of the gross weight of the aircraft. However, the thrust to weight ratio required to achieve acceptable transonic acceleration performance may be a critical design parameter and it is preferable to undertake a reasonably accurate analysis at the preliminary design phase. When no specific requirement is stated for the time of transonic acceleration it is necessary to make an appropriate assumption. For example the time for level flight acceleration over the Mach number range of 0.9 to 1.2 may be set at 1 min, which implies a mean acceleration of about $0.15g$. Clearly some latitude may be required if such an assumption proves to be unduly demanding. It must be pointed out that working to a mean acceleration requirement does not automatically ensure that there is a positive acceleration at the Mach number where drag is highest. This also must be checked subsequently.

8.4.6.2 Time for transonic acceleration

In analysing the transonic acceleration it is convenient to define it, arbitarily, as occurring between Mach numbers, M_N, of 0.9 to 1.2 and further treating it in two separate stages, namely:

i) $0.9 \leq M_N \leq 1.0$.

ii) $1.0 \leq M_N \leq 1.2$.

Equation (7.45a) gives the transonic acceleration time in terms of the two phases of acceleration, m_a and m_b defined by Eqs (7.44a) and (7.44b) respectively. Equations (7.45b), (7.44c) and (7.44d) are for the special case of transonic acceleration at 11 km altitude and a powerplant bypass ratio of 0.8. Rearrangement of Eq (7.45) with substitution from the appropriate parts of Eqs (7.44) enables the required thrust to weight ratio to be expressed, as in Eq (7.45c). This thrust to weight ratio can be related to the sea level static condition by allowing for the reduction in aircraft mass up to the start of the transonic phase, the factor k_m of Eq (7.13f) of Chapter 7.

8.4.6.3 Fuel used for transonic acceleration

The fuel used during the acceleration from $M_N = 0.9$ to $M_N = 1.2$ is given by Eqs (7.46a) and (7.46b) for the general case and the special one referred to in the previous paragraph, respectively.

8.4.7 Cruise - range considerations

8.4.7.1 General

The need to maintain a small rate of climb at the start of the cruise, as dealt with in paragraph 8.4.4.2. above, establishes the thrust to weight relationship with the wing loading for the start of cruise condition. As the cruise continues the speed will either increase to balance the drag against the thrust, or the aircraft will climb to achieve the same effect. As explained in paragraph 7.8 of Chapter 7, the analysis of the cruise performance is complicated by the number of possible flight patterns. For the first stage of the parametric analysis it is usually adequate to assume constant conditions, that is to use Eq (7.50) for low bypass ratio and fan engines and Eq (7.53) for propeller engines. The thrust to weight ratio in cruise is simply equal to the reciprocal of the lift to drag ratio and may be converted back to the equivalent take-off condition by using the (k_m/τ_m) factor of Eq (7.13f) or the relevant ceiling equations referred to in paragraph 8.4.4.2.

Experience suggests that in many cases it is most satisfactory to base the initial configuration analysis on the maximum range requirement rather than reduced range with greater payload.

8.4.7.2 Cruise wing loading

The maximum range is achieved in constant flight conditions when the lift to drag ratio has its greatest value at the given velocity/Mach number and altitude. It follows from this that there is a unique, optimum, value of wing loading at the start of the cruise. This value is given by Eqs (7.50d) or (7.50e) for low bypass ratio and fan engines and Eq (7.53e) for propeller engines.

There may, however, be an overriding restriction due to the buffet limited values of the cruise lift coefficient. This is defined in Chapter 6, paragraphs 6.2.4.6. and 6.2.5.6. Comparison may be made between the wing loadings defined by start of cruise, ceiling (paragraph 8.4.4.2 above), buffet limited lift coefficient and maximum lift to drag ratio leading to the possible revision of the altitude and speed conditions at the start of cruise to enable the best value to be achieved.

8.4.7.3 Gust sensitivity in cruise

Unless some form of ride control is to be incorporated in the design of the aircraft it is desirable to ensure that aircraft will not possess unacceptable responses to atmospheric turbulence. Equation (5.9b) of Chapter 5 suggests a criterion which may be used for this purpose. The criterion is primarily intended for passenger transport aircraft and it can be unduly severe for some military and light general aviation aircraft.

8.5 Constraints and checks

It may be necessary to impose constraints which limit the range of parameters, other than those which are directly related to performance. Some of these, such as wing span limitation, can only be imposed when the second stage of the parametric analysis is being completed and when the actual dimensions become known. However, some checks can be undertaken during the first stage of the analysis. Among these is the evaluation of the structural parameter, *SP*, referred to in Chapter 5, paragraph 5.4.2, Eq (5.8). While this is not a constraint as such it does provide a useful guide to practical limits to certain wing geometric characteristics.

8.6 Case study - short/medium haul airliner

8.6.1 General

This case study of a short/medium haul airliner has been selected to illustrate the application of the powerplant, lift, drag, mass and performance models given in Chapters 3, 6 and 7. The procedure which leads to the parametric analysis and eventually to the definition of the preliminary design follows that outlined in Chapter 1 and involves the use of all the preceding material. The particular example has been chosen as representative of one of the best documented class of aircraft thereby enabling the procedural methods to be emphasised rather than the characteristics of the design.

8.6.2 Specification *(see Chapter 1)*

The aircraft is required to have the following characteristics and performance.

a) Performance

i) A basic payload of 14,000 kg, representing 120 passengers, to be carried over a still air range of 6500 km with landing reserves only.
ii) A high density payload of 20,080 kg, representing 180 passengers.
iii) Maximum cruise Mach number (M_{NCRIT}) of 0.82.
iv) At the start of cruise at a Mach number of 0.82 and at 10 km altitude the residual rate of climb should be at least 1.5 m/s.
v) Maximum altitude at the end of cruise is to be 11.9 km.

b) Operational

i) The aircraft is to be fitted with two fan engines.
ii) The take-off field length should not exceed 2200 m under standard conditions.
iii) Landing field length should not exceed 1850 m when reverse thrust is not used, or 1700 m when reverse thrust is in operation, under standard conditions.
iv) Landing approach speed should not exceed 73 m/s (140 knots) for Air Traffic Control reasons.
v) The wing span should not exceed 34 m to facilitate handling at airport gates.

8.6.3 Configuration *(see Chapter 2)*

There is nothing in the specification to suggest that the configuration should be anything other than conventional. It will therefore be assumed, subject to later confirmation, that the following are appropriate:

i) Two underslung podded engines will be located on a low wing.
ii) The wing will be swept back by at least 15° to facilitate stowage of the landing gear in the wing root area (see Chapter 2, paragraph 2.3.2.2.)
iii) The empennage will consist of a single vertical fin with the horizontal tail located on the rear fuselage.

*8.6.4 **Propulsion** (see Chapter 3)*

Current practice suggests that for a short/medium haul airliner the fan engine bypass ratio is likely to be about four. Subsequent analysis may result in a refinement to this assumption, or alternatively a known powerplant of specific characteristics may be adopted.

For performance evaluation it will therefore be assumed that the engine thrust characteristic is that given by Eq (3.8) of Chapter 3 with bypass ratio, $R = 4$. The altitude correction factor, s, will be assumed to be 0.7 as giving a good overall representation up to the highest altitude of 11.9 km quoted in the specification. That is:

$$\text{For } 0 \leq M_N = 0.4 \qquad \tau = [1 - 0.76M_N]\,\sigma^{0.7}$$
$$0.4 \leq M_N \leq 0.9 \qquad \tau = [0.816 - 0.3M_N]\,\sigma^{0.7} \qquad (8.1)$$

When a spreadsheet approach is used it is convenient for the bypass ratio to be used as an input parameter, with the equivalent of Eq (8.1) being part of the analysis. The estimation of specific fuel consumption is not required at the first stage of parametric analysis, provided the simple approach to climb performance is considered to be adequate.

*8.6.5 **Fuselage layout and size** (see Chapter 4)*

8.6.5.1 Cabin layout

For an airliner the layout of the fuselage commences with a study of the seating and other cabin features. In this case for the 180-seat high density layout assume:

i) Seat pitch, s — 0.7 m
ii) Number of toilets — 2 pairs of 2 along length
iii) Number of galleys — 1 only (for snacks)
iv) Cross aisles — 2
v) Cabin length is nominally 65% of fuselage overall length

Then from Chapter 4, paragraphs 4.5.3.3 and 4.5.3.4 [Eq (4.1)] the cabin length in terms of the number of seats across the width, p, is:

$$0.7\left(\frac{180}{p} + 1\right) + 1 + 2 \times 0.8$$

$$= 0.7\left(\frac{180}{p} + 1\right) + 2.6$$

A 180 passenger requirement is most likely suited to a single aisle, narrow body configuration, but a twin aisle layout should be considered. Assuming initially that the cross-section is circular the layout arrangements possible are:

No. of aisles	Seats across	Cabin width	No. of rows	Cabin length	Overall dia.	Overall length	Length / dia.
1	5	3.05	36	28.5	3.25	43.85	13.49
1	6	3.55	30	24.3	3.8	37.38	9.84
2	7	4.65	26	21.5	3.9	33.0	6.73

From this table it is clear that a single aisle, six seats across, configuration has the most appropriate overall length to diameter ratio of about 9.84 with a diameter of 3.8 m, and it gives some scope for possible stretch.

8.6.5.2 Fuselage length geometry

Using typical fuselage proportions given in Table 4.1 of Chapter 4 gives:

i) Nose length 1.4 diameters
ii) Tail length 2.6 diameters
iii) Parallel length 9.84 - 1.4 - 2.6 = 5.84 diameters

and cabin length is 1.096 times the parallel length which is typical for this class of aircraft.

8.6.5.3 Fuselage cross-section and provision for freight

The depth of the fuselage cross-section is made up of:

i) Cabin height at centreline, say 2.05 m
ii) Upper structure 0.20 m

From which it follows that the floor is 2.25 m down from the upper surface, or about 40.8% of the depth above the bottom surface if the cross-section is truly circular. With a similar structure allowance at the lower surface and a further 0.2 m floor depth the clear height of the freighthold would be:

$$3.8 - 2.25 - 0.4 = 1.15 \text{ m}$$

This is marginal and it is likely that it will be necessary to accept a somewhat non-circular, slightly double-bubble, cross-section to give a freighthold height of, say, 1.3 m and an overall fuselage depth of 3.95 m.

8.6.6 Wing geometry - parametric range and high lift devices (Chapter 5)

8.6.6.1 Wing geometric properties

Reference to Tables 5.2 and 5.3 of Chapter 5 suggests the following parameter ranges for the initial investigation of this class of aircraft:

i) Wing loading 4000 to 7000 N/m^2

ii) Wing root thickness to chord ratio 0.11 to 0.15
(Sweep-back follows from this parameter to meet the required $M_N = 0.82$ critical cruise case, but with 15° minimum)

iii) Taper ratio assume 0.25 as a constant value

iv) Aspect ratio 7 to 10 (in this example this will be increased up to a value of 12 to illustrate the effect of constraints)

8.6.6.2 Leading edge high lift devices

It is necessary to make an approximate estimate of the criterion given in Eq (5.2) of Chapter 5 to determine whether leading edge devices are likely to be necessary. This is done by using typical values, either derived from comparison with existing similar types of aircraft or by use of the general data given in Chapter 5.

A typical wing loading is, say, 6000 N/m^2 and the minimum sweep back suggested by paragraph 8.6.3 above is 15°. Hence:

$$F_{LE} = \frac{M_0 g}{S \cos \Lambda_{1/4}} = 6380 \quad \textit{typically}$$

As this is greater than 5500 it may be concluded that leading edge high lift devices will be required.

8.6.6.3 Trailing edge flaps

Single- and double-slotted flap systems are used for this class of aircraft. It is always desirable to adopt the simplest approach which meets the requirements and hence it will be assumed initially that a single-slotted flap system will be adequate. However, this decision is subject to review should the subsequent analysis indicate a difficulty in meeting low speed performance conditions.

8.6.7 Basic lift characteristics (Chapter 6)

8.6.7.1 General

Since the sweepback is a variable in the synthesis process it is initially convenient to estimate the values of lift coefficient for zero sweep. Subsequent calculation in a given case must correct for the sweep effect. Initial calculation of lift coefficients is simple and when a spreadsheet approach is adopted it is convenient to input the zero sweep values.

8.6.7.2 Maximum lift coefficient (landing configuration)
Using Eq (6.2) and Table 6.1 of Chapter 6, with leading edge and single-slotted trailing edge flaps:

$$C_{LMAX} = (1.5 + 0.65 + 1.0)\ cos\Lambda_{1/4} = 3.15\ cos\Lambda_{1/4} \quad (8.2a)$$

8.6.7.3 Landing approach lift coefficient
Equation (6.3) of Chapter 6 gives:

$$C_{La} = 0.6C_{LMAX} = 1.89\ cos\Lambda_{1/4} \quad (8.2b)$$

8.6.7.4 Take-off, unstick, lift coefficient
Equation (6.4) and Table 6.1 of Chapter 6 gives:

$$C_{LUS} = 0.8(1.5 + 0.4 + 0.5)cos\Lambda_{1/4} = 1.92cos\Lambda_{1/4} \quad (8.2c)$$

8.6.7.5 Buffet limited cruise lift coefficient
Equation (6.6) of Chapter 6 suggests:

$$C_{LUSE} = 0.65cos\Lambda_{1/4} \quad (8.2d)$$

8.6.8 Drag characteristics (Chapter 6)

8.6.8.1 Zero lift drag coefficient
Tables 6.3 and 6.4 and Figure 6.1 of Chapter 6 enable the following drag parameters to be estimated:

i) Wetted area ratio, R_W 5.5
ii) Type factor, T_f 1.1
iii) Wing area factor, $(S)^{-0.1}$ 0.6 (subject to subsequent revision)
iv) There is no aerofoil laminar flow

Paragraph 8.6.7.5 above indicates that the maximum value of cruise lift coefficient is not likely to exceed 0.6, when allowance is made for sweep. Whence, using Eq (5.1a) of Chapter 5:

$$A_f = 0.95 - (0.1 \times 0.6) = 0.89$$

and

$$(M_{NCRIT})_{2D} = 0.89 - t/c$$

Using Eq (5.4a) of Chapter 5 :-

$$(M_{NCRIT})_{3D} = 0.82 = (0.89 - t/c) / (cos\Lambda_{1/4})^{1/2}$$

Hence

$$\cos\Lambda_{1/4} = [1.22\ (0.89 - t/c)]^2 \quad (8.3a)$$

which may be calculated within a spreadsheet.

Substituting the above data into Eq (6.13) of Chapter 6 the zero lift drag coefficient is:

$$C_{DZ} = 0.005\ \bar{\tau}\left[1 - 0.2M_N + 0.12\left(1.22M_N\right)^{20}\right] 5.5 \times 1.1 \times 0.6$$

$$= 0.01815\ \bar{\tau}\left[1 - 0.2M_N + 6.4\ M_N^{20}\right] \qquad (8.3b)$$

where $\bar{\tau}$ is given by Eq (6.13b) being dependent on t/c and R_W. It is close to unity here and for a Mach number of 0.3 appropriate to the take-off condition, approximately:

$$\left(C_{DZ}\right) = 0.0171\bar{\tau} \qquad (8.3c)$$

and at a Mach number of 0.82:

$$\left(C_{DZ}\right)_{Cr} = 0.0174\bar{\tau} \qquad (8.3d)$$

All the calculations are conveniently performed within a spreadsheet.

8.6.8.2 Drag coefficient due to lift

The drag coefficient due to lift as given by Eq (6.14) of Chapter 6 can be written as:

$$C_{DI} = \left[\frac{\left(1 + 0.12M_N^6\right)}{\pi} f(A)\right] C_L^2 = K_V\ C_L^2 \qquad (8.4a)$$

where

$$f(A) = \frac{1}{A}\left\{1 + \frac{(0.142 + 0.0062A)}{\left(\cos^2\Lambda_{1/4}\right)}\left(10\frac{t}{c}\right)^{0.33} + 4\ \frac{0.1\left(3N_E + 1\right)}{(4 + A)^{0.8}}\right\} \qquad (8.4b)$$

Clearly the dependence of the induced drag factor, K_V, on the various geometric and speed parameters implies that it is most conveniently calculated within a spreadsheet. For the typical Mach number conditions in this example:

i) Low speed, $M_N = 0.3$: $(K_V)_0 = 0.3183f(A)$ (8.4c)

ii) Cruise, $M_N = 0.82$: $(K_V)_{Cr} = 0.33\,f(A)$ (8.4d)

8.6.8.3 Equivalent zero lift drag at climb out
Using Eq (6.15) of Chapter 6 with $F_f = 1.0$ for a single slotted flap:

$$\Delta C_{DT} = (0.03 - 0.004)/A^{0.33}$$

and hence:

$$(C_D)_{CO} = 0.0171 + 0.026/A^{0.33} \tag{8.5}$$

which is best evaluated in the spreadsheet.

8.6.9 Mass ratios at start and end of climb and landing (see Chapter 7)

8.6.9.1 Take off - initial climb mass, M_1
It is assumed, as suggested in Chapter 7, paragraph 7.3.6, that the take-off and initial climb phases use fuel equivalent to 0.01 of the take-off mass, and thus the ratio of start of climb to take-off mass:

$$\frac{M_1}{M_0} = 0.99$$

8.6.9.2 Climb - initial cruise mass, M_{Cr1}
It is necessary to make a reasonably realistic assumption of the climb pattern, whilst at the same time keeping the analysis simple. It may well be necessary to undertake a more detailed analysis to substantiate the results, and this could be done at the second stage of the procedure. For the present purposes it will be assumed that:-

i) The aircraft will climb initially at 170 m/s equivalent airspeed, in spite of the fact that in practice Air Traffic Control may limit the start of climb to 127 m/s (see Chapter 7, paragraph 7.6.2.1). 170 m/s EAS will be maintained until a Mach number of 0.82 is reached.

ii) The remainder of the climb up to the start of cruise at 10 km altitude will be at a constant $M_N = 0.82$.

The constant EAS climb reaches $M_N = 0.82$ at between 7 and 7.5 km altitude where σ is 0.4812, say.

From Table 7.2 of Chapter 7 the mass ratios in the climb may be deduced as:

a) Sea level to 7 km altitude: $0.4812^{0.016} = 0.988$ approx.

b) 7 to 10 km altitude: $(0.3369/0.4812)^{0.02} = 0.993$ approx. and the overall ratio to give the mass at the end of the climb:

$$\frac{M_2}{M_1} = 0.988 \times 0.993 = 0.9811$$

$$\textit{therefore} \qquad \frac{M_2}{M_1} = 0.98 \qquad \textit{say}$$

Hence the ratio of the start of cruise mass, M_{Cr1}, to take-off mass, M_0, is:

$$\frac{M_{Cr1}}{M_0} = 0.99 \times 0.98 = 0.97$$

since in this case $M_{Cr1} = M_2$.

8.6.9.3 Landing - mass, M_L

The design landing mass ratio is based on the maximum design range, which in this case is 6500 km. Using Table 7.1 of Chapter 7 for short haul transports:

$$\frac{M_L}{M_0} = 0.98 - 2(6500 - 1000) \times 10^{-5} = 0.87$$

(Use of the medium/long haul formula gives the same result.)

Again this calculation may be carried out within a spreadsheet to allow for the possibility of examining the effect of the range requirement.

8.6.10 Second segment take-off climb parameters (Chapter 7)

For twin-engined aircraft, Eq (7.4a) gives the values:

$$\alpha = 2.74$$
$$\gamma = 0.02$$

These values may be input into Eq (7.4a) of Chapter 7 within the spreadsheet.

8.6.11 Structural design criteria

In order to make a first estimate of wing structural limitations it is necessary to allocate values to the design speed, V_D, and the ultimate normal manoeuvre factor, N, see paragraph 8.5 and Chapter 5, Eq (5.8). With a design critical Mach number of 0.82 at altitude it is reasonable to assume that the design limiting Mach number, M_D, will have to be 0.87 to 0.9. This is equivalent to a true airspeed of about 280 m/s at 7 km altitude where $M_N = 0.82$ is first reached, or an equivalent airspeed of about 195 m/s. Allowing for cruise flight conditions at somewhat lower altitude V_D will be assumed to be 200 m/s EAS. For this class and size of aircraft the ultimate normal manoeuvre factor is 3.75 as specified by FAR/JAR Pt.25 requirements.

8.6.12 Evaluation of the variation of thrust to weight ratio with wing loading

The procedure for undertaking the first stage of the parametric analysis is outlined in paragraph 8.2.2. The required equations are listed in paragraphs 8.4 and 8.5. As has already been stated in the proceeding paragraphs 8.6.2 to 8.6.10 above it is convenient to use a spreadsheet for the actual calculations and analysis. In the case of this example use has been made of Microsoft EXCEL© as is shown in Spreadsheet 1. The full explanation of the spreadsheet is to be found in the Appendix A 8.1.

The Parameters Box on the left of the spreadsheet contains the three variables chosen for examination in this case, and initial values are input here to commence the process. The Requirements Box lists the specified data needed for the first stage analysis. The Assumed and Initial Calculations Boxes hold the information derived in paragraphs 8.6.4 and 8.6.6 to 8.6.11 and the Preliminary Calculations use the data to evaluate numerical values which are independent of the parametric variation.

The centre of the spreadsheet covers the analysis which is undertaken in two steps:

a) Analysis of a specific configuration. The upper part is concerned with the evaluation of the thrust to weight ratios for the various performance requirements as a function of a range of wing loadings and a given wing configuration. The results are summarised and presented in chart form at the bottom left of the spreadsheet. In some cases it is possible to identify a wing loading corresponding to a minimum thrust to weight ratio which meets all the requirements. However, this is often not the case and as in the chart for this example it is necessary to make a reasoned selection. This is usually the maximum allowable wing loading as limited by the landing conditions since experience shows that in most cases it is best to use the highest possible wing loading, see paragraph 8.2. The selected "design" wing loading is then input into the analysis to obtain the corresponding output data which is shown in the Results Box located below the general analysis, and the relevant data is automatically transferred back to the opening cells to represent the derived result.

b) Analysis of alternative configurations. The lower part of the spreadsheet consists of a Summary of Results obtained by repeating the process described above for a range of wing configurations, aspect ratio and thickness to chord ratio in this case. A structural parameter, *SP*, of 16 has been selected as a nominal limiting value and the last three rows in the upper part of the box are for specific aspect ratios which give an *SP* of 16. The lower part of the box covers cases required for the second stage of the analysis, see paragraph 8.12.2. Only those results required for further consideration are shown. The results are best appreciated when presented in the form of the carpet plot located on the right of the spreadsheet. Fundamentally this shows the "design" thrust to weight ratios as a function of aspect ratio and root thickness to chord ratio. For an aspect ratio above about ten the selected wing loadings are those which give the same thrust to weight ratios for the take-off case as for the second segment climb. Otherwise they are the landing wing loadings and second segment climb thrust to weight ratios. Also shown is the structural parameter, *SP* = 16 limit, which suggests that aspect ratios much in excess of ten are likely to result in a heavy wing structure. As a matter of information the "design point" shown has been brought forward from the result of the second stage of the parametric analysis, see paragraph 8.12.2.5.

8.6.13 Comments and conclusions

It is possible to make some general comments concerning the results.

a) Although not specifically shown it was found that the start of cruise lift coefficient was below the allowed limit in all cases.

b)The second segment climb requirement determined the installed thrust in effectively all the realistic configurations investigated.

c)Increase of aspect ratio and root thickness to chord ratio results in reduced thrust to weight ratio. The former variation is to be expected as increased aspect ratio results in lower drag. The latter variation is due to the lower sweep associated with less thickness to chord ratio and hence higher allowable wing loading in the landing condition.

d)Aspect ratio much above ten is likely to result in a heavy wing structure.

The chart on the right of the spreadsheet suggests that the lowest practical thrust to weight ratio is associated with an aspect ratio of rather more than nine. However, it is necessary to undertake the second stage of the parametric analysis to see what the effect of component and fuel masses has on determining a true minimum mass optimum configuration.

8.7 Introduction to the second stage of parametric analysis

The process of parametric analysis was introduced in paragraph 8.2 which was primarily concerned with the first stage of the analysis and led to the derivation of the variation of the thrust to weight ratio with wing loading for various configurations intended to meet given performance requirements. This was undertaken in non-dimensional form and was valuable in giving a good indication of the importance of the different performance requirements and configuration parameters. However, it did not produce optimum solutions, merely yielding a set of "design" solutions which can meet the specified requirements. In order to derive an optimum solution it is necessary to move to actual masses and a dimensionalised design.

8.8 Procedure for the second stage parametric analysis and optimisation

8.8.1 Second stage parametric analysis

The second stage of the parametric analysis takes the output data of the first stage and uses them to derive a mass for each of the selected configurations. It is possible to place the total aircraft mass contributions into three categories:

a) Absolute mass, such as the required payload.

b) Masses dependent upon the total mass of the aircraft.

c) Masses dependent upon other parameters, such as the wing loading.

The actual information needed to undertake the mass calculation is dealt with in paragraph 8.9 below.

It is necessary to use an iterative process to deduce the absolute value of aircraft mass. One method of doing this is to first assume an initial value for the total mass and then to calculate a total mass which results from the assumption and which will, in general, be different to the assumed value. By repeating the process for a number of assumed total mass values and plotting the calculated values as a function of the assumed values, the value where the two are equal can be found. A more convenient approach is to use a similar facility provided by a spreadsheet, such as "Goal Seeker" on Microsoft EXCEL©.

8.8.2 Optimisation

The masses calculated for each of the "design" configurations may be compared in conjunction with any overriding constraints which may apply. The configuration which yields the lowest acceptable mass may be taken as the optimum solution, on the basis that mass is the criterion being used, see Chapter 1, paragraph 1.4.2. However, it should be pointed out that while the result achieved by the procedure is likely to be very close to the optimum value, it may not exactly coincide with it. The reasons for this lies in the assumptions needed to select the "design" values of thrust to weight ratio and wing loading discussed in paragraph 8.2.2(*j*). It is therefore desirable to investigate the effect of small changes in the variable parameters with the aim of converging to a true optimum.

An alternative approach to the optimisation is to use a mathematical optimisation technique such as multi-variate optimisation. This facility is provided on some spreadsheets, such as "Solver" in Microsoft EXCEL©. This approach should automatically find a true optimum and it is most conveniently undertaken as a "seamless" process covering both the first and second stages of the parametric analysis in one operation. The apparent disadvantage of this is that the end results of the first stage of the analysis are not explicitly needed and therefore there is no statement of the relative importance of the performance requirements. However, this may be overcome by superposing the optimisation upon a completed first stage analysis except that the listing of results for the various wing configurations is not necessary.

The case study of the short/medium haul airliner introduced at paragraph 8.6 will be continued and both methods of selecting an optimum used and compared.

8.9 Mass calculation

8.9.1 General

The mass of each "design" solution appropriate to a given wing configuration is estimated by use of the mass prediction data given in paragraph 6.4 of Chapter 6. In addition certain of the performance equations given in Chapter 7 are required for use both directly and indirectly in the calculation of the fuel mass.

8.9.2 Absolute masses (fixed independently of configuration variation)

The absolute masses are those which are determined by the requirements directly and other considerations which may be considered to be independent of the variation of the lifting surface configuration, such as fuselage geometry.

The absolute masses consist of:

a) *Payload for a given range, or the equivalent.* For the purposes of the initial design it is usually most satisfactory to use the payload associated with the maximum design range.

b) Fuselage mass, as given in Chapter 6, Eq (6.20) and Table 6.6. The information needed for this estimation is:

i) Overall fuselage dimensions.
ii) Maximum working pressure differential when this is relevant. It is determined by the highest design flight altitude.
iii) Structural design speed, V_D.
iv) The type of aircraft.

c) Operational items, as given by Chapter 6, Eq (6.21). This item can vary according to the payload-range combination and must therefore be consistent with the input used for (a) above.

8.9.3 ***Variable masses*** *(wing configuration dependent)*

The variable masses are determined by such considerations as the design parameters and performance characteristics.

a) Lifting surface mass, as given by Chapter 6, Eqs (6.22) to (6.24) and Table 6.7. For simplicity this item includes the masses of the control/stabiliser surfaces as well as that of the wing. If so required the two components, wing and stabilisers, may be separated by reference to the factor given in Table 6.10 of Chapter 6. For each set of parameters the required geometric and design data are known except for:
i) The wing area, which may be expressed in terms of take-off wing loading and mass using Eqs (6.29a) and (6.29b) of Chapter 6.
ii) The take-off mass itself which is, of course, the fundamental unknown in the calculation.

b) Powerplant mass, which is derived from Chapter 6, Eq (6.26) for the basic powerplant and Eq (6.25) and Table 6.8 for the installation factor. An assumption of the basic engine thrust to weight ratio, or power to weight ratio, Eq (6.26), enables the powerplant mass to be expressed in terms of the aircraft take-off mass, as given by Eq (6.30) of Chapter 6.

c) Systems mass, which for convenience is taken to include a nominal 4% of the all up mass as an allowance for the landing gear. It also includes such items as furnishings. Chapter 6, Eq (6.27) and Table 6.9 give typical values for this item in terms of take-off mass.

d) Fuel, which is estimated by reference to the relevant equations in Chapter 7. It consists of the following contributions:
i) Take-off and initial climb - assumed to be 1% of the total mass - see Chapter 7, paragraph 7.3.6.
ii) Climb - calculated by using the appropriate equation of Chapter 7 paragraph 7.6, dependent upon the type of climb and climb stage.
iii) Transonic acceleration for a supersonic aircraft - see Chapter 7, Eq (7.46).

iv) Cruise, in its widest sense - using the relevant equation from Chapter 7, paragraph 7.8. In evaluating the fuel required for this phase of the flight the distance covered during the climb and approach may be deducted from the total range requirement.

v) Descent, landing and landing reserve - for simplicity this is also assumed to be 1% of the take-off mass - see Chapter 7, paragraph 7.9. The ground distance covered during the descent is given by Chapter 7, Eq (7.56).

The total fuel mass is expressed solely in terms of the gross mass of the aircraft.

8.9.4 Take-off mass evaluation

The take-off mass, M_0, appropriate to any one set of design parameters is found by equating the sums of the individual masses covered by paragraphs 8.9.1 to 8.9.3 above to the take-off mass itself. The data have been formulated so that it is one of the following:

a) An actual mass.

b) A fraction of the take-off mass, M_0.

c) In the case of the lifting surfaces only, a function of wing loading and take-off mass. The wing loading is, of course, known in a particular case so this component also may be related to the mass M_0, although it is as a power function.

The overall mass equation is expressed in Chapter 6, Eqs (6.30) and (6.31). The presence of the power of M_0 arising from the lifting surface terms somewhat complicates the expression and precludes a direct solution. Some kind of numerical approach is necessary, as described in paragraph 8.8.1 above.

8.9.5 Derivation of optimum mass solution - graphical approach

A take-off mass is calculated for each of the sets of design parameters resulting from the first stage of the parametric analysis. The values of the take-off mass are conveniently presented in graphical form as a function of the variable parameters, as shown for the example covered by Spreadsheet 2, to which further reference is made in paragraph 8.12.2. Any overriding constraints or requirements may also be shown on the graph. These may include such items as:

a) Wing span limitation, which may apply to, for example a naval aircraft or a civil transport in relation to terminal limitations, see Chapter 5, paragraph 5.3.2.

b) A minimum allowable wing volume for a long range aircraft determined by the requirements of fuel stowage, see Chapter 5, paragraph 5.4.1.

c) A nominal limit of the wing structural parameter, SP, already considered at the first stage of the parametric analysis, see Chapter 5, paragraph 5.4.2.

The optimum design point is taken as that which gives the lowest take-off mass within the restriction imposed by the constraints. Having determined the mass it is possible to derive:

i) The leading geometric values, such as wing area, wing span, etc.
ii) A first level mass breakdown which may be compared with those of similar types of aircraft where these are available to ensure that there is no significant difference which cannot be accounted for.

8.9.6 Derivation of optimum mass solution - optimiser approach

8.9.6.1 General

When a mathematical optimisation technique is available, such as discussed in paragraph 8.8.2 above, it is not necessary to proceed with the graphical approach described in the previous paragraph. The optimiser may be used to move directly to the solution within the imposition of the same constraints as used for the graphical approach. The same consequences arise from the optimum mass thus derived.

8.9.6.2 Comments on the use of "Solver" in Microsoft EXCEL©.

The multi-variate optimiser "Solver" provided in the Microsoft EXCEL© spreadsheet is a powerful tool. However, some experience is desirable for its use for the type of work involved in the optimisation of conceptual design. The following points are relevant:

a) Should the optimiser fail to find a solution for any reason the previous values should be retained while the difficulty is investigated. A failure to retain the previous values can result in a need to examine and reset some parameters such as wing loading and the initial mass estimate (see Appendix A8.1, paragraph A8.2; Appendix A8.2, paragraph A8.14; and Appendix A8.3, paragraph A8.22).

b) "Solver"© may require a very large number of iterations to reach a solution. This difficulty may be overcome by relaxing the accuracy of the convergence.

c) It is possible that "Solver"© may be unable to find a solution within the limits of the input data and the constraints imposed. This can be investigated by relaxing the *constraints* in the first instance. If a solution is still not possible the requirements such as landing distance or payload/range may be relaxed until a solution can be found.

d) "Solver"© sometimes fails to find a solution with message "error in target cell or constraint cell ". If the targets or constraints have been changed an error may have been introduced. However, the message may appear when no change has been made to these cells. The explanation of this is that in seeking to find a solution "Solver"© has moved to a situation where one or more of the parameters in the problem, such as lifting surface mass, has taken a negative value and the impossibility of calculating the value of a power of such a term is the source of the apparent, but not real, error. Again the problem may be investigated by relaxation of the constraints or requirements, until a solution can be found.

8.10 Wing location and control/stabiliser surface areas

8.10.1 General

It is possible to derive sufficient further data for a preliminary drawing of the aircraft to be produced, but this involves making some additional assumptions.

8.10.2 Centre of gravity and wing location

The fore and aft position of the wing on the body is determined by the need to match its position relative to the overall centre of gravity. For initial design purposes it is adequate to assume that the overall centre of gravity is located at the quarter mean aerodynamic chord point. A knowledge of the fore and aft centre of gravity positions of the major items of mass is also required. Some of these may be readily estimated with reasonable accuracy and others determined from the fuselage layout. In the absence of more accurate information it is suggested that the following assumptions be made:

a) Wing mass: locate 0.1 of wing standard mean chord ($\bar{c}$) behind the wing quarter mean aerodynamic chord point (Δl_W). The standard mean chord is defined as the wing area divided by the wing span.

b) Fuselage mass: locate 0.45 of the fuselage length aft of its nose unless the layout indicates a more realistic position (l_{FUS}).

c) Tail mass: locate at 0.9 of the fuselage length aft of the nose (l_{TAIL}).

d) Main landing gear mass: assume initially that this item is 0.034 of the total aircraft mass and it is located at a point 1.1 times the distance of the centre of gravity aft of the nose (l_{MG}).

e) Nose landing gear mass: assume initially that this item is 0.006 of the total aircraft mass and that it is located at 0.1 of the fuselage length aft of the nose (l_{NG}).

f) Powerplant mass: the location of this item will depend upon the configuration of the aircraft. When the powerplants are located on, or within, the fuselage, the layout will give the approximate distance aft of the nose (l_{PP}). When the powerplants are wing mounted it is suggested that the following initial assumptions be made:

i) Two engines: locate at ($0.85\bar{c}/\cos\Lambda_{1/4}$) forward of the aircraft centre of gravity (Δl_{PP}), where $\Lambda_{1/4}$ is the sweep back of the wing quarter chord line.

ii) Four engines: locate at ($0.45\bar{c}/\cos\Lambda_{1/4}$) forward of the aircraft centre of gravity (Δl_{PP}).

g) System mass: unless there is evidence to the contrary locate at 0.45 of the fuselage length aft of the nose (l_{SYS}).

h) Operational items mass: as for system mass (l_{OPIT}).

i) Payload mass: locate where given by the fuselage layout, or when there are wing-mounted stores assume that they are located on the aircraft centre of gravity. For a single deck transport aircraft locate the payload at 0.45 of the fuselage length aft of the nose (l_{PAY}).

j) Fuel mass: If the fuel is wholly contained in the wing locate the fuel mass at the same point as the wing structure mass, as defined at (a) above. Otherwise use the information from the fuselage layout (l_{FUEL}).

The lengthwise location of the overall centre of gravity is then found by taking moments about the nose of the aircraft. If l is the overall length of the fuselage and l_{CG} is the distance of the centre of gravity aft of the nose, then in general:

$$M_0 l_{CG} = M_W(l_{CG} + \Delta l_W) + M_{FUS} l_{FUS} + M_{TAIL} l_{TAIL} + 0.034 M_0 l_{MG}$$

$$+ 0.006 M_0 l_{NG} + M_{PP} l_{PP} + M_{SYS} l_{SYS} + M_{OPIT} l_{OPIT} + M_{PAY} l_{PAY} + M_{FUEL} l_{FUEL} \quad (8.6a)$$

When the typical assumptions outlined above are made this becomes:

$$l_{CG}/l = [\{0.45(M_{FUS} + M_{SYS} + M_{OPIT} + 2M_{TAIL}) + 0.0006 M_0\} + \{M_{PP} l_{PP}$$

$$+ M_{FUEL} l_{FUEL} + M_{PAY} l_{PAY} + 0.1 M_W \bar{c}\}/l] / [0.963 M_0 - M_W] \quad (8.6b)$$

For the special case of a single-deck, twin-engined, transport aircraft:

$$l_{CG}/l = [\{0.45(M_{FUS} + M_{SYS} + M_{OPIT} + 2M_{TAIL} + M_{PAY}) + 0.0006 M_0\}$$

$$0.1\{(M_W + M_{FUEL}) - 8.5 M_{PP}/\cos\Lambda_{1/4}\}\bar{c}/l] / [0.963 M_0 - M_W - M_{FUEL} - M_{PP}] \quad (8.6c)$$

and for a four-engined, single-deck, transport aircraft the coefficient of M_{PP} in the numerator is replaced by $(4.5/cos\Lambda_{1/4})$.

The position of the apex of the wing on the centre of aircraft is given by:

$$l_{APEX} = l_{CG} - \bar{x}_{1/4} / c_0 \tag{8.7a}$$

where $\bar{x}_{1/4}$ is the location of the quarter mean aerodynamic chord point on the centreline chord, c_0, which is the assumed overall centre of gravity position:

$$\bar{x}_{1/4} / c_0 = \{(1 + 2\lambda)\, A tan\Lambda_{1/4}\} / 12 + 0.25 \tag{8.7b}$$

where λ is the taper ratio, and hence:

$$l_{APEX} = l_{CG} - \{(1 + 2\lambda)\, A tan\Lambda_{1/4}\}/12 + 0.25 \tag{8.7c}$$

Note that the aerodynamic mean chord, *MAC*, is derived from:

$$MAC = 4\bar{c}\,(1 + \lambda + \lambda^2) / \{3\,(1 + \lambda)^2\} \tag{8.7d}$$

where the standard mean chord, $\bar{c}$, is the reference wing area divided by the wing span.

8.10.3 Approximate estimate of the areas of the control/stabiliser surfaces

8.10.3.1 General

There exits a substantial body of empirical data relevant to the areas of both horizontal and vertical stabiliser/control surfaces for conventional, naturally stable, aircraft. There is less information on unconventional layouts, including those which are basically unstable and employ active control techniques to provide artificial stability. However, since by implication the areas of control surfaces on unstable concepts will be less than those for corresponding stable designs it is reasonable to use the existing data to give a first indication of control/stabiliser sizes.

8.10.3.2 Horizontal surfaces (typically a tailplane/elevator combination)

The horizontal tail volume coefficient is conventionally used as an empirical measure of stability and control, and is defined as:

$$\bar{V} = S_H\, l_H / S\bar{c} \tag{8.8a}$$

where S_H is the reference area of the surface and l_H the nominal distance of its centre of lift from the aircraft centre of gravity. S is wing area and $\bar{c}$ the mean wing chord. Using the assumption relating to tail location from paragraph 8.10.2(c):

$$S_H = \bar{V} S \bar{c} /(0.9l - l_{CG}) \quad (8.8b)$$

Table 8.1A gives typical values of $\bar{V}$ for various classes of aircraft, including a suggestion for foreplane layouts.

Table 8.1 Typical control/stabiliser characteristics

A Volume coefficients

Class of aircraft	Horizontal $\bar{V}$	Vertical $\bar{V}_V$
Interceptors - conventional	0.4	0.065
canard *	0.12	0.065
Single-engine general aviation	0.65	0.05
Twin-engine general aviation	0.85	0.065
Trainers, patrol, bombers	0.65	0.065
Business jets	0.70	0.065
Turboprop transports	1.00	0.08
Jet transports - subsonic	1.20	0.09
- supersonic	-	0.065
High performance sailplane	0.5	0.015 to 0.02

* May be in excess of 0.2 for very "close coupled " designs.

B Geometry

Surface	A	Λ/Λ_{WING}	λ/λ_{WING}
Horizontal tail	(0.5 to 0.6) A_{WING}	1.0	1.2
Canard	(1.0 to 1.3) A_{WING}	0.8 to 1.15	1.3
Vertical tail	0.9 to 3.0**	1.0^{x}	0.5

** 0.9 for single engine otherwise 1.2 or more, upwards of 3 for transport types.
x Usually not less than 20° on quarter chord.

8.10.3.3 Vertical surface (typically a fin/rudder combination)

In this case the volume coefficient is defined as:

$$\bar{V}_V = S_V l_V / Sb \quad (8.9a)$$

where b is the wing span, S_V the fin/rudder reference area and l_V is distance from the centre of gravity of the aircraft. Again using the assumption of paragraph 8.10.2(c):

$$S_V = \bar{V}_V Sb/(0.9l - l_{CG}) \quad (8.9b)$$

Table 8.1A also gives typical values for $\bar{V}_V$.

8.10.4 Control/stabiliser surface geometry

There can be considerable variation in the geometry of the control/stabiliser surfaces. However, the significance of many of the variations is related to a detailed analysis of the requirements and operating characteristics. Some further information is given in Addendum 2, for horizontal surfaces, and Addendum 3 for vertical surfaces. For initial design purposes it is usually satisfactory to base the design on typical values, which are related in some degree to the geometry of the wing. Table 8.1B summarises this matter.

8.10.5 Aileron geometry

The definition of aileron geometry is not necessary for the purposes of defining the overall configuration of the aircraft. However, it is convenient to consider typical aileron geometry since it does have an impact upon flap characteristics and is related to fin/rudder design. Table 8.2 summarises the typical aileron geometry for a range of aircraft and also introduces an aileron volume coefficient, $\bar{V}_A$ defined as:

$$\bar{V}_A = 0.5 S_A l_A / S\, b \quad (8.10)$$

where S_A is the total area of all ailerons
l_A is the distance between the midpoints of the ailerons, $l_A = 2y_A$, see Table 8.2
See also Addendum 3.

Table 8.2 Typical aileron characteristics

AIRCRAFT TYPE	INBOARD AILERONS			OUTBOARD AILERONS				
	b_A/b	c_A/c	y_A/b	b_A/b	c_A/c	y_A/b	Total S_A/S	Total $\bar{V}_A$
Single piston - general	-	-	-	.37	.25	.37	.08	.030
Single piston - agriculture	-	-		.37	.25	.37	.10	.037
Single prop trainers	-	-	-	.38	.25	.37	.086	.032
Jet trainers	-	-	-	.34	.27	.39	.075	.029
Twin prop general	-	-	-	.32	.26	.36	.065	.024
Twin regional turboprop	-	-	-	.33	.25	.40	.058	.023
Military turboprop transport	-	-	-	.26	.30	.41	.058	.024
Executive jets	-	-	-	.27	.25	.39	.052	.020
Jet transport - (1)	-	-	-	.21	.29	.42	.032	.013
Jet transport - (2)	.08	.23	0.19	.20	.23	.43	.04	.014
Military jet transport	-	-	-	.28	.28	.42	.05	.021
Fighter	-	-	-	.26	.26	.38	.04	.015
Attack	-	-	-	.28	.27	.37	.06	.022
Supersonic cruise (1)	No inboard controls			.80	.50	.30	.13	.039
Supersonic cruise (2)	With inboard controls			.50	.24	.30	.09	.027

NOTES: b_A is total span (both sides) of one set of ailerons
c_A is the average aileron chord
y_A is the semi span distance to the mid-point of the ailerons
S_A is the total area of all ailerons
$\bar{V}_A$ Is the total volume coefficient of all ailerons
Military transports include patrol aircraft and subsonic bombers

8.11 Overall layout of the aircraft - referee design

At this stage of the design sufficient data have been derived to enable an initial layout of the aircraft to be prepared. Effectively this represents a "referee" design which can form the basis of a detail analysis as outlined in Chapter 9. Addendum 1 contains some guidelines for landing gear layout, which is relevant at this point in the design process.

8.12 Case study - short/medium haul airliner

8.12.1 Introduction

This case study continues the parametric analysis of the example given in paragraph 8.6 and proceeds through the second stage of the analysis in order to derive an "optimum" design. Two approaches are illustrated:

a) A graphical approach which leads to a selection of the lowest mass configuration tenable within the imposed constraints.

b) An optimiser approach, where the lowest mass solution is derived directly.

In both cases use is made of a Microsoft EXCEL© spreadsheet.

8.12.2 Graphical approach

8.12.2.1 General
This is shown in Spreadsheet 2, the full explanation of the spreadsheet being given in Appendix A8.2. It is summarised below.

8.12.2.2 Data
The information in the Requirements Box has been derived directly from the original specification. It is to be noted that the ultimate manoeuvre factor, N, used at the first stage of the process to predict the structural parameter, SP, has been replaced by $\bar{N}$, as required for the calculation of lifting surface mass, see Chapter 6, Eq (6.22a).

The Climb Path Box also contains data used at the first stage of the analysis. Provision is made for the constant Mach number climb to continue to a start of cruise at altitudes in excess of 11 km, by using the suffices (1) and (2) to distinguish climb below or above 11 km respectively. In this example the cruise commences at 10 km altitude so zero has been inserted in the Sigma Cruise (2) cell.

The Parametric Box contains a summary of the results given by the first stage for a given set of the original parameters. The data have been derived directly from the Summary of Results listed at the lower centre of Spreadsheet 1. Drag coefficient data have been placed in the Input Data Box for convenience.

The Assumed Box includes information which is needed to complete the analysis, but which is either not considered as a primary parameter, such as wing taper ratio, λ, or is related to the powerplant. Nevertheless, it may be varied if so required.

The Input Data Box contains the overall fuselage dimensions derived in paragraph 8.6.5, the mass coefficients C_1 to C_5, see paragraph 6.4 of Chapter 6, the drag coefficient data referred to above, and the assumed location of the major components of mass as discussed in paragraph 8.10.2.

8.12.2.3 Calculations

The Calculations Box uses the data contained in the previously mentioned Boxes to derive the mass components for the particular configuration under examination. These include:

a) Climb, cruise and approach fuel masses, which are given as a fraction of the total mass, M_0.

b) Powerplant and systems masses, also a fraction of M_0 and together with the fuel make up (*Kappa* M_0).

c) Fuselage mass, payload and operation items mass which have a definite value and are collectively referred to as the absolute or fixed mass.

(d) Lifting surface mass which is a complex function of both M_0 and the known wing loading.

It is to be noted that the last cell in the Calculations Box is the evaluation of X_2, the rate of climb correction factor, for conditions above 11 km altitude, see Chapter 7, Eq (7.26). Evaluation of X_2 includes division by the end of climb relative density Sigma (2) and as this is zero in this case X_2 cannot be evaluated. For the same reason the cells relating to this second phase of the constant Mach number climb contain zeros.

The calculations are brought together in the Analysis Box. To derive a value of the total mass, M_0, it is first necessary to make an estimate of the value of M_0, namely [(M_0) est 1]. This is done by assuming a typical value of $0.12M_0$, for the lifting surface mass. The value so derived is directly input to the adjacent cell as [(M_0) est 2] followed by cell (KM_0) and the formula for the lifting surface mass. [(M_0) calc] is the sum of the fixed mass, (KM_0) and the lifting surface mass and will, in general be different to [(M_0) est 2] on which it is based. The "Goal Seeker" tool in EXCEL© is then used to reconcile the two values by minimising the difference between them as shown in the error cell.

The first numerical column in the Summary Box is a listing of the masses of the main components. These data are used with the location data from the Input Box to derive the overall centre of gravity position, wing location and control/stabiliser surfaces areas in the lower row of the Analysis Box. The Summary Box also contains the details of the implied wing geometry, including an estimate of the volume available for fuel stowage in the wing, and for convenience the control/stabiliser sizes.

8.12.2.4 Results

The total set of results for the various configurations under examination is given in the Summary of Results table, the Case Summary being used as a convenience for listing. These results are shown as a carpet graph in the associated figure. In this example there are two constraints to be imposed:

a) The wing span must not exceed 34 m. The results data have been used to derive the aspect ratio corresponding to 34 m span for each of three thickness to chord ratios, as shown in the upper right-hand figure on the spreadsheet, the values being added to the carpet plot.

b) The fuel should all be contained in the wing so that the available fuel volume must be at least as great as that required. The conditions which give this have been analysed in the lower of the two graphs on the right-hand side of the spreadsheet, and again the results transferred to the carpet plot.

The remaining numerical groups on the spreadsheet were used in preparing the graphical data, see Appendix A8.2.

8.12.2.5 Comments on results

It is instructive to compare the carpet plot of the results with the comparable one derived at the conclusion of the first stage of the analysis. The thickness to chord ratio trend is reversed. Aspect ratio trends are similar up to a value approaching ten, beyond which the mass of the aircraft increases in spite of the lower required thrust to weight ratios. As it happens these higher aspect ratios are ruled out in this example by the 34 m wing span constraint. The tendency of increase of total mass at high aspect ratio is reasonably consistent with the selection of a structural parameter value of 16 as a check in this case.

8.12.2.6 Optimum design point

The optimum mass is chosen as the lowest value allowed within the constraints. Inspection of the carpet plot suggests that this will occur at a point along the available fuel volume condition at an aspect ratio of between nine and ten, with a thickness to chord ratio somewhat less than 0.13. A larger scale plot of the results enables actual values of 9.1 for aspect ratio and 0.128 for thickness to chord ratio to be deduced. The corresponding results are the ones shown in the Summary and give a take-off mass of 72,757 kg with a wing span of 33 m.

It will be noted that the deduced wing area of 119.73 m^2. implies that the value of $S^{\wedge -0.1}$ is almost 0.62, rather than the value of 0.6 originally assumed. If a correction is made for this, using the same aspect ratio and thickness to chord ratio, the new mass is found to be 73,892 kg. Unless the whole analysis is repeated it cannot be confirmed that this will still be the optimum wing geometry and the result does indicate a slightly deficient fuel volume, but it is expected that it will be very near to the optimum. See paragraph 8.12.3.4.

8.12.3 Optimiser approach

8.12.3.1 General

This is shown in Spreadsheet 3 and is effectively a combination of the first part of Spreadsheet 1 with Spreadsheet 2 referred to in the previous paragraph. The explanation of the spreadsheet is given in Appendix A8.3. It is possible to derive a somewhat simpler spreadsheet if the first stage of the analysis is combined directly with the optimiser approach, as for example in Addendum 5, paragraph AD5.4.

8.12.3.2 Data

The information contained under the left-hand Parameters, Requirement, Assumed, Initial Calculated Values and Preliminary Calculations is identical to that of Spreadsheet 1, as are the Analysis, Summary and Result Boxes, and the diagram of thrust to weight ratio as a function of wing loading. The lower, right-hand, Requirements, Parameters, Input Data, Calculations, Climb Path and Assumed Boxes are identical to those of Spreadsheet 2.

8.12.3.3 Analysis

While the Analysis Box also appears to be the same as that of Spreadsheet 2 it differs in one important respect. The [(M_0) calc] cell is used in conjunction with EXCEL© "Solver" to determine a minimum value of M_0 by varying the wing loading, aspect ratio and thickness to chord ratio. In addition constraints are imposed to ensure that:

i) The M_0 value on the [(M_0) est 2] cell and [(M_0) calc] are identical.
ii) The wing span does not exceed 34 m.
iii) The wing loading is equal to or less than that determined by the landing condition [$(Mg/S)_0$ ld].
iv) The wing loading is equal to or greater than that determined by the gust sensitivity condition [$(Mg/S)_0$ gt].

As is the case with Spreadsheet 2 the resulting value of M_0 is used to evaluate the centre of gravity position, component mass breakdown and control/stabiliser surface areas, all of which are given in the final Summary Box.

8.12.3.4 Result

Using the optimiser approach yields only one result which, providing all the imposed conditions have been met, is the overall minimum mass. The aspect ratio is seen to be 9.102 and the thickness to chord ratio 0.1280.The corresponding total mass is 72,749 kg. The fuel volume constraint has been exactly met. These are based on $S\wedge^{-0.1}$ equal to 0.6.

When using the optimiser approach it is a simple matter to correct the value of the parameter $S\wedge^{-0.1}$ and re-optimise to obtain an exact solution. When this is done $S\wedge^{-0.1}$ has a value of 0.6187. The aspect ratio increases slightly to 9.158 and the thickness to chord ratio to 0.1288. The corresponding mass is 73,764 kg and wing span 33.38 m.

8.12.4 Comparison of the results obtained from the two approaches

If $S\wedge^{-0.1}$ is taken to be 0.6 the results derived from both approaches are, as is to be expected, virtually identical. The mass discrepancy is only some 8 kg, or about 0.01%, and the aspect ratio and thickness to chord ratio are the same. However, when the corrected value of $S\wedge^{-0.1}$ is used the graphical approach, which was not re-optimised, has a mass some 218 kg higher than that of the true optimum. Even this is only a 0.3% error. That the graphical approach did not quite meet the fuel volume requirement is reflected in the slightly higher aspect ratio and thickness to chord ratio given by the optimiser method.

8.12.5 Outline layout of aircraft

Figure 8.2 is an outline layout of the aircraft which has been produced by the use of the derived data and the information contained in paragraph 8.10 above. Additional data used were:

- i) Dihedral angle assumed to be 3°, see Chapter 5, paragraph 5.3.7.
- ii) Horizontal tail and fin geometry used Table 8.1B.
- iii) Fuselage nose and tail proportions from paragraph 8.6.5.2.
- iv) For an individual powerplant thrust of about 105,000 N, static, there are several existing powerplants the dimensions of which have been used to give a typical shape.

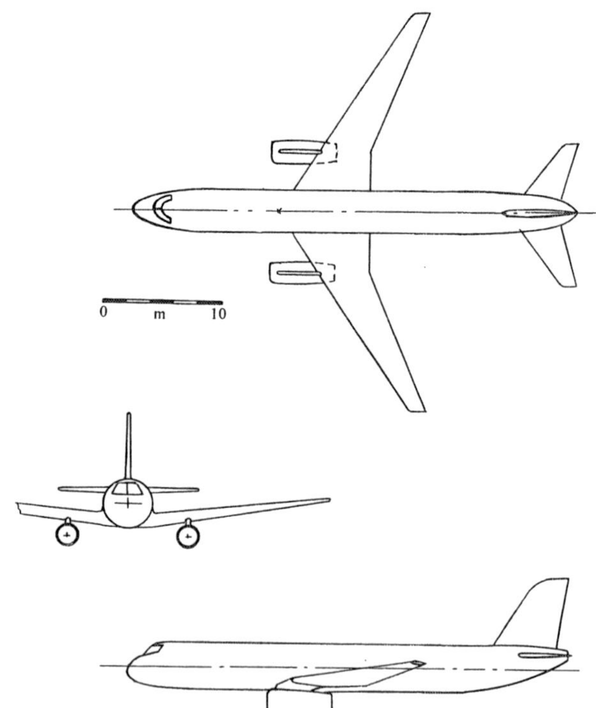

Figure 8.2 Short/medium haul airliner

Appendix A8.1 Case study - short/medium haul airliner

First stage of the parametric analysis - explanation of SPREADSHEET 1

(Subsonic flight - Fan engine 1: Microsoft EXCEL©)

A8.1 Introduction

The following listing contains a detailed explanation of the formulation of each cell of the spreadsheet so that the procedure may be followed completely. However, the spreadsheet may be used by making reference only to paragraphs A8.2 to A8.4 and A8.8.

A8.2 Input parameters

In the example shown in Spreadsheet 1 the parameters chosen for variation are limited to three, as shown in the Parameters Box, namely:

Row 3 Cell A Aspect ratio (initial entry only - see cell O47) - A
B Root thickness to chord ratio (initial entry only - see cell P47) - t/c
C Bypass ratio of engine - R

If subsequent overall optimisation is undertaken using EXCEL Solver© it is more convenient to operate from the output of the first stage of the analysis, Rows 45 to 47, and to use cells O47 and P47 only to input the parameters initially. Other items outlined in paragraphs A8.3 to A8.5 may also be varied as parameters if so required.

A8.3 Requirements box

The requirements necessary to undertake the first stage of the parametric analysis and which are given in paragraph 8.6.2 are listed in the spreadsheet as follows:-

Row 3 Cell D Factored take off field length - ToL
E Factored landing length (no reverse thrust) - LL
F Maximum allowable approach speed - Va max
G Design speed (structural - see paragraph 8.6.11) - Vd
H Cruise critical Mach number - (Mn) crit
I Actual cruise Mach number - (Mn) cr
J Rel. density at start of cruise altitude (if below 11 km) -Sigma cr 1
K Rel. density at second cruise altitude (if above 11 km) - Sigma cr 2
(This is used to handle the situation when the cruising or climbing flight covers two altitude bands - see Ch. 7, paragraph 7.6.3).
L Maximum range requirement - Max Rge

A8.4 Assumptions and basic data

It is necessary to make some assumptions and to input basic data from relevant sections of the notes. In the example those shown in the Assumed Box are:

Row 7 Cell A Ratio of start of climb to take-off mass (see para 8.6.9.1 and Ch.7, para 7.3.6) - M1/Mo

B Wing area parameter in zero lift drag term (Ch.6, Figure 6.1) - (S^-0.1)

C Wetted area ratio (Ch.6, Table 6.3) - Rw

D Type factor (Ch.6, Table 6.4) - Type Fac

E Relative air density at end of equivalent air speed phase of climb (see para 8.6.9.2) - Cl EAS sig.

F Second segment climb factor, α (Ch.7, Eq (7.4a)) - SS alpha

G Second segment climb factor, γ, (Ch.7, Eq (7.4a)) - SS gamma

H Take-off flap factor (Ch.6, Eq (6.15)) - Flap Fac

I Speed of sound at initial cruise altitude - a Cruise

J Ultimate normal manoeuvre factor (para 8.6.11) - N

K Powerplant altitude dependency power, s, (para 8.6.4) - PP Fac-s

L Proportion of wing laminar flow, c_l, (para 8.6.8.1) - Lam Chrd

M Gust sensitivity switch (para 8.4.7.3, - 1 on, 0 off) - Gust Sen

N Take-off Mach number - TO Mn

A8.5 Initial calculation and input data

Some of the required data are independent of the parametric analysis and it is convenient to calculate these before undertaking the analysis associated with the variable parameters. Certain other items are suitable for direct input. These items are in the Initial Calculated Values Box:

Row 11 Cell A Ratio of design landing mass to take-off mass (para 8.6.9.3 and Ch.7, Table 7.1 - uses cell L3) - Ml/Mo

B Ratio of initial cruise mass to take-off mass (see para 8.6.9.2 and Ch.7, para 7.6.1, uses cells J3, A7 and E7) - Mcr/Mo

C Zero sweep maximum lift coefficient (para 8.6.7.2 and Ch.6, Eq (6.2) and Table 6.1) - (Cl max) o

D Zero sweep unstick lift coefficient (para 8.6.7.4 and Ch.6, Eq (6.4) and Table 6.1) - (Clus) o

E Zero sweep approach lift coefficient (para 8.6.7.3 and Ch.6, Eq (6.3)) - (Cla) o

F Zero sweep usable lift coefficient in cruise (para 8.6.7.5 and Ch.6, Eq (6.6)) - (Cl use) o

G Low speed zero lift drag coefficient (para 8.6.8.1 and Ch.6, Eq (6.13), uses cells B3, B7, C7, D7, L7 and N7) - (Cdz)ls

H Rel. density at start of cruise (para 8.6.2) - Sigma Cr

A8.6 Calculations

Some further calculations must be completed to provide data for use in the parametric analysis. The set of calculations depends upon the information from the requirements, assumptions, input parameters and initial calculations covered by paragraphs A8.2 to A8.5 above. They are shown in the Preliminary Calculations Box.

Row 15 Cell A Cosine of wing quarter chord sweep (see para 8.6.8.1 and Ch.5, Eq (5.4a), uses cells B3 and H3) - cos delta

B Wing quarter chord sweep (uses cell A15) - Delta deg.

C Wave drag factor (the term to the power of 20 in Ch.6, Eq (6.13), uses cells A15, B3, and I3) - Wave Dr F

D Zero lift drag coefficient in cruise (see para 8.6.8.1 and Ch.6, Eq (6.13), uses cells C15, G11, I3 and N7) - (Cdz) cr

E Equivalent zero lift drag coefficient in climb out (see para 8.6.8.3. and Ch.6, Eq (15), uses cells A3, G11 and H7) - (Cd)co

F Low speed induced drag factor (see para 8.6.8.2 and Ch.6, Eq (6.14), uses cells A3, B3 and A15) - (Kv)o

G Cruise induced drag factor (see para 8.6.8.2 and Ch.6, Eq (6.14), uses cells F15 and I3) - (Kv)cr

H Thrust factor at start of cruise conditions (see para 8.6.4 and Ch.3, Eq (3.8b), uses cells C3, E7 and K7) - Tau (Cl Mn1)

I Thrust factor at start of second cruise phase (see para 8.6.4 and Ch.3, Eq (3.8b), uses cells C3, I3 and K7) - Tau(Cl Mn2)

Row 18 Cell A Maximum lift coefficient (see Ch.6, Eq (6.2), uses cells A15 and C11) - Cl max

B Unstick lift coefficient (see Ch.6, Eq (6.4), uses cells A15 and D11) - Cl/us

C Approach lift coefficient (see Ch.6, Eq (6.3), uses cells A15 and E11) - Cl/a

D Usable lift coefficient in cruise (see Ch.6, Eq (6.3), uses cells A15 and F11) - Cl/use

E Approach speed from landing distance requirement (see Ch.7, Eq (7.6g), uses cell E3) - Va calc

F Actual approach speed, being the lowest of Va calc (cell E18) and Va max (cell F3) - Va

G Take-off wing loading defined by limiting approach speed (see para 8.3.3 and Ch.7, para 7.4, uses cells A11, C18 and F18) - (Mg/S) o ld

H Take-off wing loading to meet the gust sensitivity requirements (see para 8.3.7.3 and Ch.5, Eq (5.9b), uses cells A3, A15,G3, I3, and H7; the latter may be used to negate this) - (Mg/S) o gt

I Actual landing length with selected value of approach speed (see para 8.4.3 and Ch.7, para 7.4 and Eq (7.6), uses cell F18) - Cor LL.

A8.7 Parametric analysis

The parametric analysis consists of evaluating the variation of the take-off wing loadings, $(Mg/S)_0$ with the installed static thrust to weight ratio, $(T/Mg)_0$. For the particular example the relevant performance conditions are:

a) Take-off (factored runway length).

b) Accelerate-stop (runway length).

c) Second segment climb (hot and high environment with one engine failed).

d) Start of cruise with a residual rate of climb of 1.5 m/s. The analysis allows for the cruise to commence below 11 km altitude (cells S16 to 25) or above 11 km (cells T16 to 25).

e) Reversed thrust landing, which is determined by the maximum thrust to weight ratio derived from the other conditions.

In the spreadsheet the analysis is arranged as follows:

Column Q — Row 4 to 13: Range of wing loading associated with cases (*a*), (*b*) and (*c*). - (Mg/S)o. Row 13 is a particular case derived from the analysis, see also cell Q37. The default value is the wing loading based on the landing condition (cell G18) but this may be overridden when appropriate - (see para A8.8.1).

Columns R and S — Row 4 to 13: Thrust to weight ratio for take-off, case (*a*), using Ch.7, Eq (7.1b). Column R is a first approximation based on selection of an initial value of $(T/Mg)_0$ from Ch.3, Table 3.2. Column S is a corrected value using the result from column R.(uses cells B18, D3, R4 and Q4, etc.).

Column U — Row 4 to 13: Thrust to weight ratio for the accelerate-stop, case (*b*), using Ch.7, Eq (7.3b). (Uses cells B18, D3, and Q4 etc.)

Columns W, X, Y — Row 4 to 13: Derived thrust to weight ratio for the second segment climb case (*c*) using Ch.7, Eq (7.4a). Column W is the appropriate climb out thrust factor based on a

speed of 1.1 times the unstick speed (uses Ch.3, Eq (3.8a) with cells B18, C3 and O4, etc) - Tau co.
Column X is the appropriate lift to drag ratio (uses cells B18 and E15) - (Cd) co/(Cl) us.
Column Y is the required thrust to weight ratio (uses cells, E15, F7, F15, G7, W4 and X4, etc.).

Column Q Row 16 to 25: Range of wing loading associated with cases (*d*) and (*e*) - (Mg/S)o. (See under column Q rows 4-13 above.)

Columns R and S Row 16 to 25: Thrust to weight ratio for the start of climb case (*d*) using Ch.7, Eq (7.23b) for cruise below 11 km altitude. Column R evaluates the factor Q_M (Ch.7, Eq (7.22b), uses cells D15, I3, and Q16, etc.) - Qm. Column S is the required sea level static thrust to weight ratio (uses cells A7, D15, E7, G15, H15, I3, J3, K7 and R16, etc.) - (T/Mg)o-1).

Column T Row 16 to 25: Thrust to weight ratio for start of climb case (*d*) using Ch.7, Eq (7.26c) for cruise above 11 km altitude (uses cells A7, D15, G15, I3, I15, J3, K3, and R16, etc.) - (T/Mg)o - 2.

Column U Row 16 to 25: Lift coefficient at start of cruise (uses cells B11, H11, I7 and Q16, etc.) - Cl case-1.

Column V Row 16 to 25: Drag coefficient at start of cruise (uses cells D15, G15 and U16, etc.) - Cd-1.

Column W Row 16 to 25: Corresponding lift/drag ratio (uses cells U16, V 16 etc) - (L/D)-1.

Column X Row 16 to 25: Product of cruise Mach number and lift/drag ratio (uses cells I3 and W16, etc.) - Mn* L/D.

Columns Y and Z Row 16 to 25: Reverse thrust landing case (*e*) using Ch.7, Eq (7.6e). Column Y is the highest value of (T/Mg)o from the previous cases (cells S4, U4, X 4, S16 and T16, etc.). Column Z is the landing length (uses cells A11, F18 and Y16, etc.) - L length.

A8.8 Results

Results are presented at various stages of the analysis:

A8.8.1 Summary of thrust to weight ratios for a given configuration.

Columns Q to X, rows 28 to 42, present a summary of the thrust to weight ratios needed to meet the performance requirements for a given configuration. In the case of this example it is for a given aspect ratio and thickness to chord ratio. Rows 23 to 36 repeat the results obtained for the selected range of wing loadings. Row 37 is for a particular value of wing loading to which reference will subsequently be made. Rows 38 and 39 together with rows 41 and 42 are the wing loadings derived from the gust sensitivity condition, cell H18, and landing condition, cell G18, respectively. Arbitrary thrust to weight ratios which are anticipated to be at the extremities of the likely range are associated with these wing loadings for convenience of graphical presentation.

This set of results is converted into the chart format on the lower left-hand side so that the critical performance conditions may be identified. From this chart it is possible to select a "design" value of the wing loading and corresponding thrust to weight ratio. The selection of this "design" value is not always straightforward. In some cases there may be a clear minimum thrust to weight ratio, but this by no means always the case, as indicated on the chart. Here the choice is made by selection of the highest value of wing loading allowed, which is in fact the landing condition. Only at aspect ratios in excess of ten does the take-off case became critical and indicate a somewhat lower value of wing loading than the landing case, see paragraph A8.8.3 below. Whichever criterion is used for selection the chosen value of wing loading is input to cell Q37 (and hence cell Q13). The associated "design" thrust to weight ratio is given in cell Y37.

A8.8.2 Result of analysis of given configuration

The particular set of results for a given configuration is shown in the Result Box columns O to Z and rows 45 to 47. Cell Y47 is the Structural Parameter for the given configuration (see para 8.5 and Ch.5, Eq (5.8)). It makes use of cells A3, B3, A15 and J7.

A8.8.3 Analysis of various configurations - Summary of Results

The most convenient way to investigate alternative configurations is to change the value of the parameters in the Results Box, columns O and P, row 47. The changes are automatically repeated in the Parameters Box, columns A and B, row 3.The process of selecting "design" wing loading and thrust to weight ratio values is repeated, using the chart, see paragraph A8.8.1 above, and the revised results appear in the Results Box when the adjusted wing loading has been placed in cell Q37.

Below the Results Box in columns O to X and rows 51 to 68 there is the Summary of Results appropriate to the various configurations. This table contains only the data needed for further analysis and is produced by direct input from the Results Box, this being simpler than transferring the data automatically. Columns O to X, rows 70 to 72, include supplementary configurations chosen to give the Structural Parameter, *SP*, a value of 16, which is taken to be the nominal upper limit. This set of results was obtained simply by altering the values of the aspect ratio in cell O47 until the desired *SP* resulted in cell Y47. Rows 75 to 77 and 79 to 84 contain data needed for the second stage analysis - see Appendix A8.2, paragraph A8.17.

It is convenient to present the overall results in graphical form, preferably in carpet format. This is shown in the chart on the right-hand sidc. Thc tabulated values above the chart have been derived directly from the overall Summary of Results and are presented in the form necessary to produce the carpet format. The chart is effectively a series of overlays of each set of tabulated data.

It may be deduced that aspect ratios in excess of about ten are likely to result in an unduly heavy wing structure, but this is a matter to be checked at the next stage of the parametric analysis.

A8.9 Conclusions

The evidence of the first stage of the parametric analysis is that the lowest thrust to weight ratio, and hence the lowest relative powerplant mass, will result when the aspect ratio is rather more than nine. Only the second stage of the analysis can show if this is near to the optimum and define the associated value of the thickness to chord ratio.

Appendix A8.2 Case study - short/medium haul airliner

Second stage of the parametric analysis - graphical approach - explanation of SPREADSHEET 2

(Subsonic flight - fan engine 2: Microsoft EXCEL©)

A8.10 Introduction

The following listing contains an explanation of the source or formulation of the data in each cell of Spreadsheet 2 so that the procedure may be followed completely. The spreadsheet may be used by making reference to the Requirements, Parameters, Climb Path, Assumed, Input and Analysis Boxes, together with the Summary of Results.

The spreadsheet is a direct continuation of Spreadsheet 1 and Appendix A8.1. When it is necessary to refer to cells from this previous spreadsheet, they are given the prefix 1 for clarity.

A8.11 Initial data

The initial data are shown in the four boxes at the top of the spreadsheet and the majority of it is extracted directly from the first stage of the analysis or the requirements.

The Requirement Box consists of:

Row 3 Cell	A	Design range, 1L3 - Range
	B	Number of passengers, specification - PAX
	C	Payload from specification - Payload
	D	Cruise Mach number, 1I3 - (Mn) cr

E Initial cruise altitude from specification - In CrAlt.
F Initial cruise relative density, 1J3 - In Cr sigma
G Final cruise altitude, specification - Fin Cr Alt
H Final Cruise relative density, specification - Fin Cr sig
I Structural design speed derived from specification (see para 8.6.11) - Vd
J Effective factored wing normal acceleration factor (see para 8.6.11 and Ch.6, Eq (6.22a)) - N bar

The Parameters Box contains:
Row 7 Cell A Aspect Rati, 1A3 or 1047 - A
B Thickness to chord ratio, 1B3 or 1P47 - t/c
C Cosine of wing quarter chord sweep, 1A15 - cos delta
D Bypass ratio, 1C3 - R
E Take-off wing loading, 1Q37 or 1Q47 - (Mg/S)o
F Static thrust to weight ratio, 1R47 - (T/Mg)o
G Cruise lift to drag ratio, 1S47 - L/D
H Start of climb mass ratio, 1A7 - MI/Mo
I Wing quarter chord sweep, radians, A(cos C7) - Delta rad
J Structural parameter - SP

The Climb Path Box covers:
Row 3 Cell L Constant EAS climb speed (see para 8.6.9.2) - ClEAS
M Altitude at end of constant EAS climb (see para 8.6.9.2) - CLEAS H2
N Density ratio at end of constant EAS climb, 1E7 - ClEASsig
O Constant Mach number climb condition (see para 8.6.9.2) - Climb Mn
P Density ratio at start of cruise (if below 11 km) 1J3 - Sig cr 1
Q Density ratio at start of cruise (if above 11 km), 1K3 - Sig cr 2 (0 if climb starts below 11 km.)
R Altitude at end of climb (first phase), specification - End cr l

The contents of the Assumed Box are:
Row 7 Cell L Wing taper ratio (see para 8.6.6.1) - Lambda
M Basic powerplant thrust to weight ratio (see Ch.6, Eq (6.26(h)) - (T/Mg) eng
N Allowance factor for operational items mass (see Ch.6, para 6.4.2.3.) - Op It Fac
O Approach fuel allowance mass ratio (see Ch.6, para 6.4.3) - App Fuel/Mo
P Assumed horizontal tail volume coefficient (see Table 8.1A) - V bar
Q Assumed vertical tail volume coefficient (see Table 8.1A) - Vv bar
R Powerplant altitude dependency power, 1K7 - PPFac- s

A8.12 Additional input data

The information in the Input Data Box is:

Row 11 Cell A Fuselage length (see para 8.6.5.1) - Fus L
- B Fuselage width (see para 8.6.5.3) - Fus B
- C Fuselage depth (see para 8.6.5.3) - Fus H
- D Lifting surface mass coefficient (see Ch.6, Table 6.7) - c1
- E Fuselage mass coefficient (see Ch.6, Table 6.6) - c2
- F Engine installation mass factor (see Ch.6, Table 6.8) - c3
- G Systems mass ratio (see Ch.6, Table 6.9) - c4
- H Lifting surface mass to wing mass ratio (see Ch.6, Table 6.10) - c5
- I Low speed zero lift drag coefficient, 1G11 or 1U47 - (Cdz)ls
- J Cruise zero lift drag coefficient, 1D15 or 1V47 - (Cdz)cr
- K Incompressible drag factor, 1F15 or 1W47 - (Kv)o
- L Cruise induced drag factor, 1G15 or 1X47 - (Kv)cr
- M Incremental position of wing mass (see para 8.10.2(*a*)) - Del lw
- N Position of fuselage mass (see para 8.10.2(*b*)) - l Fus
- O Position of tail mass (see para 8.10.2(*c*)) - l Tail
- P Incremental position of powerplant mass (see para 8.10.2(*f*)) - Del lPP
- Q Position of systems mass (see para 8.10.2(*g*)) - l SYS
- R Position of payload mass (see para 8.10.2(*i*)) - l PAY
- S Position of mass of operational items (see para 8.10.2 (*h*)) - l OP IT
- T Increment position of fuel mass (see para 8.10.2 (*j*)) - Del l Fuel

A8.13 Calculations

The calculations which lead to the estimation of the mass of a given configuration are shown in rows 15 and 19.

Row 15 Cell A Wing loading factor in lifting surface mass (see Ch.6, Eq (6.32) uses cell E7) - (S/Mo) ^ 0.45
- B Passenger cabin relative differential pressure (based on a cabin altitude of 6000 ft, uses cell H3) - p bar
- C Required cruise thrust to weight ratio (inverse of lift to drag ratio, uses cell G7) - Req (T/Mg).
- D Available cruise thrust to weight ratio (Ch.3, Eq (3.8b) uses cells D3, D7, F3, F7 and R7) - Av (T/Mg)
- E Ratio of available to required cruise thrust to weight ratios (cell D15/cell C15) - Av/Req
- F Design cruise specific fuel consumption (Ch.3, Eq (3.12d), uses cells D3 and F3) - (c) des

	G	Off design cruise specific fuel consumption (Ch.3, Eq (3.12b), uses cells E15 and F15) - (c) od
	H	Ratio of constant equivalent airspeed climb to datum value (Ch.7, Eq (7.15c), uses cells E7, H7, I11, K11 and L3) - Z
	I	Thrust factor for constant equivalent airspeed climb (Ch.3, Eq (3.8b) uses cells D7 and L3) - Tau Cl EAS
	J	Mean constant equivalent airspeed rate of climb (Ch.7, Eqs (7.15a), (7.15b), (7.15d) and (7.16d), uses cells D7, F7, H7, H15, I11, I15, K11, L3, N3 and R7) - (Vv) EAS
	K	Distance covered during constant equivalent airspeed climb (Ch.7, Eq (7.21), uses cell, J15, L3, M3 and N3) - Dist EAS
	L	Factor in constant Mach number climb equation (see Ch.7, Eq (7.22b), uses cells B19, D3, E7, H7 and J11) - Qm
	M	Thrust factor for constant Mach number climb (first phase) (Ch.3, Eq (3.8b), uses cells D7, N3, 03 and R7) - Tau Mn1
	N	Mean constant Mach number rate of climb (first phase) (Ch.7, Eq (7.22a), uses cells B19, F7, H7, J11, L11, L15, M15, N3, O3, P3 and R7) - (Vv) Mn1
	O	Distance covered during constant Mach number climb (first phase) (see Ch.7, Eq (7.25), uses cells F3, M3, N3, O3 and R3) - Dist Mn1.
	P	Thrust factor for constant Mach number climb (second phase) (Ch.3, Eq (3.8b) uses cells D7, O3 and R7) - Tau Mn2
	Q	Mean constant Mach number rate of climb (second phase) (Ch.7, Eq (7.26a), uses cells B19, D19, F7, H7, J11, L11, L15, O3, P15, Q3, R7 and T19) - (Vv) Mn2
	R	Distance covered during constant Mach number climb (second phase) (Ch.7, Eq (7.29), uses cells E3, O3, Q3 and Q15) - (Dist Mn2)
	S	As cell R15 but checks against final climb altitude density ratio (uses cells Q3 and R15) - Dist Mn2
	T	Distance covered during descent (Ch.7, Eq (7.56), uses cell G3) - App Dist
Row 19 Cell	A	Datum specific consumption, static sea level (Ch.3, Eq (3.12a), uses cell D7) - Cl EAS (c) o.
	B	Fuel mass ratio for constant equivalent airspeed climb (Ch.7, Eq (7.17), uses cells A19, F7, I15, J15, L3, M3, N3, and R7) - Cl EAS Wf/(Mg)o.
	C	Specific fuel consumption in constant Mach number climb (first phase), datum value (Ch.3, Eq (3.12a) uses cells F15 and P3) - Cl Mn1 (c) 1

D Fuel mass ratio for constant Mach number climb (first phase) (Ch.7, Eq (7.24) uses cells C19, E3, F7, M3, M15, N3, N15, P3 and R7) - Cl Mn1 Wf/(Mg)o

E Fuel mass ratio for constant Mach number climb (second phase) (Ch.7, Eq (7.28) uses cells E3, F7, F15, P15, Q3 and Q15) - Cl Mn2 Wf/(Mg)o

F Start of cruise mass ratio; all up mass less take-off and total of climb fuel ratios (uses cells B19, D19, E19 and H7) - Mcr/Mo

G Fuselage mass (Ch.6, Eq (20a) uses cells A11, B11, B15, C11 and E11) - M fus

H Factor in lifting surface mass evaluation (Ch.6, Eq (6.24b) uses cells A7, B7, C7, D15, I3, J3 and L7) - c1 bar

I Installed powerplant mass ratio (Ch.6, Eq (6.25) uses cells F7, F11 and M7) - Mpp/Mo

J Systems mass ratio (transfer of cell G11) - Msys/Mo

K Mass of operational items (uses cells B3 and N11) - M op it

L "Fixed" mass component; sum of fuselage, payload and operational items masses (uses cells C3, G19 and K19) - M fixed

M Net range requirement less distance covered in climbs and descents (uses cells A3, K15, O15, S15 and T15) - Net range

N Required logarithm (to base 10) of cruise fuel ratio (Ch.7, Eq (7.50b) uses cells D3, G7, G15 and M19) - log 10 (Mc1/Mc2)

O Required cruise mass ratio, antilog of cell N19 - Mc 1/Mc2

P Ratio of end of cruise mass to take-off value (uses cells F19 and O19) - Mc2/Mo

Q Ratio of fuel mass to take-off values (uses cells O7 and P19) - Mf/Mo

R Ratio of all masses directly proportional to total mass (sum of cells I19, J19 and Q19) - KappaMo

S Location of quarter mean aerodynamic chord point on wing centreline chord (Ch. 8, Eq (8.7b), uses cells A7, I7 and L7) - x^0.25/co

T Correction factor in constant Mach number climb calculation (second phase) (Ch.7, Eq (7.26b) uses cells B19, D19, F7, H7, J11, L11, L15, P15 and Q3) - X2

A8.14 Analysis

The Analysis Box uses the previous calculations to derive the mass, centre of gravity, wing position and tail surface areas.

Row 23 Cell F First estimate of total mass, based on the typical assumption of lifting surface mass being $0.12M_0$ (cell L19 divided by 0.88 less cell R19) - (Mo) est 1

G Direct numerical transfer of value from cell F23 - (Mo) est 2

H Mass of items proportional to M_0, based on cell G23 (uses cell R19) - Kappa* Mo

I Mass of lifting surfaces, based on cell G23 (Ch.6, Eq (6.32) uses cells A15 and H19) - Mlift sur

J Calculated mass; sum of cells H23, I23 and J23 - Mo calc

K Difference between calculated and estimated mass, (cell J23 - cell G23) - error

The EXCEL© Goal Seek tool is then used on cell K23 to bring it to a nominal minimum value by adjusting the estimated mass, cell G23. The result is the total mass for that particular configuration. The value is used to summarise the component masses in the Summary Box, see paragraph A8.15.

Row 26 Cell F Position of overall centre of gravity aft of the nose (Ch.8, Eq (8.6c), uses cells A11, M11 to T11, N22 to N28, N30 to N32) - lCG

G Position of wing root chord leading edge (apex) aft of nose, cell F26 less cell S19 - l WG APX

H Tail arm; cell O11 less cell F26 - 1 TLARM

I Area of horizontal tail surface (Ch.8, Eq (8.8b), uses cells H26, P7, Q22 and Q24) - S Hor Tail

J Area of vertical tail surface (Ch. 8, Eq (8.9b), uses cells H26, Q7, Q22 and Q23) - S Vert Tail

A8.15 Summary of one configuration

The details of the design of a given configuration are conveniently located in the Summary Box.

Column N Row 22 Wing mass (cell I23 divided by cell H11) - M wing

23 Fuselage mass; cell G19 - M fus

24 Tail surfaces mass; cell I23 less cell N22 - M tail

25 Landing gear mass; assumed to be $0.04M_0$ (0.04 cell J23) - M gear

26 Powerplant mass; product of cells I19 and J23 - M power p

27 Systems mass; based on cell J19 less 0.04 (uses also cell J23) - M sys

28 Operational items mass; cell K19 - M opit

29 Operating empty mass; sum of cells N22 to N28 - M oew

30 Mass of payload; cell C3 - M pay

31 Mass of fuel; product of cells J23 and Q19 - M fuel

	32	Overall mass; sum of cells N29, N30 and N31 and is equal to cell J23 - Mo
Column Q Row	22	Wing area; weight of aircraft divided by wing loading (uses cells E7 and J23) - Wing area
	23	Wing span; square root of product of wing area and aspect ratio (uses cells A7 and Q22) - Wing span
	24	Wing standard mean chord; ratio of wing area to wing span (uses cell Q22 and Q23) - Mean chord
	25	Aspect ratio, cell A7 - A
	26	Wing quarter chord sweep, in degrees, uses cell C7 - Del 0.25
	27	Wing taper ratio; cell L7 - Lambda
	28	Wing thickness to chord ratio; cell B7 - t/c
	31	Fuel mass as given by the available volume in the wing (Ch. 5, Eq (5.7), uses cells Q22, Q23, Q25, Q27 and Q28) - Fuel mass avail
Column T Row	22	Position of wing apex from nose; cell G26 - Wg Apex
	23	Horizontal tail area; cell I26 - S Hor Tail
	24	Vertical tail area; cell J26 - S Vert Tail
	26	Wing area tp the power of -0.1 - S^$^{-0.1}$
	31	Total static thrust from cells F7 and N32 - Stat Thr

A8.16 Overall summary of results

The overall summary of the results for the variations of the aspect ratio and thickness to chord ratio investigated are shown in the Summary of Results, columns F to N, rows 38 to 55. Row 34 contains a particular set of Case Results for ease of handling the data. For each set of parameters the following information is shown:-

Column	H	Wing loading (numerical value from cell E7) - (Mg/S)o
	I	Static thrust to weight ratio (numerical value from cell F7) - (T/Mg)o
	J	Mass (numerical value from cells J23 or N32) - Mo
	K	Fuel mass required (numerical value from cell N31) - Fuel mass
	L	Available fuel mass (numerical value from cell Q31) - Fuel Mass Avail
	M	Structural parameter (numerical value from cell Y47 of the stage 1 Spreadsheet 1) - SP
	N	Wing span (numerical value from cell Q23) - Wing span

The values of the total mass M_0 are used to construct the carpet plot which shows them as a function of aspect ratio and thickness to chord ratio. The data shown in rows 36 to 55 and columns P to U are used to produce the graph.

A8.17 Constraints

There are two constraints to be imposed, namely a wing span limit of 34 m and the requirement to carry all the fuel in the wing.

A8.17.1 Wing span limit

The wing span data, column N of the Summary of Results, have been extracted and presented in a convenient form for analysis in rows 59 to 67 and columns P to U. The data have been used to produce the graphs immediately below, from which the conditions appropriate to 34 m span have been read off. These have been then used in the overall procedure to produce the three additional mass values shown in the Summary of Results in rows 57 to 59 of column J. They have been converted to the form required for the carpet plot in row 60 to 64, columns V and W, and are shown as the constraint line on the plot.

A8.17.2 Fuel volume requirement

The ratios of the fuel mass required to that available have been extracted from the Summary of Results and are shown as a function of aspect ratio and thickness to chord ratio in rows 92 to 108 and columns Q to V. The values are plotted in the graph below this table and for each aspect ratio the thickness to chord ratio appropriate to a fuel mass ratio of unity have been read off. These are shown in rows 61 to 66 at the bottom of the Summary of Results, together with the masses appropriate to them. The required and available fuel volumes are also shown as a check, the small discrepancies being acceptable within the accuracy of the process. In order to show this information on the carpet plot it is necessary to adjust the aspect ratio values for the fact that the thickness to chord ratios are not the standard values used in the basic parametric analysis. This has been done assuming a linear variation of aspect ratio with thickness to chord ratio in the range of $0.11 \leq t/c \leq 0.13$. Thus for example for the initial value appropriate to $A = 7$, the thickness to chord ratio is 40% of the difference between 0.11 and 0.12. Since for the purposes of the carpet plot aspect ratio has been given an increment of two for each increase of 0.02 in thickness to chord ratio, this amounts to an effective aspect ratio increment of 0.8. The results of this process of correction are shown in rows 36 to 42 of columns V and W. The corresponding constraint boundary is shown on the carpet plot.

A8.18 Overall optimum

The overall optimum configuration is considered to be that which gives the lowest mass within the imposed constraints. This is seen to be the lowest point on the available fuel volume line. A larger graph indicates an aspect ratio of 9.1 and a thickness to chord ratio of 0.128. Thus, for example, $A = 9.1$ and $t/c = 0.13$ may be input at cells O47 and P47 of the stage 1 Spreadsheet 1, and the results read into the present spreadsheet to check the comparability of the constraints. This design point is shown on both the spreadsheets. The predicted mass is 72,757 kg.

Appendix A8.3 Case study - short/medium haul airliner

Second stage of the parametric analysis - optimiser approach - explanation of SPREADSHEET 3

(Subsonic flight - fan engine Optimiser: Microsoft EXCEL©)

A8.19 Introduction

The procedure employed for the optimiser approach is shown in Spreadsheet 3. In many respects it consists of data which are directly comparable to those of Spreadsheet 1 and Spreadsheet 2, but is aimed at deriving an optimum design direct from the initial inputs. In order to demonstrate the comparability some of the information is duplicated which is not necessary when this procedure is used.(See, for example, Addendum 5, paragraph 5.4.)

A8.20 First stage parametric analysis

The data above and including row 47 of Spreadsheet 3 are identical in all respects to those of the corresponding cells of Spreadsheet 1, Appendix A8.1, to which reference should be made for explanation. However, since the optimiser approach bypasses the analysis of parametric variation there is no Summary of Results or carpet plot of the results. The graph which shows the relative importance of the different performance requirements for a given configuration in terms of thrust to weight ratio and wing loading is retained but it is only for illustration.

A8.21 Second stage parametric analysis

The data below row 47 are identical to those of the upper part of Spreadsheet 2, Appendix A8.2, although inevitably the cell identification is different. The correlations using the prefix 2 to identify the cell notation from Spreadsheet 2 are as follows:

Requirement Box: Row 52 columns K to T - see 2Row 3 columns A to J (cells K52, M52, O52 are an unnecessary repetition from cells 2L3, 2I3 and 2J3 respectively)

Parameters Box: Row 56 columns K to S - see 2Row 7 columns A to I (cells K56 to R56 are unnecessary repetition from cells 2O47, 2P47, 2A15, 2C3, 2Q47, 2R47 and 2A47 respectively)

Climb Path Box: Row 52 columns V to AB - see 2Row 3 columns L to R (cells X52, Z52, AA52 are unnecessary repetition from cells 2E7, 2J3and 2K3 respectively)

Assumed Box: Row 56 columns V to AB - see 2Row 7 columns L to R (cell R56 is unnecessary repetition from cell 2K7)

Additional Input Data: Row 60 columns K to AD - see 2Row 11 columns A to T (cells I60, J60, K60, L60 are unnecessary repetition from cells 2U47, 2V47, 2W47 and 2X47 respectively)

Calculation Box: Row 64 columns K to AD and Row 69 columns K to AD - see 2Row 15 columns A to T and 2Row 19 columns A to T

A8.22 Analysis

The data in the Analysis Box are contained at Row 73 columns P to U and Row 77 columns P to T. The corresponding data in Spreadsheet 2 are shown at 2Row 23 column F to K and 2Row 26 columns F to J. Although the two Boxes are superficially the same, in the present case the EXCEL© Solver is used on cell T73, [(Mo) calc], to obtain a minimum by altering:

Cell Q13 - Wing loading chosen for design condition
Cell O47 - Aspect ratio
Cell P47 - Thickness to chord ratio
Cell Q73 - Estimate of total mass - [(Mo) est 2]

Subject to the constraints:-

Cell AA73 < = 34 - Wing span limitation
Cell AA81 > = Cell X81 - Fuel volume requirement
Cell Q13 < = Cell G18 - Wing loading limited by maximum landing value
Cell Q13> = Cell H18 - Wing loading at least as great as value for gust sensitivity
Cell T73 = Cell X82 - Compatibility of total mass values

Providing a solution is found which satisfies all the conditions the resulting mass is a true minimum within the overall assumptions made.

A8.23 Result

In this case the minimum mass is found to correspond to an aspect ratio of 9.102 and thickness to chord ratio of 0.128. The mass is 72,749 kg which is 8 kg less than that found from the somewhat less precise graphical approach, but the configuration is identical for all practical purposes. A more accurate result is obtained by using cells B7 and B8 to give a consistent value of $S^{\wedge -0.1}$. The optimised mass becomes 73,764 kg, aspect ratio 9.159 and thickness to chord ratio 0.12879.

A8.24 Summary of results

The Summary of Results is shown in columns: X rows 72 to 82; AA, rows 72 to 78 and 81; and AD, rows 72 to 81 - see Spreadsheet 2, columns: 2N, rows 22 to 32; 2Q, rows 22 to 28 and 31; 2T, rows 22 to 31 respectively.

Chapter 9

Analysis of concept design

9.1 Introduction

The output of the parametric analysis described in Chapter 8 is an initial design concept for which the leading characteristics and outline general arrangement are defined. The achievement of this result depended upon the many assumptions needed to derive the relatively simple models used for lift, drag, mass, powerplant characteristics and, to some extent, performance. Essentially the procedure thus far has been one of synthesis, and it is now possible to analyse the design concept with the aim of verifying, or adjusting, the assumptions and refining the concept generally. As with all design work the analysis is an iterative process. Providing the initial assumptions prove to be reasonably justified the process will converge to a finalised design of which the characteristics can be established with an acceptable degree of accuracy. It may be, however, that certain of the assumptions are not justified by subsequent analysis and it is not impossible for the design to diverge and prove to be untenable. Figure 9.1 is a simple illustration of how this may occur. It is particularly liable to be a problem with advanced concepts where it is inevitable that significant extrapolation is necessary to establish the assumed input data. Should divergence occur it is necessary to re-think the whole concept, possibly investigating alternative overall configurations or reviewing the dominant performance requirements.

It is clear that as the analysis proceeds it is necessary to have access to a considerable source of design data. There are numerous references which contain such data. A selected list is contained in the bibliography located in Appendix A9.1 to this Chapter, to which reference is made in the following paragraphs. In general it is necessary to supplement such data by ad hoc investigations relevant to the particular design concept, such as wind tunnel testing or dedicated computational fluid dynamics studies. Further consideration of these is outside the scope of this work.

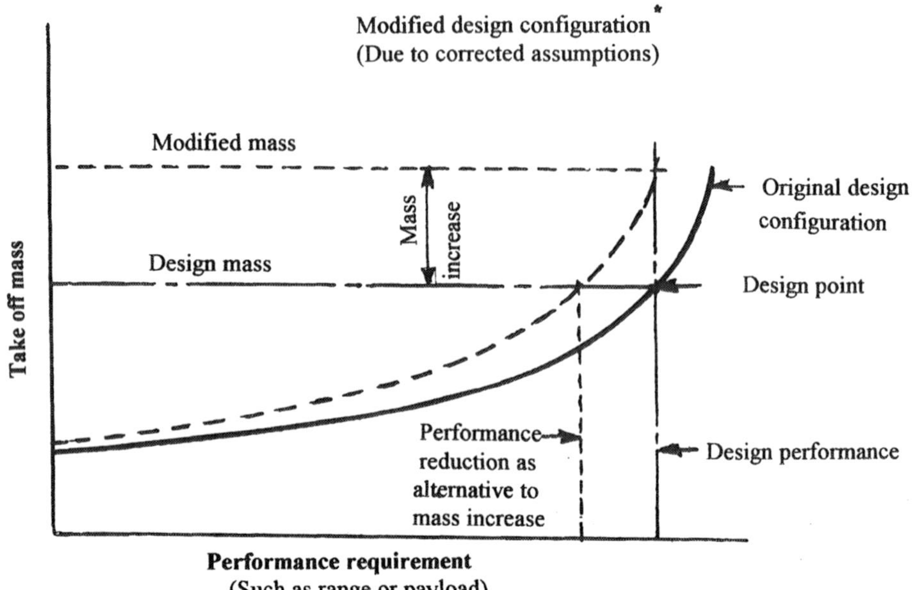

New curve may become asymptotic
at a mass less than the design value

Figure 9.1 Influence of design assumptions on mass growth

9.2 Powerplant

9.2.1 Selection of engine

The parametric analysis has to be based on a model powerplant which may be a relatively simple representation such as suggested in Chapter 3 or a more accurate set of data derived as typical of a given type of engine. In either case until the overall mass of the design concept has been established it is not possible to specify the actual static thrust or power of the powerplant. Once this information is available it is possible to define the characteristics of the required engine and, preferably, select an existing unit from those already available. This may not always be possible and an alternative approach is to base the subsequent design analysis on the performance characteristics of an engine in the correct category which has the potential for development to an appropriate size. Although it is sometimes inevitable, the combination of a completely new powerplant and new aircraft almost invariably leads to unanticipated development problems. It is thus best avoided.

9.2.2 Engine data

In order to undertake the analysis of the design concept it is highly desirable to be in possession of specific engine mass data as well as a complete performance map. The latter has to include the variation of thrust and fuel consumption as a function of flight speed, altitude and engine setting (throttle condition). Unfortunately this amount of data may not always be readily available and it may become necessary to produce a model of the engine based on a number of point performance characteristics. Some guidance can be found in Reference D1.

9.2.3 Powerplant location

9.2.3.1 General

Some thought as to the location of the powerplants is necessary in defining the overall aircraft configuration, or configurations, as discussed in Chapter 2, para 2.2.2. The development of the design concept now enables the question of powerplant location to be considered more precisely. It is convenient to discuss the issues involved as separate items but there is inevitable interaction between them. Chapter 4, paragraph 4.2.4 contains some reference to powerplant location.

9.2.3.2 Versatility and access

As a general rule engines mounted in pods or nacelles on the outside of the aircraft are more versatile than those buried within the airframe. It is usually possible to accommodate different engines without too much difficulty, thereby more readily handling upgrades or specific customer requirements. Podded and nacelle-mounted engines are also preferable for maintenance since access is generally more direct and need not involve significant structural penalties. Also air intakes and exhausts are usually shorter and can be more efficient.

9.2.3.3 Installation clearance

As the engine dimensions are defined it is necessary to check installation clearances, such as propeller tip spacing from adjacent structure or the ground and underwing nacelle ground clearance on a low wing aircraft.

9.2.3.4 Safety

Safety issues are mainly concerned with the aircraft itself although there may be some matters where ground operations influence powerplant location. In particular powerplants must be positioned so that in emergency situations they do not pose a threat to the occupants of the aircraft or other safety critical components. The major concerns are the consequences of an engine becoming detached during an emergency alighting situation or the failure of rotating components such as fan and turbine blades or discs and propeller blades. The engines must be located such that these component failures do not present a hazard to other engines, primary structures or vital systems.

9.2.3.5 Acoustic considerations

The further aft on an aircraft the engine is located the less is the noise impact upon the occupants, but this is not usually a major consideration. Some engine positions, for example those above the wing or fuselage, reduce the noise as perceived at the ground but at the expense of powerplant efficiency. Acoustic fatigue damage to airframe components located near the exhaust is likely to be a more important consideration. It is especially significant for engines which employ reheat when the exhaust outlets should be behind all major airframe components.

9.2.3.6 Stability and control

Lateral stability and control requirements for a multi-engined aircraft may well be dominated by the need to maintain the path of the aircraft subsequent to the failure of the most critical engine. Thus from this point of view wing-located powerplants should be as far inboard as allowed by other considerations. See also Addendum 3.

9.2.3.7 Powerplant/airframe efficiency

Clearly the aim must be to produce the most efficient overall aircraft. This often means a compromise between engine requirements and those of the overall aerodynamics. Thus, for example, it may be desirable on a short take-off propeller driven aircraft to space the engines across the wing to enhance slipstream effect upon low speed lift. Ground clearance may suggest that large diameter power units should be mounted over the top of the wing but this can lead to a considerable loss of efficiency of lift development. In the case of buried engines the intake arrangements may have a significant effect upon powerplant efficiency and must be given careful consideration, see Chapter 2, paragraph 2.2.2.2 and Chapter 4, paragraph 4.2.4.

9.3 Structure

9.3.1 Introduction

At this stage of the conceptual design it is essential to give specific thought to the arrangement of the primary airframe structure. There are two main considerations:

a) The overall layout of the aircraft, as defined by aerodynamic, powerplant, payload and similar considerations, must be compatible with an efficient structure. To ensure this it is necessary to have some understanding of the way in which a structure reacts to the imposed loads as well as the nature of that loading. All major load paths should be continuous and the effect of the inevitable cutouts must be reduced to a minimum. This is further considered in paragraph 9.3.2.

b) The aerodynamic configuration, particularly that of the wing, must not be such as to imply an unacceptably high structure weight. In practice this issue has already been addressed, at least to some extent, by the introduction of the structural parameter, *SP*, in Chapter 5, paragraph 5.4.2. As the design develops this matter must be investigated in more detail.

The selection of the material for the airframe is significant but need not have a big effect upon the concept of the overall structural layout.

9.3.2 Structural function and configuration (see References E1 and E3)

9.3.2.1 Lifting surfaces (see Chapter 5, paragraph 5.4)
The wing may be considered to be typical of the lifting surfaces although not all the considerations of its structural design apply to the secondary surfaces.

The main structural role of the wing is to transmit the lift loads inwards to its attachments to the fuselage. Primarily this requires it to act as a spanwise beam but there is also a need for adequate strength and stiffness in torsion along the same direction. Sometimes the spanwise bending and torsion requirements may be met by a single beam or spar located near to the maximum chordwise depth of the aerofoil and associated with a shear carrying leading edge, or 'D' nose, terminating at the spar. More usually there are two or more spars located across the chord, and together with the upper and lower covers between their chordwise extremities they form a box beam which also meets the torsion requirement. The forward and aft extremities of this beam are determined by the auxiliary lifting surfaces and ailerons. Even when there is no high lift device on the leading edge, aerofoil depth and systems requirements result in the front spar being located at least 10% back on the chord. The rearmost spar is often located at 60-70% of the chord, leaving room for the trailing edge surfaces.

The box beam also conveniently acts as integral tankage for the fuel and access panels are required for inspection purposes. However, these may usually be relatively small. In the case of moderate to high aspect ratio wings the box beam should be continuous across the whole span. Thus large cutouts for such items as landing gear stowage should be outside its boundaries. When the powerplant is buried within the fuselage and the wing is of relatively low aspect ratio, as for many combat aircraft, it is often necessary to use several spanwise spars and to pass the bending loads round the fuselage by means of a series of ring frames to which the spars are connected.

9.3.2.2 Fuselages (see Chapter 4, paragraphs 4.2.3 and 4.3.2)
It is possible to distinguish between primarily pressurised fuselages and those where there is either no pressurisation or its extent is very limited, as in the case of the crew compartment of a combat aircraft. Essentially the fuselage is a fore and aft beam.

A pressurised fuselage invariably consists of a semi-monocoque, shell-type construction where the outer skins are reinforced by ring frames and longitudinal stiffeners. It is important to minimise the number and size of cutouts but these are inevitable for doors and windows, etc. Relatively substantial structural members must be provided for the attachment of the lifting surfaces and landing gear.

The nominally unpressurised type of fuselage is frequently associated with aircraft where the payload occupies a relatively small part of the volume. Crew access and vision, powerplant installations and such items usually dominate the layout. Large cutouts are often required for air intakes, access to the engine and possibly weapons bays. While a semi-monocoque construction may well be used it normally has to be associated with a number of substantial fore and aft members, or longerons, which provide reinforcement at the edges of the cutouts. The longeron members should be as continuous along the fuselage length as is possible. They may change direction at appropriate positions, but should not be stepped.

9.4 Landing gear

Although specific consideration of landing gear detail layout has been deliberately avoided thus far in the design process, it can have an important impact on the overall concept of the aircraft. The main forms of landing gear layout are discussed in Chapter 2, paragraphs 2.2.2.5 and 2.2.3.7, and it is now appropriate to establish the geometry more precisely. Some relevant information is to be found in References *F*1 to *F*3 and Addendum 1. The main points to be investigated are:

i) Overall layout geometry, including the number of main gear units.
ii) Number and size of the wheels/tyres.
iii) Location of suitable stowage volume when the units are retracted. This volume should be located in regions of secondary structural importance.
iv) Location, or provision, of suitable load attachment points.

Clearly these latter two issues are closely related to the layout of the structural members discussed in the previous paragraph.

9.5 Fuselage layout

There is an extensive discussion of fuselage layout in Chapter 4 to which reference should be made.

Although a preliminary overall layout of the fuselage is a prerequisite for the parametric analysis it is only essential to specify the overall dimensions. A more precise and detailed layout of the fuselage may now be undertaken. Items which have to be investigated include:

i) Influence of the wing junction with the fuselage (Chapter 2, paragraph 2.2.2.3 and Chapter 4, paragraph 4.4.1).
ii) Landing gear attachment and stowage (see paragraph 9.4 above and Chapter 4, paragraph 4.4.3).
iii) Layout of the cockpit region, windscreen geometry, etc. (Chapter 4, paragraph 4.5.1).
iv) Details of the powerplant installation where this interacts with the fuselage. This especially includes the design of the air intakes, engine maintenance access and removal, exhaust locations (see paragraph 9.2.2, Chapter 2, paragraph 2.2.4 and Chapter 4, paragraph 4.2.4).
v) Payload details, for example doors, windows and facilities on a passenger aircraft (Chapter 4, paragraph 4.5.3), freight loading access and doors (Chapter 4, paragraph 4.5.4), or weapons (Chapter 4, paragraph 4.5.2).
vi) Special detail considerations. One example of these is the local geometry at the rear end of a combat aircraft having twin side-by-side engines.

9.6 Operating empty mass

9.6.1 General

The operating empty mass of the aircraft is of fundamental importance in the design as it has a major effect upon both the performance and total mass. Hence it is essential to predict it as accurately as possible. This is best done by analysing the component masses of the aircraft using the best data available and comparing them with values for comparable aircraft where this information is to hand. The process requires a more detailed approach than that given at the end of the parametric analysis, for example the systems component should be dealt with by looking at each major item within it.

9.6.2 Mass prediction techniques

There are several levels of mass prediction and which one should be used in a given case is dependent upon the detail of the data available.

a) Empirical comparisons. In some circumstances it is possible to make acceptable predictions by direct comparison with other, similar, designs. Indeed as suggested above it is always desirable to check predictions against known data. Hence an important aspect of mass prediction is the collection and collation of data. Experience plays a major role in the interpretation of mass data since no two designs are exactly similar and the basis of data presentation is not standard. Because of these variations it is often best to base comparisons with other aircraft on the operating empty mass rather than the mass of specific components.

b) Empirical formulae. This is an extension of the direct comparison technique and involves the use of formulae derived directly by a statistical analysis of available information from many aircraft. In some cases an attempt is made to analyse the data by interpreting it in the context of parameters known to be of importance although often it is simply related to the overall mass of the aircraft. Generally available mass prediction methods are usually based on this approach and the lifting surface mass formula presented in Chapter 6, paragraph 6.4 is one such example.

c) Theoretically derived formulae. Sometimes, but by no means always, it is more satisfactory to derive a prediction formula by using a theoretical approach. Such an approach is normally based on a simplified design of the component, the degree of simplification being related to the detail of the data anticipated to be available. An attempt to include too much detail may result in a formula which is too complex to use and in any case it is inevitable that many details of a real design will be unknown at the stage of a design where such a formula is of value. Therefore it is necessary to use some empirical data in the derivation of the formula. This approach is especially useful when a variation of some of the design parameters of a component is under investigation. Certain of the mass formulae quoted in Addendum 4, to which further reference is made in paragraph 9.6.3, are of this type.

d) Prediction methods. On the whole prediction methods enable more accurate results to be obtained than is possible simply by the use of formulae. Their disadvantage is the need for more extensive information and they are less suitable for simple parametric investigations. Nevertheless they are well suited to computational techniques. In the limit this method is effectively a mass estimation of the detailed component.

9.6.3 Mass prediction data

Useful information relevant to the prediction of mass may be found in References I3 to I9 among others.

Addendum 4 presents some suggested techniques for application to the mass prediction of wings and fuselages, and includes data on other structural components, systems and equipment.

9.7 Aerodynamic analysis - performance

*9.7.1 **Aerofoil and high lift devices/roll control** (Chapter 5, paragraph 5.2.2)*

The analysis of the lifting characteristics of the wing demands a more specific consideration of both the basic aerofoil section and the geometry of the high lift devices than was possible at the conceptual design stage dealt with in Chapter 5 and Chapter 6. Information can be found in References C1, 2, 3, 5, 7 and 8 as well as those in section I.

The output of the parametric analysis includes some aerofoil detail, such as the root thickness to chord ratio and the desired critical Mach number characteristics. This information may be used in the selection of a suitable aerofoil and its basic lift characteristics checked against those assumed in Chapter 6. The performance of the high lift devices is critical in that it has a direct bearing upon the maximum allowable wing loading in the landing configuration, and by implication the take-off wing loading. It is therefore important to define precisely the type and geometry of these auxiliary surfaces.

While leading edge high lift devices may be located across the greater part of the wing span, it may be necessary to accept some interruption, as for example in the region of wing-located powerplants. Such effects should be minimised and carefully considered as the details of the design are established.

Generally the spanwise extent of the trailing edge high lift devices is limited by the need to provide ailerons for roll control although in some special circumstances these may be "drooped" to augment the lift. This spanwise limitation is reflected in the flap lift increments suggested in Chapter 5, Table 5.1 and Chapter 6, paragraph 6.2. Outboard ailerons are usually needed for low speed roll control and typically occupy 20-25% of the wing semi-span, see Chapter 8, Table 8.3.

Outer ailerons on high aspect ratio sweptback wings may introduce aeroelastic difficulties when used at high speed, and to overcome this problem alternative high speed roll control may be provided in the form of inboard ailerons or differential longitudinal controls such as tailerons. In the case of inboard ailerons there will be an effect upon the extent of the trailing edge flaps although this may be minimised by locating the controls immediately behind powerplants. Within some limits the effectiveness of trailing edge devices may be adjusted by variation of their chordwise extent.

Because the landing wing loading is critically dependent upon the maximum lift coefficient of the wing it is necessary to achieve the value assumed for the parametric analysis or the whole process must be repeated.

9.7.2 Drag

The zero lift drag prediction made for the parametric analysis was based primarily on empirical data derived from typical types of aircraft. Many aircraft possess features which are specific to the particular design and hence it is important to analyse the drag characteristics once the basic aircraft layout has been established. One technique which is widely used is to estimate the drag of each individual component of the aircraft and summate the results with additional allowance for interference effects. Relevant data may be found in, for example, References C2, 4, 5, 6, 7 and 8 and all those in section I. Some dedicated computer programs are also available to facilitate drag prediction.

9.7.3 Performance

The drag characteristics have a significant impact upon performance. Thus when better substantiated drag data are available it is important to return to all aspects of the performance evaluation, utilising also upgraded powerplant and mass data as covered in paragraphs 9.2.1, 9.2.2 and 9.6.3 above. Any significant departure of the performance from that predicted during the parametric analysis can result in the need to review the characteristics of the chosen design configuration. It may be necessary to change the configuration to meet a specified performance condition, such as sustained manoeuvre, or to allow for a different fuel load ratio if, say, cruise lift to drag ratio is altered.

9.8 Aerodynamic analysis - stability and control

In spite of the developments in active control techniques many aircraft are still designed to be inherently stable, albeit with reduced margins in some cases. Even when instability is deliberately introduced into certain modes it is essential to have adequate control power and response. Analysis of the conceptual design enables the inertial and aerodynamic data to be evaluated in order to assess the stability and control characteristics. When artificial stability is accepted the required control characteristics must be established.

The design of the control and stabiliser surfaces requires ensuring that there is:

i) Ability to provide the forces/moments needed to maintain the trim of the aircraft in all flight conditions.

ii) Ability to provide control forces/moments to meet prescribed handling criteria, including situations in which a partial failure has occurred.

iii) Stability in both the static and dynamic sense, or alternatively acceptable artificial damping or stability in all critical modes.

The background to these issues may be found in the references quoted in section H of the Bibliography as well as some of those in section I.

The design of the control and stabiliser surfaces is complex because of the number of separate conditions which have to be met. Some simplifications may be made to facilitate the process during initial design work and Addenda 2 and 3 outline design methods for the longitudinal and lateral requirements respectively. The output from these methods may then be used for a full analysis.

9.9 Cost

Aspects of cost have been discussed in Chapter 1, paragraph 1.3. As stated there the ultimate optimum must be based on some aspect of cost and hence a vital part of the analysis of the conceptual design is a prediction of the costs associated with its procurement and operation. Some cost data and cost estimating techniques may be found in References I5 and I6, Part VIII. While these references give a useful insight into the factors involved many aspects of cost are dependent upon the characteristics of individual manufacturers and operators as well as market forces. For these reasons it is frequently most satisfactory to base costs on a comparative rather than absolute basis, individual designs being compared against a common set of assumptions.

9.10 Design refinement

The consequence of the analysis of the conceptual design is the iterative refinement of its characteristics until the analytically derived values coincide with the requirements and input assumptions.

Appendix A9.1

Bibliography - Selection of relevant literature generally available

Where appropriate the information given follows the format:-

Author Title Publisher Date
References marked: * include data on mass prediction
x include data on cost

A General data

There are numerous sources of general information on aircraft, powerplants and other associated equipment. These include books and periodicals. A convenient source of data which is regularly updated can be found in the Janes publications especially:

1. All the Worlds Aircraft; Janes Information Group Updated yearly
2. Aero Engines; Janes Information Group Updated regularly
3. Air Launched Weapons; Janes Information Group Updated regularly

B Aircraft flight and performance

	Author	Title	Publisher	Year
1.	Anderson J.D.	Introduction to Flight	McGraw-Hill	1989
2.	Barnard R.H. and Philpott D.R.	Aircraft Flight	Longmans	1989
3.	Hale F.J.	Introduction to Aircraft Performance, Selection and Design	Wiley	1984
4.	Mair W. A. and Birdsall D. L.	Aircraft Performance	Cambridge University Press	1996
5.	Shevell R.S.	Fundamentals of Flight	Prentice-Hall	1989

C Aerodynamics

	Author	Title	Publisher	Year
1.	Abbott I.H. and Van Doenhoff	Theory of Wing Sections	Dover	1959
2.	Anderson J.D.	Fundamentals of Aerodynamics	McGraw-Hill	1991
3.	Hoerner S.F.	Fluid Dynamic Lift	Hoerner	1985
4.	Hoerner S.F.	Fluid Dynamic Drag	Hoerner	1965
5.	Houghton E.L. and Carpenter P.W.	Aerodynamics for Engineering Students	Arnold	1993
6.	Kuckemann D.	The Aerodynamic Design of Aircraft	Pergamon	1978
7.	McCormick B.W.	Aerodynamics, Aeronautics and Flight Mechanics	Wiley	1979
8.		Aerodynamics and Performance Data Sheets	Engineering Society Data Unit-ESDU	

D Powerplant

	Author	Title	Publisher	Year
1.	Mattingley J.D., Heiser W.H and Daley D.H.	Aircraft Engine Design	AIAA	1987
2.	Oates G.C.	Aircraft Propulsion Systems	AIAA	1989
3.	Seddon J. and Goldsmith E.L.	Intake Aerodynamics	AIAA	1985
4.	Goldsmith E.L. and Seddon J.	Practical Intake Aerodynamic Design	AIAA	1993
5.	Thomson W.	Thrust for Flight	Longmans	1992
6.		The Jet Engine	Rolls-Royce	1986

See also References I2, I4, I5 and I6, PartVI

E Structural layout

1. Cutler J.	Understanding Aircraft Structures	Blackwell	1992
2. Niu M.C.Y.	Airframe Structure Design	Conmilit	1988
3. Niu M.C.Y.	Airframe Stress Analysis and Sizing	Conmilit	1997

F Landing gear

1. Conway H.G.	Landing Gear Design	Chapman	1958
2. Currey N.S.	Aircraft Landing Gear Design	AIAA	1988
3. ICAO	Aerodrome Design Manual DOC 9157-AN-901, 2nd edition	ICAO	1982

See also Reference I6, Part IV

G Equipment and cockpit layout

1. Chant C.	Modern Aircraft Armament	Stephens	1988
2. Coombes L.F.E.	The Aircraft Cockpit	Stephens	1990
3. Middleton D.H.	Avionics Systems	Longmans	1989

See also Reference I4, Part III

H Stability and control

1. Babister A.W.	Aircraft Dynamic Stability and Response	Pergamon	1980
2 Cook M.V.	Flight Dynamics Principles	Arnold	1997
3. Etkin B.	Dynamics of Flight; Stability and Control	Wiley	1982
4. Irving F.G.	Introduction to Longitudinal Static Stability	Pergamon	1966
5. Perkins C.D. and Hage R.E.	Aircraft Performance, Stability and Control	Wiley	1957

See also References C8 and I6, Part VII

I Aircraft overall design

1. Fielding J.P.	Introduction to Aircraft Design	Cambridge University Press	1999
2. Huenecke K.	Modern Combat Aircraft Design	Airlife	1987

x*3.	Jenkinson L.D, Simpkin P. and Rhodes D.	Civil Jet Aircraft Design	Arnold	1999
*4.	Nicolai L.M.	Fundamentals of Aircraft Design	METS	1975
x*5.	Raymer D.F.	Aircraft Design - A Conceptual Approach	AIAA	1999
x*6.	Roskam J.	Airplane Design: Part I Preliminary Sizing Part II Configuration, Design and Integration Part III Cockpit layout Part IV Layout of Landing Gear and Systems Part V Component Weight Estimation Part VI Aerodynamic, Thrust and Power Calculations Part VII Stability, Control and Performance Part VIII Cost Estimation	Roskam	1985
7.	Stiles R.A, Brandt, S.A. Bertin J.J and Whitford, R.	Introduction to Aeronautics - A Design Perspective	AIAA	1998
*8.	Stinton D.	The Anatomy of the Aeroplane	Blackwell	1966
*9.	Stinton D.	The Design of the Aeroplane	Blackwell	1983
*10.	Torenbeek E.	Synthesis of Subsonic Aircraft Design	Delft University	1982
11	Whitford R.	Design for Air Combat	Janes	1987

J Airworthiness requirements

1. Federal Aviation Administration: Federal Aviation Regulations
   FAR 23 - Utility, Aerobatic and Commuter Category Airplanes
   FAR 25 - Transport Category Airplanes
2. Joint Airworthiness Authorities - Joint Aviation Requirements
   JAR 22 - Sailplanes and Powered Sailplanes
   JAR 23 - Utility, Aerobatic and Commuter Category Aeroplanes
   JAR 25 - Large Aeroplanes
3. U.K. Ministry of Defence - Design and Airworthiness Requirements for Service Aircraft, Def. Stan 00-970, Vol 1. Aeroplanes

K Atmospheric properties

1.	Properties of the Atmosphere - ESDU 77021	Engineering Society Data Unit	1977

Addendum 1

Landing gear considerations

Notation (for this Addendum only)

ACN	Aircraft Classification Number (see also PCN)
D	Tyre diameter
$ESWL$	Equivalent Single Wheel Load
F	Ratio of allowable load of multiple tyre unit to $ESWL$
k	Factor in value of F
LCG	Airfield Load Classification Group
LCN	Aircraft Load Classification Number, related to LCG
M	Aircraft mass
P	Maximum dynamic load during landing
p	Tyre pressure
PCN	Pavement Classification Number (see also ACN)
v	Vertical velocity of descent at touch down
W	Tyre width
δ_s	Vertical axle travel at landing (effectively shock absorber travel)
δ_T	Tyre deflection
η	Effective shock absorber efficiency, related to δ_S
θ	Critical angle in definition of lateral ground stability
λ	Overall aircraft reaction factor during landing, related to δ_S

The references referred to may be found in Chapter 9, Appendix A9.1.

AD1.1 Introduction

Some aspects of the general layout of the landing gear are discussed in Chapter 2, paragraphs 2.2.5 and 2.3.7. Chapter 9, paragraph 9.4 summarises the landing gear considerations which have to be addressed at the initial layout stage. The following paragraphs amplify this. A comprehensive reference which covers all aspects of landing gear is Reference *F*2.

AD1.2 Overall layout of landing gear

The advantages of the conventional nose wheel type landing gear layout are such that it should be adopted unless there are compelling reasons otherwise. While a tailwheel arrangement may be somewhat lighter it suffers from a major disadvantage of lack of ground stability during braking, as well as high drag on take-off and a tendency to "balloon" on landing. It is only suitable for small general aviation aircraft and even then is not preferred. A bicycle layout having main gear units disposed more or less equally ahead and aft of the centre of gravity can confer certain overall fuselage layout advantages. However, it has major disadvantages due to the critical performance during landing and braking, a difficulty in lifting the aircraft nose for take-off and the probable need for outriggers.

See Chapter 2, paragraphs 2.2.2.5 and 2.2.3.7 and Figures 2.1, 2.2, 2.4, 2.14-2.20, 2.26 and 2.27.

AD1.3 Ground operating conditions

AD1.3.1 General comments

The specified ground operation condition has a major impact on the layout and design of the landing gear and in some cases the layout of the aircraft as well. It determines the tyre pressure, number of wheel/tyre units and their geometric configuration.

AD1.3.2 Airfield types

It is necessary to distinguish between different classes of airfield operating surfaces:

a)Unpaved airfields where the surface may be grass or sand. Sometimes metal reinforcing matting is used to improve the quality of the surface.

b) Rigid paved runways which usually consist of concrete slabs laid on a prepared sub-base. The sub-base quality is of importance, as well as is the thickness of the slabs.

c)Flexible paved runways where asphalt is laid on to the base and sub-base. Again the quality of the base and subbase is of great importance.

The use of unpaved airfield surfaces is limited to relatively light aircraft, usually those in the general aviation category, and to military tactical operations. In the latter case the aircraft may be large and heavy and the need to restrict tyre pressure has a dominating effect upon the design of the landing gear and some parts of the aircraft.

For convenience of classification the sub-base quality of paved runways is placed into one of four categories ranging through high stiffness, medium, low to ultra-low stiffness, Grades A to D respectively. This enables pavement load carrying capacity to be defined.

AD1.3.3 Unpaved surfaces - tyre pressure

A number of units are used for defining tyre pressure:

i) lb/in^2
ii) kg/cm^2 (1 kg/cm^2 is 14.22 lb/in^2)
iii) Mpa (1 Mpa is 145lb/in^2)

For convenience this note will principally use the Imperial unit of lb/in^2, tyre sizes and related items being stated in inches. Tyre pressure is the dominant consideration for unpaved surfaces and typical allowable maximum values are:

1) Loose desert sand 30 lb/in^2
2) Hard desert sand 60 lb/in^2
3) Wet, boggy ground 40 lb/in^2
4) Hard, firm ground 60 to 70 lb/in^2
5) Metal reinforced grass 90 lb/in^2

However, there is another consideration, especially in the case of large aircraft. This arises because of the possibility of damage to the surface when an aircraft passes over it. This is covered by the "Flotation" requirement. Flotation is the number of passes an aircraft can be allowed to make over a surface of specified characteristics. The subject is fully covered in Reference F2 and can include paved as well as unpaved surfaces.

Apart from tyre pressure the important aircraft characteristic is the "Equivalent Single Wheel Load", *ESWL*, which relates to a measure of the surface strength, as opposed to local bearing capability. The surface characteristic for unpaved surfaces are quoted in terms of California Bearing Ratio (*CBR*). There are numerous methods of analysis. In some circumstances the "Flotation" requirement may be replaced by the more general methods covered in the following paragraph.

AD1.3.4 Paved surfaces - ACN/PCN method

The most commonly accepted means of comparing aircraft requirements and airfield capability is the Aircraft Classification Number/Pavement Classification Number method, or *ACN/PCN*. It is described in Reference F3. This method results in a simple and accurate means of defining either the runway requirement for a given aircraft status or the allowable aircraft characteristics for a given runway surface.

a) Pavement Classification Number - PCN. This is derived by analysis/testing of a given airfield. The *PCN* includes the following information:

i) Type of surface (rigid or flexible)
ii) Category of subsoil (A to D as defined in paragraph AD1.3.2)
iii) The *PCN* value

(For aircraft of less than 5700 kg, the *PCN* is the aircraft mass, kg, divided by the tyre pressure, MPa.)

b) Aircraft Classification Number - ACN. The *ACN* is the Derived Single Wheel Load, kg, divided by 500. The Derived Single Wheel Load is not the same as the Equivalent Single Wheel Load mentioned in paragraph AD1.3.3. The latter is a function of the geometry of multiwheel units and the total tyre contact area, while the former also depends upon the type and grade of runway. Thus, for a given aircraft mass and tyre pressure there is a specific single wheel load, but for the same aircraft condition there are eight values of Derived Single Wheel Load and eight different *ACN* values.

The aircraft *ACN* for a given pavement type and subsoil category may be compared with the *PCN* value for a runway having those characteristics and should be no greater than the *PCN* to allow unlimited operations.

The analytical complexity of the *ACN/PCN* method is such that it is not very convenient for use as a synthesis tool for initial design purposes, although as soon as a gear/wheel/tyre configuration has been established the *ACN* values may be readily evaluated by use of the program contained in Reference F3.

AD1.3.5 Paved surfaces - LCN/LCG method

The Load Classification Number, *LCN*, method was used extensively over a long period of time and in the light of experience was modified to the Load Classification Number/Load Classification Group method which was the forerunner of the *ACN/PCN* method. In the updated version of this technique pavements are placed in a Load Classification Group (*LCG*) category and the Load Classification Number (*LCN*) of an aircraft is evaluated for comparison with the *LCG*. The *LCN* is based upon the concept of the Equivalent Single Wheel Load and tyre pressure. As the *ESWL* is independent of the pavement surface the method is much simpler than the *ACN/PCN* method for initial work.

As suggested by the name the *ESWL* is the datum load on an isolated single wheel, to which the load carried by a multi-tyre/wheel unit can be related.

The following *LCG* groups are related to aircraft *LCN* values as shown:

Runways *LCG*	Type of Runway	Aircraft *LCN* range
I	Best civil and military pavements	101-120
II	Good quality pavements	76-100
III	Moderate pavements	51-75
IV	Lower grade asphalt	31-50
V	Reinforced soil	16-30
VI	Soil in good condition	11-15
VII	Boggy soil and loose sand	Below 11

For comparison with the *LCG* groups, the *LCN* is approximately defined by:

$$LCN = 5.7 \times 10^{-4}(ESWL)^{0.91}\, p^{0.34} \tag{AD1.1a}$$

where *ESWL* is the equivalent single wheel load (lbs) and *p* is the tyre pressure (lb/in^2). If *LCN* is a defined requirement, Eq (AD1.1a) may be inverted to give:

$$ESWL = 3710(LCN)^{1.1}/p^{0.37} \tag{AD1.1b}$$

Equation (AD1.1) may be used in various ways. For example given a landing gear layout with tyre dimensions, spacing (see paragraph AD1.5) and distribution of aircraft weight the *ESWL* may be calculated and used to derive the *LCN*. However, if this amount of data is known the *ACN/PCN* technique described in the previous paragraph is preferable. On the other hand if the *LCN* is specified the allowable combinations of *ESWL* and tyre pressure may be evaluated. These may then be used in conjunction with the information given in paragraph AD1.5 to define various wheel/tyre configurations which are acceptable.

AD1.3.6 Comparison of ACN/PCN and LCN/LCG methods

General correlation between the two methods described in paragraphs AD1.3.4 and AD1.3.5 respectively is not straightforward. However, it is of interest to look at a particular case, that of the A300-B4 airliner. In this case the relevant data are:

Take-off mass:- *ACN* in the range 45 to 70 for rigid pavements
46 to 80 for flexible pavements
(The higher figures are for the lowest grade sub-base).
LCN 70 approx.

From this it might be concluded that in this case the *LCN* method gives a value comparable with the *ACN* for the lowest grade rigid runways, but does not cover the case of the corresponding flexible ones. Nevertheless, as a first indication of likely characteristics the *LCN/LCG* method gives a good insight into the situation, especially for rigid pavements on which it is based.

AD1.4 Tyre characteristics

AD1.4.1 Tyre sizes

There is a wide range of standard tyre sizes ranging in diameter, *D*, from as low as 8 in. (0.2 m) up to 64 in. (1.63 m).

The tyre width/diameter ratio (*W/D*) varies from as low as 0.2 to about 0.4 with the tendency for the higher values to be associated with greater diameters. A typical value of (*W/D*) is 0.32 and this may be assumed for initial design work.

AD1.4.2 Tyre load capacity

The rated static load capacity of a tyre is the product of the maximum allowable pressure and the ground contact area. The ground contact area is typically 0.3 (*WD*) which is about $0.1D^2$ for normal (*W/D*) values. The tyre pressure is determined by the strength of the carcass which is defined in terms of the ply rating. This leads to:

$$\text{Tyre static load rating} = 0.5\,p\left(\frac{W}{D}\right)^{1.5} D^2 \text{ lb} \qquad \text{approximately} \qquad \text{(AD1.2a)}$$

p in lb/in^2 and D in inches

or

$$= 5000p\left(\frac{W}{D}\right)^{1.5} D^2 \text{ kg} \qquad \text{(AD1.2b)}$$

p in kg/cm^2 and D in m

AD1.5 Multiple tyre units

The equivalent single wheel load of multiple tyre units is a complex function of the geometry of the tyre layout and total contact area. Thus it is a function, by implication, of tyre size and pressure. In typical multiple tyre layouts the following dimensions are representative.

Wheel effective track (pair of single tyres)	$0.8D$
Wheel effective track (pair of twin tyres)	$1.1D$
Wheel base (bogie)	$1.2D$

For these particular cases the ratio of the actual load capacity of the unit, as a ratio of the equivalent single wheel load is:

$$F=\left[1+k\left(D/100\right)\right] \tag{AD1.3}$$

where D is in inches

$k = 1.1$ for a pair of side-by-side single tyres
4.0 for a four-wheel bogie
4.8 for a four-wheel bogie, each wheel having twin tyres

For given tyre dimensions Eq (AD1.3) may be used with Eq (AD1.1) to determine either the corresponding *LCN/LCG* or the allowable pressure.

AD1.6 Shock absorber characteristics

AD1.6.1 General

The vertical energy of landing has to be absorbed, and subsequently dissipated, in the shock absorber units. The stroke of the shock absorber determines the overall length of the landing gear leg and hence has a major influence on certain aspects of the landing gear layout.

AD1.6.2 Axle travel

For a given vertical velocity requirement the axle travel needed to absorb the energy is independent of aircraft mass but does depend upon the efficiency of the shock absorber and size of the tyre. The relevant energy equation for the condition when lift is equal to weight at touch down is:

$$\Delta M v^2/2 = P(\eta \delta_S + 0.47\delta_T) \tag{AD1.4a}$$

where ΔM is the proportion of the mass carried by a given leg unit in static conditions
P is the maximum dynamic load during touch down

η is the efficiency of the shock strut (effectively the non-dimensional area under the dynamic load-deflection curve)
δ_S is the axle travel, vertically
δ_T is the tyre deflection
v is the specified touch down vertical velocity

The 0.47 factor represents the shock absorption efficiency of the tyre.
Equation (AD1.4a) may be written in the form:-

$$\left(\eta\delta_S + 0.47\,\delta_T\right) = \frac{\Delta M v^2}{2P} = \frac{v^2}{2\lambda g} \qquad \text{(AD1.4b)}$$

where λ is the ratio of the dynamic to static load and is also known as the reaction factor, and

$$\eta\,\delta_S = \frac{v^2}{2\lambda g} - 0.47\,\delta_T \qquad \text{(AD1.4c)}$$

The tyre deflection, δ_T, may be estimated with acceptable accuracy from a knowledge of the tyre dimensions, say $\delta_T = 0.5W$. Thus for a given vertical velocity ($\eta\delta_S$) is effectively a function of ($1/\lambda$). Relevant values of η and λ are discussed below.

AD1.6.3 Design vertical velocity

Table AD1.1 Design values of vertical velocity

Type of aircraft	Limit Vertical Velocity	
	ft/s	m/s
Very light, low wing loading	7	2.13
General aviation	7-10	2.13-3.05
Transport types	10	3.05
Military combat, land based	12	3.66
Trainers, civil	13	3.96
Trainers, military	13-14	3.96-4.27
Naval, aircraft carrier operation	20 or more	6.1 and up

The design vertical velocity, v, is specified in requirements for a given class of aircraft. It is usual to calculate the axle travel based on the limit (actual design) value of v and to make a small additional allowance, say 10 -12%, for the ultimate condition which is normally 1.2 times the design value. Required limit values are summarised in Table AD1.1.

AD1.6.4 Shock absorber efficiency

The shock absorber efficiency, η, is 0.5 for a simple spring although in practice friction effects increase this somewhat. A much higher value, up to 0.85, can be obtained by combining the spring and damping functions, the most common type used being an air spring and oil damper (oleo-pneumatic). In practice an efficiency of 0.8 is commonly used, but this does imply a rapid rise of force with initial deflection. Where this gives rise to the possibility of critical fatigue conditions, as on a transport aircraft which must undergo a large amount of taxying, the shock absorber is designed to have low load with initial deflection. In this case the effective efficiency may be as low as 0.6.

AD1.6.5 Reaction factor

The reaction factor is chosen within the imposed geometrical constraints to minimise the impact of the loads developed during ground operation. The usual range of values is 1.2 to 3, the higher valves being associated with smaller aircraft where layout considerations restrict the available stroke and the aircraft tends to be designed to higher load factors generally. Transport aircraft utilise values at the lower end of the range. Reaction factors above 3, possibly as much as 5, are found on naval aircraft and for some helicopter emergency landing conditions. It should be noted that if the reaction factor is much above 3 the tyres are designed by the dynamic, rather than static, load capacity.

AD1.7 Landing gear structure and kinematics

AD1.7.1 Types of landing gear structure

There are two basic forms of landing gear structure.

AD1.7.1.1 Telescopic

With this arrangement the main structural member, or leg, incorporates the shock absorber as illustrated in Figure AD1.1. The wheels may be located on a single axle or a bogie as shown. The top of the leg is attached to a suitable point on the airframe, usually by a long pivot or trunnion. The trunnion may lie at any angle in a horizontal plane but is often lateral for fore/aft retracting units and lengthwise for sideways retracting units. Inclinations from these two basic directions are sometimes used to facilitate retraction kinematics and stowage. An additional fore and aft member, or drag strut, is needed to complete the structure of a fore/aft retracting unit although sometimes this may be replaced by a direct connection to a suitable point on the airframe. Likewise a side strut is needed for sideways retracting units. Drag and side struts usually have to fold about some point along their length.

In order to use a telescopic arrangement it is necessary for the distance between the axle in the fully extended position and the top trunnion to be at least 2.5 times the shock absorber stroke.

The leg must be more or less vertical in the front elevation as any inclination can result in unacceptable tyre side friction forces which may prevent smooth closure of the shock absorber.

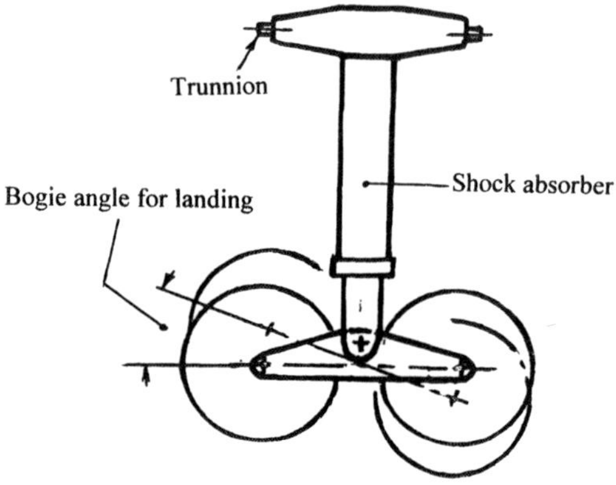

Figure AD1.1 Telescopic landing gear layout

AD1.7.1.2 Lever suspension (trailing arm)

When the geometry of the aircraft is such that there is inadequate space for a telescopic unit it is necessary to use a lever suspension arrangement. This situation usually arises on small aircraft. Although heavier and more complex than a telescopic unit, the lever suspension does have advantages as well as its compact layout. These advantages include: a good ride on the ground due to lower effective shock absorber friction; lower aircraft pitching; more scope in fore and aft location of wheel relative to the structure; easy access to shock absorber for maintenance; relative ease of shortening unit for retraction and the possibility of having the structure of the unit inclined outwards to increase track while still maintaining vertical axle travel.

A lever suspension unit is illustrated in Figure AD1.2. It can be seen that the axle is located at one end of the lever, the other end being pivoted to the bottom of the basic structure. The shock absorber is a separate unit placed between the lever and the structure.

The load in the shock absorber is increased by the lever effect but the stroke is correspondingly shortened. This may facilitate shock absorber design on smaller, less highly loaded aircraft. The lever geometry should normally be arranged to preclude any undue fore and aft movement of the wheel during load application. Drag or side struts are required in the same way as telescopic units.

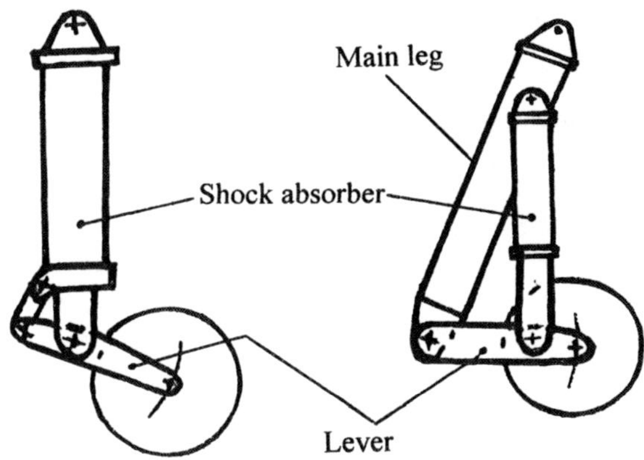

Figure AD1.2 Lever suspension landing gear layout

In addition to the more usual trailing lever configuration some designs utilise lateral beams for the lever units, as illustrated by Figure AD1.3.

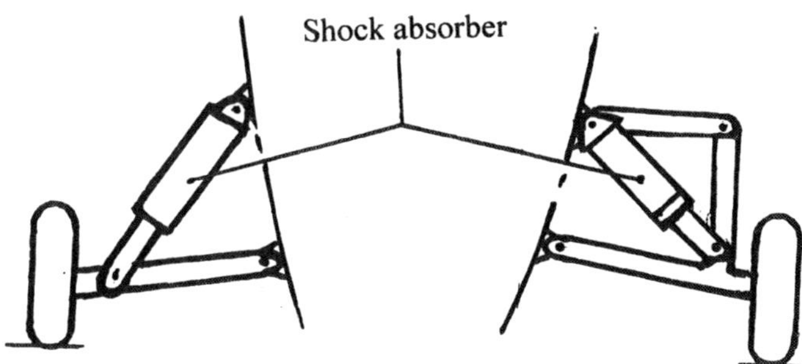

Figure AD1.3 Lateral lever suspension gear arrangements

AD1.7.2 Retraction kinematics

The choice of location of the landing gear units in their extended positions is discussed in paragraph AD1.8. It is necessary to identify suitable stowage volumes in the retracted positions of the units and the kinematics is concerned with the transfer between the extended and retracted positions. There are numerous ways of achieving this, see for example Reference F1.

The following points may be of help in determining the kinematics:

i) It is possible to use the shock absorber stroke to change the configuration of the unit by "pre-shortening" it during retraction. This may be done mechanically or hydraulically.

ii) It is possible to rotate a bogie unit during retraction to change its orientation relative to the main leg.

iii) It is possible to rotate the wheels of a telescopic unit about the shock absorber axis. This may be especially helpful with fore and aft retraction.

iv) In the case of lever suspension units which employ lateral beams, the wheels may be rotated about a fore and aft axis to provide compact stowage.

v) The use of an inclined, or oblique, fore and aft hinge, enables the unit to be moved fore or aft during retraction but it also implies some rotation.

AD1.8 Layout of the landing gear units of the aircraft

AD1.8.1 Preliminary considerations

In order to undertake the layout of the landing gear at the project definition stage it is necessary to know:

i) Mass of the aircraft in the take-off (ramp) and landing configurations.

ii) Centre of gravity positions, especially the maximum forward and maximum aft/high positions.

iii) Aircraft body attitude at take-off and landing as determined by the rotation and flare conditions respectively.

iv) Ground surface/runway requirements.

v) Initial estimate of mainwheel reaction factor and axle travel, see paragraph AD1.6.

AD1.8.2 Tyre configuration and pressure

When the ground surface requirements are anything other than a fully constructed runway it is possible to select a maximum allowable tyre pressure from paragraph AD1.3.2. In some cases operation from a poorly prepared surface may only be required at reduced mass and this should be allowed for by factoring the pressure up in proportion to the design take-off mass. When the *LCN/LCG* value is specified it is necessary to select a number of tyre pressures and equivalent single wheel load combinations for investigation, as outlined in paragraph AD1.3.3.

Although the nosewheel loading must be investigated the major issue is concerned with the main gear units. For initial investigation it is suggested that it is assumed that in the most aft centre of gravity condition at take-off mass some 93% of the mass is carried by the main gear units. This may be somewhat conservative and is subject to subsequent checking. This mass must be distributed between the two main units, or for a large aircraft between three or four. It is worth noting here that arrangements with three main gear units have usually resulted from a need to handle increased mass during development, rather than as an initial design consideration.

When the load for each main gear unit has been determined it is possible to use the typical data given in paragraph AD1.5 to relate it to equivalent single wheel load for various tyre configurations. This leads also to the load on each tyre so that for a given tyre pressure a typical tyre size may be determined using Eq (AD1.2). Thus for each of the selected tyre pressures and equivalent single wheel-load combinations a practical tyre arrangement and size may be established. From these it is possible to identify one or two of the most satisfactory for further investigation. It is usually preferable to select the simplest tyre configuration which meets the requirements, although consideration must be given to possible development of the aircraft, allowing for a future mass increase of, say, 25 to 30%. One way of facilitating this is to select tyres with potential for increased load carrying capacity by increasing the ply rating. However, this alone must result in a need for operation from runways of higher load capability.

AD1.8.3 Layout in side elevation - landing condition

Figure AD1.4 illustrates this phase of the layout procedure.

i) The attitude of the aircraft at the end of the landing flare enables the ground line to be located relative to the aircraft datum (see paragraph AD1.8.1 (iii)). A small tail down clearance of, say 0.15 m (6 inches), should be allowed for or a bumper provided.

ii) At the point of touch down the landing gear shock absorber is fully extended. In the case of a bogie arrangement it is also usual to arrange for the rear set of wheels to be trimmed down by an angle of, say, about 30° relative to the static geometry.

iii) At the point of ground contact the wheels on a single axle or the centre of a bogie set should be located such that the angle between the contact point and the most adverse aft/high centre of gravity position is some 4° behind the perpendicular from the ground line, through the centre of gravity. (In the case of a tailwheel layout the corresponding angle should be about 17° forward). This ensures that there is a positive nose down moment at touch down due to impact loads, but any greater angle may result in severe loads on the nosewheel during braking, and also a large force to raise the nose at take-off.

iv) It is now possible to make a provisional identification of suitable airframe points for the attachment of the main gear units and possible stowage volume in the retracted condition. The length between the axle or centre of a bogie and the attachment point may be determined. This leads to a decision as to whether a telescopic unit is possible or not. As mentioned in paragraph AD1.7.1.2 a form of lever suspension gear may be selected if the distance available is insufficient for a telescopic unit or if it confers advantages for stowage or wheel track, see paragraph AD1.8.5.

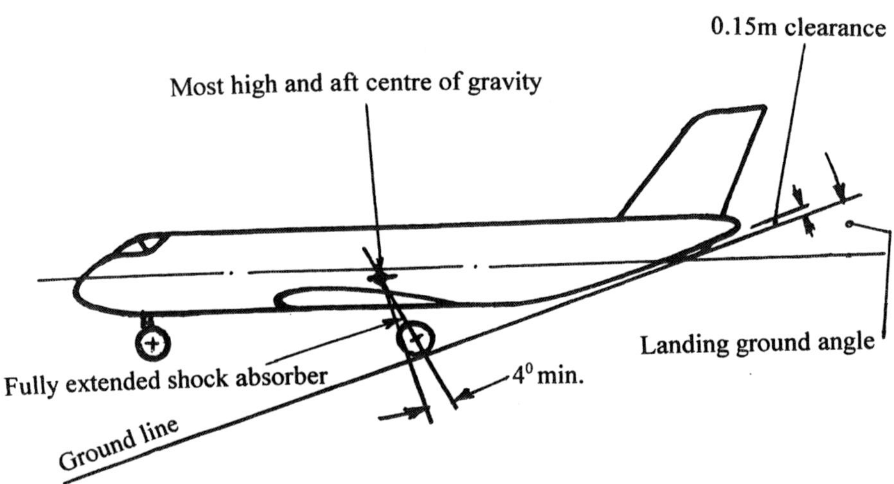

Figure AD1.4 Layout in side elevation - landing

AD1.8.4 Layout in side elevation - static condition (take-off)

The layout in the static condition, which determines the nosewheel fore and aft location is shown in Figure AD1.5.

i) Estimate the mainwheel tyre deflection and shock strut closure with the aircraft in its nominal static/take-off state. In the absence of other information the tyre deflection may be assumed to be about $0.25W$. The detail design of the shock absorber may be adapted to give a specific static closure but again for initial purposes it is sufficient to assume that the closure is equivalent to half the total axle travel.

ii) The static ground line may be located at the centre of the main wheel unit and is usually nominally parallel to the aircraft datum, especially on a transport aircraft as it gives a level floor. Slightly nose down attitude may be acceptable and a nose up attitude is sometimes used to assist in the take-off. However, any significant departure from a level fuselage situation can result in a drag penalty during the take-off.

iii) The fore and aft position of the main wheel units, as established in paragraph AD1.8.3 (iii) must now be checked to ensure that the centre of gravity is forward of the main gear centre point in all cases.

iv) The nosewheel location is determined by consideration of the full range of centre of gravity position. It is normally considered that the nosewheel ideally reacts about 10% of the mass in the static condition, but this must vary with centre of gravity position:

1) Extreme forward centre of gravity; the nosewheel load should not exceed about 15% of the total, otherwise the unit becomes unacceptably heavy and there is a possibility of instability during ground manoeuvres.

2) Extreme aft centre of gravity; the nosewheel load should not be less than about 6% of the total or the steering function may be inadequate.

A suitable position for the nosewheel must be established within these limits and, if necessary, minor changes made to the fore and aft location of the mainwheels. The position should have regard for suitable attachment structure and stowage volume for retraction.

v) The maximum load on the nosewheel may be determined by the case of touching down simultaneously on all the wheel units with a forward centre of gravity. More likely, however, it will arise during dynamic braking of the mainwheels. As a guide the nose gear reaction factor should be similar to that of the mainwheel, but the axle travel may need to be somewhat greater, say 20% higher, for initial work.

vi) In the majority of cases the nosewheel tyre size may be based upon its greatest share of the reaction of the mass of the aircraft, the cases stated in paragraph (v) above being dynamic ones which only become critical when the reaction factor is high, say above about 2.5. The tyre pressure and equivalent single wheel load must be consistent with those of the main gear in terms of *LCN/LCG*. Twin side-by-side nose tyres are usual. Single tyres are used on small aircraft and very occasionally two pairs of twin wheels on large aircraft intended for poor ground operations.

vii) It is usual, but not inevitable, for the nosewheel to be steered and also given some trail to assist in ground manoeuvres.

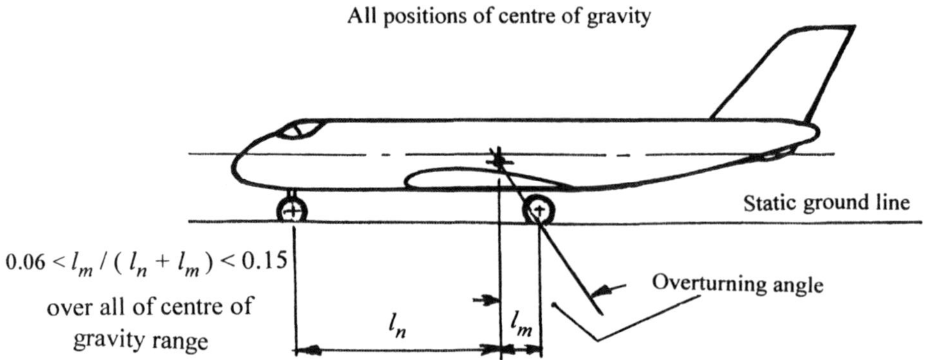

Figure AD1.5 Layout in side elevation - take-off

AD1.8.5 Layout in plan elevation

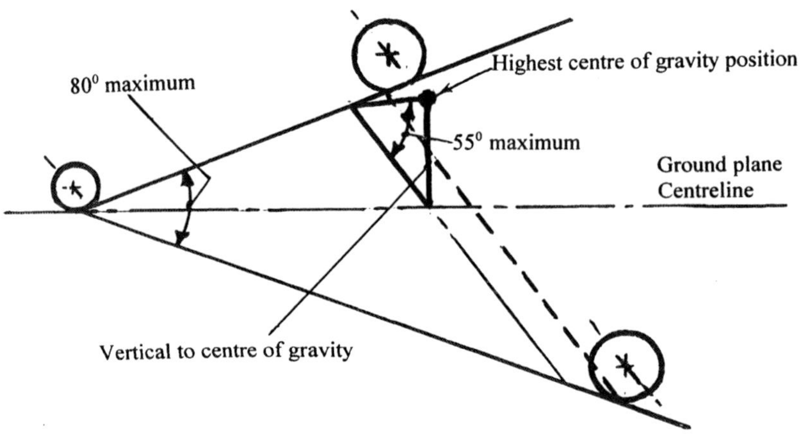

Figure AD1.6 Layout in the plan elevation

The layout of the landing gear in the plan elevation is shown in Figure AD1.6 and is primarily concerned with determining the track of the main gear. Unless the centre of gravity is high and the track narrow, the lateral positioning of the wheel units is not as critical as that in the fore and aft sense. Structural/stowage volume considerations or such matters as turning on a runway of given width may become the deciding factors. However, there is a requirement to ensure that the track is not so small as to give rise to the possibility of lateral instability while turning or landing in cross-wind conditions. As shown in Figure AD1.6, the critical design criterion is the angle, θ, which is formed by the intersection of a line in the vertical plane from the most adverse centre of gravity position to the ground at a point on a second line produced by joining the nose and mainwheels. The most adverse centre of gravity is the highest and most forward. Ideally the angle θ should not be greater than about 55° although up to 60° may be acceptable when the vertical stiffness of the shock absorbers is adequate.

AD1.8.6 Ground clearances

Once the positions and provisional travels of all the gear units have been established it is necessary to check that there is adequate ground clearance of all parts of the aircraft in all operational conditions. This includes such items as propellers and control surfaces, especially outer ailerons when the aircraft banks near to the ground. Often 5° bank with the maximum down aileron is used for checking. Flaps and underwing engines may also be affected as may an elevator on a low-mounted horizontal tailplane. As a generalisation the minimum clearance of fixed items should be about 0.18 m (7 inches), and on moving items 0.25 m (10 inches).

AD1.9 Comments on unusual main landing gear layouts

AD1.9.1 Six-wheeled bogies

The necessity of spreading the main landing gear load over as wide an area as possible has sometimes resulted in the adoption of six-wheel bogie units with three axles. See Chapter 2, paragraph 2.2.2.5.

One arrangement is that used on the Tupolev Tu 154 and Boeing 777 and illustrated in Chapter 2, Figure 2.19. Here the axles are located one behind the other as a simple extension of a four-wheel bogie. It might be anticipated that with this arrangement the value of the load factor, F, defined in paragraph AD1.5, Eq (AD1.3) would be about 1.4 times that of the equivalent four-wheel bogie. The long wheelbase of such a bogie layout may well result in the need to steer the bogie for ground manoeuvring. The complexity of this has to be compared with that of an additional main gear unit. In the case of the Boeing 777 the rear pair of wheels are steered in conjunction with the nosewheel.

Another approach was used for the Lockheed C 5A, Chapter 2, Figure 2.14, which has four main gear units. In this case the two rearmost axles on each bogie are in line laterally and spaced apart to spread the load. Load carrying capacity is thus more nearly equivalent to three, separate, twin-wheel units. Other wheel/tyre arrangements are possible but so far have not found application.

AD1.9.2 Multiple in-line main gear units

The Antonev AN 124, illustrated in Chapter 2, Figure 2.20, has twenty main wheels arranged in ten pairs, five in line on each side just below the bottom of the fuselage.

Each pair of wheels is located on an independent leg and the front two pairs on each side are steerable in conjunction with the nose gear. This latter consists of two separate, side-by-side units each with two wheels. The main tyres are 50 in. dia and 20 in. width. Although complex in terms of the number of units the concept does result in a layout which achieves the required spreading of the load without undue penalty on the overall design.

Addendum 2

Longitudinal control and stability surfaces

Notation (for this Addendum only)

The datum point for fore and aft dimensions is the leading edge of the aerodynamic mean chord of the wing (*MAC*), referred to as *c* in this Addendum). See also Figure AD2.1.

C_D	Overall drag coefficient
C_{D0}	Zero lift drag coefficient
C_L	Overall lift coefficient
C_{M0}	Pitching moment coefficient at zero lift (less stabiliser)
F_1, F_2	Functions in definition of phugoid motion [Eq (AD2.5)]
G	Overall lift correction factor - stick fixed [Eq (AD2.1b)]
$\overline{G}$	Overall lift correction factor - stick free [Eq (AD2.2c)]
$H_G c$	Location aft of leading edge of mean aerodynamic chord of most aft wheel position
$H_0 c$	Location aft of leading edge of mean aerodynamic chord of wing-body aerodynamic centre
K_n	Stick-fixed static margin
K_n^1	Stick-free static margin
M	Aircraft mass
S	Wing area
S_T	Horizontal stabiliser/control area
T	Thrust
V	Velocity
a_1	Wing-body lift curve slope
a'_{1T}	Canard lift curve slope allowing for wing upwash (paragraph AD2.3.3.3a)
a_{1T}	Horizontal surface lift curve slope due to incidence

a'_1 Wing lift curve slope in presence of a close coupled canard (paragraph AD2.3.3.3b)
$\bar{a}_{1T}$ $a_{1T} + b_1 a_2 / b_2$
a_2 Horizontal surface lift curve slope due to elevator deflection
b_1 Elevator hinge moment coefficient due to incidence
b_2 Elevator hinge moment coefficient due to deflection
c Aerodynamic mean chord of wing
g Gravitational acceleration
hc Position of centre of gravity aft of leading edge of aerodynamic mean chord
h_T Distance of thrust line above ground
k_B Pitch radius of gyration
l_H Location of horizontal surface lift aft of centre of gravity
$l^1{}_H$ Location of horizontal surface lift aft of leading edge of aerodynamic mean chord (negative for a canard configuration)
l_η, l^1_η Values corresponding to l_H and l'_H for a tailless design
$m_{\dot\theta}$ Derivative of pitching moment of wing body due to pitch velocity (usually negligibly small except for a tailless design)
q Dynamic pressure
z_T Distance of thrust line below centre of gravity
α Wing-body aerodynamic angle of incidence
α_B Body angle of incidence relative to airflow (α_B - α_W)
α_T Angle of incidence of horizontal surface relative to α_B
α_W Body angle at no lift condition with flaps in the appropriate position, it is usually negative.
ϵ Downwash angle at tail due to wing airflow
η Elevator deflection
ζ_L Damping ratio in phugoid motion
ζ_S Damping ratio in short period motion
ρ Air density
ω, ν Terms in definition of ζ_L, [Eq (AD2.5c)]
ω_S Frequency of short period motion

AD2.1 Introduction

In order to achieve a relatively straightforward conclusion to the project synthesis process described in Chapter 8 the sizes of the longitudinal, or horizontal, control and stabilising surfaces were based on historical data. This Addendum introduces a more refined procedure which examines the roles of these surfaces in order to yield a more accurate configuration.

The equations given are, for the most part, standard except that some simplification has been introduced to facilitate initial design as opposed to analysis. They are given in a form which enables general application when appropriate adjustments are made for specific configurations. The basic equations deal with what can be regarded as a conventional tailplane/elevator arrangement where the tailplane has either fixed incidence or has an incidence adjusted only for specific flight phases.

The roles of the horizontal control/stabiliser surfaces are presented in an order which facilitates the design process and suggestions are given as to how this should be undertaken. Where possible the equations are arranged to give directly the ratio of the horizontal surface area to that of the wing.

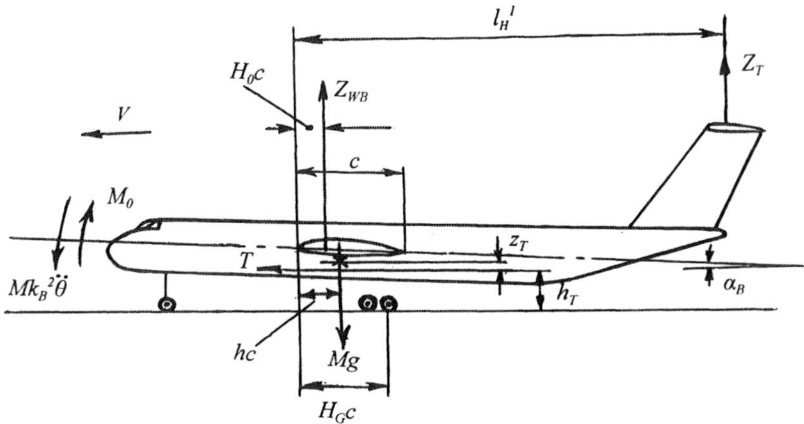

Figure AD2.1 Explanation of the notation

AD2.2 Design conditions

AD2.2.1 Trim

Regardless of the layout of the longitudinal surfaces and the philosophy of the design of the flying control system it must be possible to trim the aircraft in all steady flight conditions. For preliminary design work the conditions which are likely to be most important, and therefore need investigation, are the nominal cruise and the approach to landing with the high lift devices fully deployed.

The requirement for the aircraft to be in trim may be expressed as:

$$\frac{S_T}{S}=\frac{G\left[\frac{Tz_T}{c}+qSC_{MO}\right]+Mg(h-H_0)}{qS\left[\frac{Mga_{IT}}{qSa_I}\left(1-\frac{d\varepsilon}{d\alpha}\right)+\left\{a_{IT}(\alpha_T+\alpha_W)+a_2\eta\right\}\right]\frac{l_H}{c}} \tag{AD2.1a}$$

The factor G in the numerator is the allowance for the contribution of the horizontal stabilising area to the overall lift and for initial work may be assigned a value of 1.1. For more precise analysis it must be replaced by:

$$G=\left[1+\frac{a_{IT}}{a_I}\frac{S_T}{S}\left(1-\frac{d\varepsilon}{d\alpha}\right)\right] \tag{AD2.1b}$$

The range of centre of gravity positions, defined by hc, must be examined.

For a design with trailing edge control surfaces it is usual to establish a value of the tail setting angle, α_T, so that in cruise the variation of η through the centre of gravity range is as near zero as possible. Likewise on the approach to land either α_T is assumed fixed by the cruise case, or a new value is found to ensure that 35% of elevator control remains for manoeuvre.

When the whole surface is used for both control and trim the $\alpha_2\eta$ term is not present but the same principles apply in calculating the values of α_T.

AD2.2.2 Static stability

Unless the aircraft employs an advanced flying control system with the capability to provide artificial stability it is necessary to ensure that the geometry and size of the auxiliary horizontal surfaces are adequate to ensure natural stability. When the control surfaces are irreversibly connected to the control system it is only necessary to ensure stick-fixed stability, otherwise stick-free conditions must also be investigated. The degree of artificial stability in an unstable system may be measured in terms of a negative value of the stability margin, K_n.

The relevant equation for the stick-fixed case is:

$$\frac{S_T}{S}=\frac{G\left[K_n-H_0+h\right]}{\left[\frac{a_{IT}}{a_I}\left(1-\frac{d\varepsilon}{d\alpha}\right)\right]\frac{l_H}{c}} \tag{AD2.2a}$$

K_n is the stick-fixed static margin defined as $(-dC_M/dC_L)$.
When natural stability is required K_n must be positive, say at least 0.05 to 0.1 for initial calculations. K_n will take a negative value if artificial stability is assumed and in this case Eq (AD2.2a) is not really relevant since the problem now becomes one of adequate control. The clear implication is that in this case the value of (S_T/S) may be less than that given by using $K_n = 0$ in Eq (AD2.2a) and the equation may be used inversely to estimate the degree of static instability for a given surface size.

In the stick- free condition G, K_n and a_{IT} of Eq (AD2.2a) have to be replaced by $\bar{G}, K^1{}_n$ and $\bar{a}_{IT}$ where:

$$\bar{a}_{IT} = a_{IT} + a_2 \frac{b_1}{b_2} \tag{AD2.2b}$$

and

$$\overline{G} = \left[1 + \frac{\overline{a_{IT}}}{a_1} \frac{S_T}{S} \left(1 - \frac{d\varepsilon}{d\alpha}\right)\right] \tag{AD2.2c}$$

Because $\overline{a_{IT}}$ includes control hinge moment characteristics, b_1 and b_2, it is often best to delay investigation of stick-free stability until the design has become more established and, to some extent, the values can be adjusted without change to the basic size and geometry of the horizontal surface.

AD2.2.3 Rotation at take-off

There must be sufficient control power to enable the nose of the aircraft to be lifted towards the end of the take-off run. While the analysis of the actual condition is complex an approximate equation of sufficient accuracy for initial work may be derived by making appropriate simplifications:

$$\frac{S_T}{S} = \frac{\left[T\frac{h_T}{c} - q\,S\left(C_{M0} - a_1\,a_W\left(H_G - H_0\right)\right)\right] + Mg\left[H_G - h + \frac{0.1\,k_B^2}{l_H^1 c}\right]}{q\,S\left[a_{IT}\left(a_T + a_W \cdot \frac{d\varepsilon}{d\alpha}\right) + a_2\eta\right]\left(H_G - \frac{l_H^1}{c}\right)} \tag{AD2.3}$$

If conventional trailing edge controls are used then the important term is $a_2\eta$ and α_T is assumed constant, its value determined by other considerations. On the other hand if the surface as a whole is used for control $a_{1T}\alpha_T$ is the important term and unless a trailing edge surface is provided for trim, $a_2\eta$ is zero.

The critical case is the forward centre of gravity position where h has its lowest value. It is suggested that the dynamic pressure, q, should be based on 1.1 times the stalling speed in the take-off condition.

AD2.2.4 Dynamic stability

AD2.2.4.1 General

The two primary modes of dynamic stability are the short period oscillation and the long period phugoid. The former is largely a pitching oscillation of the aircraft about its centre of gravity, while the latter is a vertical oscillation of the centre of gravity relative to the nominal flight path with an implied variation of forward speed.

AD2.2.4.2 Short period oscillation

Satisfactory handling qualities are determined by a combination of both frequency and damping in the short period mode. The optimum depends on the type of aircraft and also the stage of the flight. Too much damping at low frequency may be undesirable. For small, agile, aircraft it is unlikely that a damping ratio below 0.4 will be acceptable, but for large, transport, types the figure may be reduced to as low as 0.2. However, as a check for initial design work it is suggested that the higher of these figures be used generally, with the rider that should the case prove to be critical some lower value may be acceptable for a large aircraft.

A somewhat simplified equation which enables the damping ratio to be estimated is:

$$\zeta_S = \frac{\left[\frac{a_1}{4} + \frac{a_{1T}}{4}\left(\frac{l_H}{k_B}\right)^2 (1 + d\varepsilon / d\alpha)\left(\frac{S_T}{S}\right)\right]}{\left[\frac{Ma_1 c}{2\rho S k_B^2}\left\{(H_0 - h) + \left[\frac{1}{G}\frac{a_{1T}}{a_1}\frac{l_H}{c}\left(1 - \frac{d\varepsilon}{d\alpha}\right) + \frac{\rho S a_{1T} l_H^2}{2Mc} - \rho\frac{l_H^2}{c}\frac{S^2}{S_T} m_{\dot{\theta}}\right]\left(\frac{S_T}{S}\right)\right\}\right]^{1/2}} \quad \text{(AD2.4a)}$$

$m_{\dot{\theta}}$ is likely to be negligibly small except for a tailless aircraft.

The undamped short period frequency is given by:

$$\omega_S = \rho \frac{SV}{2\pi M}\left[\frac{Ma_l c}{2\rho S k_B^2}\left\{(H_o - h) + \left[\frac{1}{G}\frac{a_{lT}}{a_l}\frac{\ell_H}{c}\left(1 - \frac{d\varepsilon}{d\alpha}\right) + \rho\frac{Sa_{lT}l_H^2}{2Mc} - \rho\frac{l_H^2 S^2}{cS_T}m_{\dot{\theta}}\right]\left(\frac{S_T}{S}\right)\right\}\right]^{1/2} \quad Hz$$

(AD2.4b)

AD2.2.4.3 Phugoid oscillation

Simple estimation of the characteristics of the phugoid mode is difficult. A very approximate approach is:

$$\zeta_L = F_l C_{D0} + F_2 C_L^{\,2} \qquad \text{(AD2.5a)}$$

where:

$$F_l = 0.5 + \frac{\omega}{2(2\omega + a_l \nu)} \qquad \text{(AD2.5b)}$$

$$F_2 = \left[\frac{3\omega - a_l\nu}{4\omega + 2a_{ll}\nu}\right]\left(\frac{C_D - C_{D0}}{C_L^2}\right) - \frac{a_1}{2}\,\frac{\left\{\omega - \nu^2\left\{1 + \frac{d\varepsilon}{d\alpha}\right\}\right\}}{(2\omega + a_l\nu)^2}$$

$$\omega = \frac{Mc}{\rho S k_B^2}\frac{a_{lT}}{2}K_n$$

$$\nu = \frac{a_{lT}}{2}\frac{S_T}{S}\left(\frac{l_H}{k_B}\right)^2 \qquad \text{(AD2.5c)}$$

In addition to appearing in the definition of ν, the (S_T/S) parameter is also included in ω by virtue of the K_n term. The algebraic complexity is such that there is little point in endeavouring to extract the (S_T/S) parameter explicitly. Having determined the area of the horizontal surface from other considerations it can be checked in this equation. Ideally the damping ratio should be about 0.2, but this may be difficult to achieve and a low, positive, damping is likely to be acceptable.

AD2.3 Design procedure

AD2.3.1 Conventional tail layouts

The suggested procedure for conventional tailplane/elevator layouts is:

i) Evaluate the cruise trim condition. Calculate the variation of tail setting angle as a function of (S_T/S) and centre of gravity (h) with the elevator angle zero, using Eq (AD2.1a), and plot the results. A range of 0.15 to 0.3 for (S_T/S) is suggested for investigation. There will be a mean value of α_T which minimises the actual value of η to trim, see Figure AD2.2.

ii) Evaluate the landing trim condition, Figure AD2.3, again using Eq (AD2.1a). This may need to be done in two stages:

A) Using the mean value of α_T derived from (i) find the elevator angles to trim as a function of (S_T/S). Providing these do not exceed about 67% of the available angle for any of the range of values of (S_T/S) and (h), no further action is needed.

B) If the elevator angles to trim significantly exceed 67% of the available movement, then new values of α_T should be calculated, assuming the elevator angle is 67% of that available. From this a revised, landing, value of α_T can be identified, although it may require exclusion of lower values of (S_T/S).This implies a variable incidence tailplane unless the situation is marginal and some penalty in cruise is acceptable.

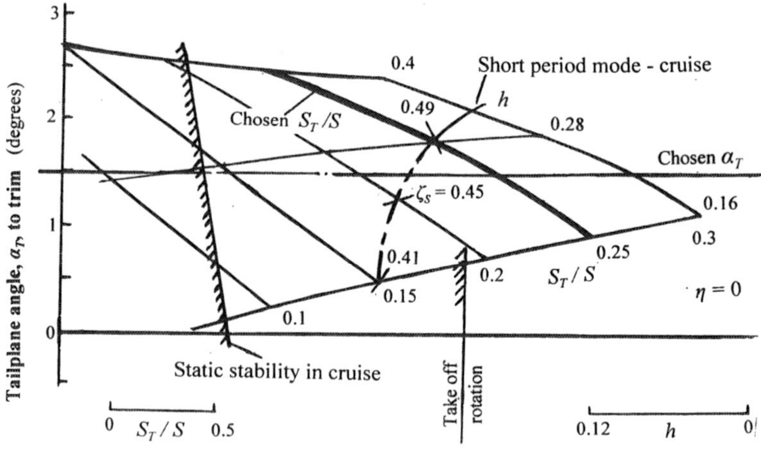

Figure AD2.2 Cruise conditions

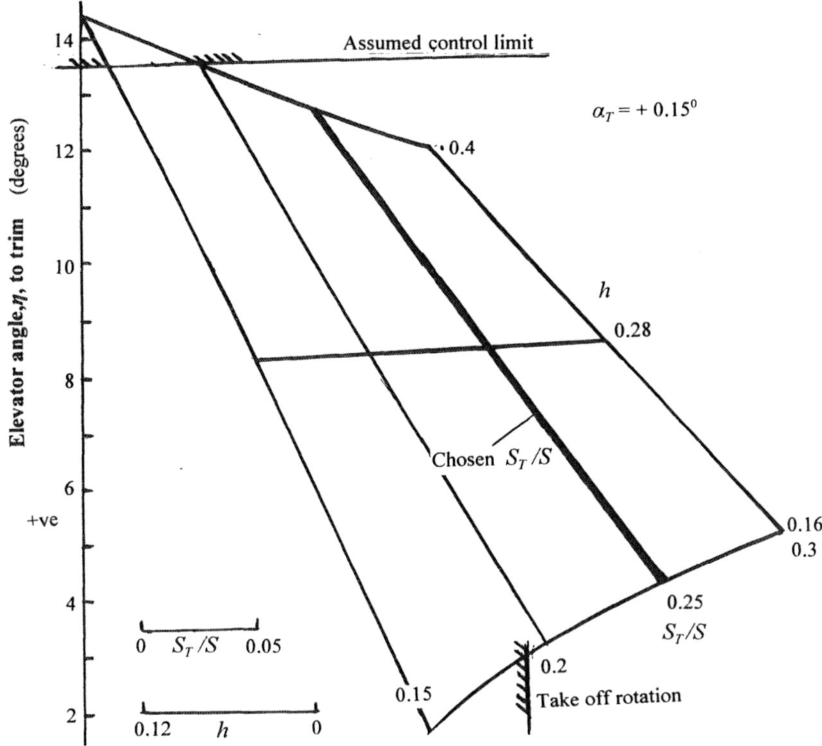

Figure AD2.3 Approach to landing

iii) Calculate the values of (S_T/S) as a function of (h) needed to give the required static stability, using Eq (AD2.2a). The static stability boundary may then be marked on the cruise trim diagram, Figure AD2.2.

iv) Calculate the value of (S_T/S) required for rotation at take-off at forward centre of gravity and with the available up elevator angle, using Eq (AD2.3). The result can also be marked on the cruise trim diagram, Figure AD2.2.

v) Calculate the values of short period damping ratio as a function of (S_T/S) and (h), using Eq (AD2.4a) for various aircraft inertial conditions. Identify the minimum value for a given value of (h) and the corresponding value of (S_T/S). The minimum values may be plotted on the cruise trim diagram, and may place a limit on the lowest acceptable value of (S_T/S).

vi) Select the lowest value of (S_T/S) which meets conditions (iii), (iv) and (v), and identify this on the landing trim diagram. Check that it enables the chosen cruise and landing values of α_T to be achieved consistent with the allowable elevator angles, and if necessary increase (S_T/S) until the requirements are met.

vii) Use the selected value of (S_T/S) to check the damping in the phugoid mode, using Eq (AD2.5a). The value of (S_T/S) can also be used to calculate the undamped short period frequency using Eq (AD2.4b) and in association with the comparable damping ratio it can be checked against any relevant handling criteria.

AD2.3.2 All moving tail configuration

The procedure is essentially the same as in paragraph AD2.3.1. Step (i) is identical except that there is no need to select a mean value of α_T, since it is variable The important aspect is to ensure that it does not depart significantly from the neutral setting and this may effectively eliminate some values of (S_T/S).

Step (ii) is different in that now it is necessary to evaluate the values of α_T to give trim on the approach to land. It, therefore, resembles Step (ii, B) but with η zero unless a trailing edge surface is used for trimming. In this latter case the maximum angle of the trim surface may be used. If any of the calculated values of α_T lie outside the available range again a restriction on the choice of (S_T/S) is implied.

Steps (iii) to (v) are carried out as before, the various boundaries being positioned on the cruise trim diagram The selected value of (S_T/S) should be checked in the approach trim diagram and modified if necessary before checking the phugoid damping ratio, Step (vii).

AD2.3.3 Foreplane design configurations

AD2.3.3.1 General

The equations are stated in a form which enables them to be used for foreplane layouts. It is necessary to distinguish between 'long-coupled' and 'close-coupled' configurations.

AD2.3.3.2 Long-coupled canard layouts

Somewhat arbitrarily a long-coupled canard is defined as an arrangement where the foreplane moment arm is of the order of four wing mean chords or higher. When this is the case for initial design work it is reasonable to assume that the interference effects are small:

i) Wing upwash at the canard may be neglected.

ii) Canard wash at the wing is a combination of downwash inboard and upwash due to canard tip vortices further out. The overall effect is likely to be small and so, for simplicity, the canard wash effect may be neglected. All that is required to modify the design equations is to delete the wing downwash term $(d\varepsilon/d\alpha)$ which is no longer relevant.

Equations (AD2.1) to (AD2.5) may then be used with the following observations:

a) l_H *and* l^1_H *become negative by definition* since longitudinal dimensions are taken as positive aft of the leading edge of the wing mean aerodynamic chord.

b) It is possible that h may also be negative.

c) The term $(d\varepsilon/d\alpha)$ *is set to zero.*

AD2.3.3.3 Short-coupled canard layout (close-coupled)

Correctly designed a short-coupled canard can result in a greater overall lift than that of the sum of canard and wing in isolation. To achieve this the canard moment arm usually has to be between about 0.8 and two wing mean chords and the canard located vertically higher than the wing. Two effects contribute to this:

i) Favourable upwash from the wing which can be regarded as effectively increasing the value of the canard lift curve slope, a_{1T}, to a higher value, say a^1_{1T}.

ii) Favourable canard effect upon the wing, possibly due to enhanced flow due to the canard tip vortices. This may be considered as effectively increasing wing lift curve slope, a_1 to a higher value, a^1_1.

An approximate estimation of the increment in overall lift coefficient in these circumstances is:

$$\Delta C_{L_{MAX}} = \frac{S_T}{S}\left[0.217\left(\frac{l_H}{c}\right)^2 - 1.45\left(\frac{l_H}{c}\right) + 2.03\right] \qquad \text{(AD2.6a)}$$

for $0.08 \leq \frac{l_H}{c} \leq 3.0$, but note that ΔC_{LMAX} is negative for $\left(\frac{l_H}{c}\right)$ greater than 2.

On the assumption that this effect is equally caused by favourable wing upwash on the canard and favourable canard vortex effect on the wing, it may be allowed for by introducing the following corrections:

a) Replace canard lift curve slope a_{1T} *by:*

$$a^1_{1T} = \left[1 + \left\{0.11\left(\frac{l_H}{c}\right)^2 - 0.73\left(\frac{l_H}{c}\right) + 1.02\right\}\right] a_{1T} \qquad \text{(AD2.6b)}$$

b) Likewise the wing lift curve slope, a_l, should be replaced by:

$$a^1_l = \left[1 + \left\{0.11\left(\frac{l_H}{c}\right)^2 - 0.73\left(\frac{l_H}{c}\right) + 1.02\right\}\frac{S_T}{S}\right]a_l \qquad \text{(AD2.6c)}$$

It should recognised that in practice the canard-wing interference effects are likely to be a non-linear function of angle of attack.

AD2.3.4 Tailless configurations

The absence of an auxiliary horizontal surface in a tailless design implies a need to reorganise the basic equations. It is assumed that longitudinal control is provided by an elevator, or equivalent, located at the trailing edge of the wing. This elevator has a lift curve slope, a_2, based on the wing area S. When operated it produces an increment of wing lift located at a distance l^1_η aft of the reference leading edge of the aerodyanamic means chord. Downwash effects are not relevant and effectively a_{lT} is zero. The volume coefficient effectively becomes:

$$V_T = \frac{l_\eta}{c}$$

where

$$l_\eta = l^1_\eta - hc \qquad \text{(AD2.7)}$$

The relevant equations now take the following forms:

i) Trim, refer to Eq (AD2.1a)

$$\left[\frac{Tz_T}{c} + qSC_{M0}\right] + Mg(h - H_0) - qS\left[a_2\,\eta\left(\frac{l_\eta}{c}\right) + (h - H_0)\right] = 0$$

(AD2.8a)

ii) Static stability, stick fixed, refer to Eq (AD2.2a):

$$K_n = (H_0 - h) \qquad \text{(AD2.8b)}$$

iii) Static stability, stick free, refer to Eqs (AD2.2b) and (AD2.2c):

$$\overline{a_{1T}} = a_2 \frac{b_1}{b_2} \qquad since\ a_{1T} = 0 \tag{AD2.8c}$$

$$\therefore K^1_\eta = (H_0 - h) + \frac{\ell_\eta}{c} \frac{\overline{a_{1T}}}{a_1} \tag{AD2.8d}$$

iv) Rotation at take-off, refer to Eq (AD2.3):

$$a_2\eta\left[H_G - \frac{\ell'_\eta}{c}\right] = C_{M0} - \frac{Th_T}{qSc} - a_1\,\alpha_w\left(H_G - h\right) - \frac{Mg}{qS}\left(H_G - h + \frac{0.1k_B^2}{\ell'_\eta c}\right) \tag{AD2.8e}$$

v) Dynamic stability, short period mode, see Eqs (AD2.4a) and (AD2.4b):

$$Frequency\,;\ \omega_S = \frac{\rho SV}{2\pi M}\left[\frac{Ma_1c}{2\rho Sk_B^2}\left\{(H_0 - h) - \rho\frac{S\ell_\eta^2}{c}m_{\dot\theta}\right\}\right]^{\frac{1}{2}} \tag{AD2.8f}$$

Damping ratio; ζ_S

$$= \left[\frac{a_1}{4} - \left(\frac{\ell_\eta}{k_B}\right)^2\left(\frac{m_{\dot\theta}}{2}\right)\right] \Big/ \left[\frac{Ma_1c}{2\rho Sk_B^2}\left\{(H_0 - h) - \frac{\rho S\ell_\eta^{\,2}}{c}m_{\dot\theta}\right\}\right]^{\frac{1}{2}} \tag{AD2.8g}$$

vi) Phugoid mode, refer to Eq (5)

It is questionable as to whether the simplified expression for the long period motion is applicable to the tailless configuration. However, on the basis that it is the equations take the form:

$$\zeta_L = F_1 C_{D0} + F_2 C_L^2 \quad \Big\}$$

as previously

$$F_2 = \left[\frac{3\omega - a_1 \nu}{4\omega + 2a_1 \nu}\right]\left[\frac{C_D - C_{D0}}{C_L^2}\right] - \frac{a_1(\omega - \nu)}{2(2\omega + a_1 \nu)^2} \qquad \text{(AD2.8h)}$$

$$\omega = \frac{Mc}{\rho S k_B^2} \cdot \frac{a_1}{2}(H_0 - h)$$

$$\nu = -\left(\frac{\ell_\eta}{k_B}\right)^2 m_{\dot{\theta}}$$

in this case.

Addendum 3

Lateral control and stability surfaces

Notation (for this Addendum only)

a_{1W}	Wing lift curve slope, three-dimensional
A	Wing aspect ratio
$\overline{A}$	Roll moment of inertia
b	Wing span
B	Coefficient in lateral stability terms [Eq (AD3.8c)]
$(c_f/c)_A$	Ratio of aileron chord to wing chord
$(c_f/c)_R$	Ratio of rudder chord to fin chord
C	Coefficient in lateral stability terms [Eq (AD3.8d)]
C_L	Overall lift coefficient
CR	Cross-wind ratio
$\overline{F}$	Term in rudder effectiveness [Eq (AD3.3a)]
i_A, i_C	Non-dimensional rolling and yawing moment of inertia coefficients, obtained by dividing the actual moment of inertia by $[M(b/2)^2]$
J	Non-dimensional damped frequency of motion in pure lateral mode, Eq (AD3.10c)
K_l	Coefficient in roll equation [Eq (AD3.13b)]
l_F	Distance of fin sideforce aft of aircraft centre of gravity
L_i, N_i and Y_i	Rolling, yawing and sideforce coefficients respectively due to roll ($i = p$), yaw ($i = r$), sideforce ($i = v$), rudder deflection ($i = \zeta$), aileron deflection ($i = \xi$) and engine failure ($i = E$). Obtained from absolute values by dividing by the following expressions

	For L_V, N_V, Y_P and Y_r:	$\rho V S b/2$
	Y_V:	$\rho V S/2$
	L_P, N_P, L_V and N_r:	$\rho V S b^2/2$
	L_ζ and Y_ζ:	$\rho V^2 S/2$
	L_ξ, N_ξ, L_ζ, N_ζ, and N_E:	$\rho V^2 S b/2$
M	Aircraft mass	
M_N	Mach number	
p	Rate of roll	
r	Rate of yaw	
R	Damping coefficient in pure lateral motion [Eq (AD3.10b)]	
S	Wing area	
S_F	Fin area	
t_l	Time taken to roll through a defined back angle, ϕ_l	
v	Sideslip velocity	
V	Forward velocity	
β	Sideslip angle, v/V	
β_M	$(1 - M_N^2)^{1/2}$	
Γ	Dihedral angle	
ζ	Rudder angle	
ζ_D	Damping ratio in short period lateral mode [Eq (AD3.8a)]	
η	Location of spanwise point as a fraction of semi-span	
λ	Wing taper ratio (tip/root chords)	
λ_l	Coefficient in lateral stability terms [Eq (AD3.8b)]	
$\Lambda_{1/2}$	Sweep of wing half chord line	
μ	Non-dimensional relative density, $2M/\rho Sb$	
ξ	Aileron angle	
ρ	Air density	
τ	Non-dimensional time; equal to $[t\,(\rho SV/M)]$, where t is real time	
ϕ_l	Bank angle achieved in time t_l	
$\Phi_{\xi i}$, $\Phi_{\xi 0}$	Correction terms in aileron equation [Eq (AD3.14b)]	
ω	Frequency of short period lateral mode [Eq (AD3.8e)]	

AD3.1 Introduction

Historical data are provided in Chapter 8, Tables 8.2 and 8.3, as a means of making a first approximation to the geometry of the lateral control and stabilisation surfaces which in a conventional design may be identified as a fin/rudder combination and ailerons. The requirements which determine the size and characteristics of these surfaces are complex, especially for multi-engined aircraft where the handling behaviour subsequent to an engine failure is almost certainly a dominant issue. The information contained in this Addendum represents an attempt to resolve this by presenting simplified equations which, nevertheless, should enable a reasonable initial design to be achieved. Because of the variation of specific requirements for different aircraft: civil or military, combat, transport or general aviation, it is necessary to consult the relevant airworthiness documents to obtain the input data. See, for example, References J, Chapter 9, Appendix A9.1.

AD3.2 Preliminary calculations

AD3.2.1 Initial aircraft data

It is presumed that the following data are known from the initial overall synthesis process described in Chapter 8, subject to refinement as the design proceeds:

- i) Mass distributions, dimensions and centres of gravity from which the required moments of inertia may be derived.
- ii) Wing geometry, including a first definition of the trailing edge flap system which gives an indication of possible aileron span. Dihedral of the wing may not have been assumed and it is therefore covered in the subsequent analysis in this Addendum.
- iii) Fuselage geometry.
- iv) Horizontal tail geometry, possibly as a result of the design procedure outlined in Addendum 2. This will include a proposed vertical location relative to the fin, which may require further consideration.
- v) Powerplant nacelle geometry, where relevant, together with engine thrust characteristics.
- vi) Speed and altitude conditions appropriate to take-off /initial climb, cruise and the approach to landing.

AD3.2.2 Requirements

The requirements appropriate to the given class of aircraft must be consulted to provide:

a) Handling requirements consequent upon engine failure where this is relevant. They may include:

- i) Trim requirements.
- ii) Corrective action, including time delay.
- iii) Allowable heading change during recovery (directional).
- iv) Residual heading change capability.
- v) Allowable bank angle during recovery (lateral).
- vi) Residual bank/roll capability.

b)Directional static stability.

c)Lateral static stability.

d)Directional/lateral dynamic stability - Dutch roll and spiral modes.

e)Roll performance - lateral control.

f)Handling in cross-winds.

g)Spin recovery, where relevant.

AD3.2.3 Evaluation of basic aircraft characteristics

The basic inertial and aerodynamic characteristics of the aircraft are needed. In particular these include:

a)Inertial characteristics: mass, fore and aft and vertical centre of gravity range; moments of inertia in roll and yaw and, if thought to be significant, the corresponding product of inertia (*A*, *C* and *E* respectively).

b)Aerodynamic characteristics: the contributions to the relevant aerodynamic derivatives of the components of the aircraft excluding fin/rudder, aileron and wing dihedral effects. In general the relevant aerodynamic derivatives are:

i) Rolling coefficients, L_i, due to rate of roll, p, rate of yaw, r, and sideslip velocity, v.

ii) Yawing coefficients, N_i, due to above effects.

iii) Sideslip coefficients, Y_i, due to above effects.

In addition to the above deflection of the aileron, ξ, rudder, ζ, and engine failure, *E*, add contributions to the three basic sets of coefficients, L_i, N_i and Y_i as relevant.

The calculations of the coefficients requires a knowledge of such items as wing planform, including sweep and dihedral, fuselage geometry and wing/fuselage interaction effects. Deployment of flaps is significant in some cases. Clearly fin/rear fuselage/tailplane and rudder characteristics determine the fin/rudder contributions but as yet these data are not known. It is convenient to retain the dihedral effect as a function of dihedral angle, see paragraph AD3.3.3.

Evaluations of these derivatives, apart from initial predictions for the fin/rudder and aileron effects are outside the scope of this note. A number of sources of information may be referred to. For example the Engineering Sciences Data Unit, ESDU, Aerodynamic series (Chapter 9, Appendix A9.1, Reference C8) includes a comprehensive coverage of the subject.

It is also necessary to know the non-dimensional yawing moment, N_E, due to failure of the most critically located powerplant. Because thrust is dependent upon speed, N_E will have different values for various speed conditions.

AD3.3 Design procedure

AD3.3.1 Initial sizing of fin/rudder

AD3.3.1.1 General

It is possible to derive a logical procedure to enable the sizes (areas) of the fin and rudder to be established. Such a procedure is outlined in the following paragraphs.

AD3.3.1.2 Heading after engine failure

Subsequent to an engine failure at a prescribed speed, usually $1.4V_{SI}$, where V_{SI} is the stalling speed in the take-off configuration, the aircraft must be able to maintain a straight heading with a bank angle of not more than 5°, This leads to a numerical inequality:

$$\left(\frac{l_F}{b}\right) \geq \frac{115N_E}{C_L} \tag{AD3.1}$$

where
- l_F is the arm of the fin sideforce behind the overall centre of gravity
- b is the wing span
- N_E is the engine failed yawing moment coefficient at the prescribed speed
- C_L is the corresponding aircraft lift coefficient.

In many cases an approximate value of l_F is already known from the design synthesis procedure and will be found to be considerably more than the minimum value given by Eq (AD3.1). This, however, may not be the case for a tailless design.

AD3.3.1.3 Fin and rudder contributions to N_V and N_ζ

The fin/rudder contributions to the most important derivatives, N_V and N_ζ may now be estimated conservatively, assuming a typical fin geometry, by:

Due to fin: $$(Y_V)_F = -2.4(S_F/S) \quad \text{approx} . \tag{AD3.2a}$$

where S_F is the fin reference area and S is the wing area.

It follows that due to the fin:

$$(N_V)_F = 2.4\left(\frac{l_F}{b}\right)\left(\frac{S_F}{S}\right) \qquad \text{(AD3.2b)}$$

The ratio (S_F/S) is retained as a main parameter in the analysis.

The sideforce coefficient due to rudder deflection is approximately:

$$Y_\zeta = \bar{F}\left[\left(\frac{c_f}{c}\right)_R^{0.47} + 0.08\right]\left(\frac{S_F}{S}\right) \qquad \text{(AD3.3a)}$$

where $(c_f/c)_R$ is the mean ratio of the rudder chord to fin chord.

$\bar{F}$ is dependent upon the extent of the rudder along the fin and the tailplane position vertically. For a full span rudder, approximately:

Body mounted tailplane $\bar{F} = 2.3$

High mounted, "T" tail $\bar{F} = 3.0$

Again it follows that the corresponding yawing moment coefficient is:

$$N_\zeta = -\bar{F}\left[\left(\frac{c_f}{c}\right)_R^{0.47} + 0.08\right]\left(\frac{l_F}{b}\right)\left(\frac{S_F}{S}\right) \qquad \text{(AD3.3b)}$$

AD3.3.1.4 Rudder angle to trim after engine failure

The rudder angle required to trim the aircraft after an engine failure at the prescribed speed, in this case usually $1.2V_{S1}$, must not exceed 75% of the total angle available. It follows from this that approximately as an absolute value:

$$N_\zeta \geq 1.35 N_E / \zeta_{MAX} \qquad \text{(AD3.4)}$$

where N_E is the engine failed moment coefficient at the prescribed speed and where the assumed effective linear equivalent rudder angle is ζ_{MAX} (a maximum value of 0.35 rad is typical for a simple rudder).

AD3.3.1.5 Inability of rudder to cause fin stall

In no circumstances must full rudder deflection result in a dynamic sideslip angle which would cause the fin to stall. The allowable angle may be increased with the aid of a dorsal fin. A dynamic fin stall angle of 0.5 radians is feasible and this leads to:

$$-N_\zeta \leq 0.5 N_V / \zeta_{MAX} \qquad \text{(AD3.5)}$$

where N_V is the total aircraft value.

AD3.3.1.6 Heading hold in cross-wind

It must be possible to hold the aircraft on a given heading in the presence of a cross-wind. This implies a sideslip angle of (CR) where (CR) is the ratio of the cross-wind velocity to the forward speed of the aircraft. It may also be a requirement to introduce a heading change against the cross-wind. The usual design cross-wind ratio does not exceed 0.2, although slow flying aircraft may be an exception, and the heading change, if required, 0.175 rad. It is possible to write:

i) No heading change:

$$-N_\zeta \geq 0.20 N_V / \zeta_{MAX} \qquad \text{(AD3.6a)}$$

ii) 0.175 rad heading change:

$$-N_\zeta \geq 0.375 N_V / \zeta_{MAX} \qquad \text{(AD3.6b)}$$

AD3.3.1.7 Change of heading against failed engine

When the critical engine has failed at the same speed as that of paragraph AD3.3.1.2 it must be possible to change heading against the failed engine, typically by 0.262 radians. Assuming that the ailerons are used to offset any rolling tendency due to the failed engine, this implies that:

$$-N_\zeta \geq \left(N_E + 0.262 N_V\right) / \zeta_{MAX} \qquad \text{(AD3.7)}$$

AD3.3.1.8 Determination of design values

The inequalities of Eqs (AD3.4) to (AD3.7) may be used to derive acceptable values for (S_F/S) and $(c_f/c)_R$. The most expedient way of doing this is:-

i) Evaluate the total value of N_V as the sum of the value less fin/rudder defined at paragraph AD3.2.3 (b), and the additional contribution of Eq (AD3.2a), for a range of values of (S_F/S) - typically say 0.10 to 0.3, but possibly less when there is no engine failed requirement.

ii) When engine failure is a requirement calculate, from Eq (AD3.7):

$$\left(N_E + 0.262 N_V\right) / \zeta_{MAX}$$

iii) Compare the values of $(-N_\zeta)$ obtained from Eqs (AD3.4) and (AD3.6) and select the highest for evaluation, for example that given by Eq (AD3.6b).

iv) Calculate $0.5 N_V / \zeta_{MAX}$.

v) Inspection of the results enables the minimum value of (S_F/S) to be found which satisfies the inequalities. When there is no engine failed case this follows from the cross-wind condition, Eq (AD3.6). However, when there is an engine failure case a value of (S_F/S) must be chosen to ensure that $(-N_\zeta)$ lies between the highest value given by Eqs (AD3.4), (AD3.6) or (AD3.7) but not greater than $(0.5N_V/\zeta_{MAX})$ as given by Eq (AD3.5). Each acceptable value of (S_F/S) implies a value of N_ζ.

vi) For several values of (S_F/S) at and above the minimum determined at the previous step evaluate N_ζ in terms of $(c_f/c)_R$. Typical values of $(c_f/c)_R$ are usually in the range 0.2 to 0.5.

vii) Inspection of the results enables the value of $(c_f/c)_R$ consistent with each value of (S_F/S) to be established. In practice it will usually be found that the lowest acceptable value of (S_F/S) also gives an acceptable, that is practical, value of $(c_f/c)_R$. The value of investigating larger values of (S_F/S) is that it provides information readily available should the initially selected value of (S_F/S) prove to be insufficient when other conditions, such as dynamic stability, are investigated.

Directional static stability requires that the derivative N_V should be positive and this should have been covered by the above procedure.

AD3.3.2 Geometry of fin/rudder

Once initial values of (S_F/S) and $(c_f/c)_R$ have been established it is possible to check the fin geometry, using the guidelines of Table 8.2 in Chapter 8. A more precise value of the fin arm can be established and the fin contributions to Y_V, N_V, and N_ζ calculated more accurately than given by Eqs (AD3.2) and (AD3.3) above. However, it is sometimes prudent to leave these refinements until other requirements have been investigated.

AD3.3.3 Lateral static stability

Lateral stability may be simply equated to the value of the derivative L_V being negative. The overall value of L_V can be obtained by adding the fin/rudder contribution corresponding to the fin/rudder geometry derived in paragraph AD3.3.1 to that of the rest of the aircraft, paragraph AD3.2.3(b). This latter contribution includes a term which is a function of dihedral angle, Γ. At this stage a value of Γ may be deduced to ensure that L_V is negative. However, L_V should be only just negative - see paragraph AD3.3.5(b).

AD3.3.4 Overall aerodynamic derivatives

Tentative values of (S_F/S), $(c_f/c)_R$ and Γ are now established and it is possible to evaluate the full values of all the aerodynamic derivatives with the exception of those resulting from aileron deflection.

AD3.3.5 Lateral/directional dynamic stability

There are two primary modes of dynamic stability, namely the short period motion known as the "Dutch Roll" and a long period motion which represents a spiral mode.

a) Short period mode. It is necessary for the short period lateral/directional motion to be positively damped and some requirements give specific values for damping ratio. Automatic yaw damping by means of rudder motion is often used to obtain a satisfactory value. The period of the motion is also of some concern and must be such as not to cause unacceptable handling qualities. As a guide the effective damping ratio in cruise should not be less than about 0.08 for larger, less manoeuvrable aircraft, about 0.2 for smaller highly manoeuvrable types and as much as 0.4 for combat aircraft in combat conditions.

The frequency of the motion should not be less than 0.5 Hz for larger aircraft and 1 Hz for small aircraft. The product of the damping ratio and frequency is also of some concern and should not be less than about 0.15 for flight conditions where rapid manoeuvre is not required, but 0.35 where it is.

The damping ratio in the short period mode is approximately given by:

$$\zeta_D = (B + \lambda_1)/2\{C + (B + \lambda_1)\lambda_1\}^{1/2} \quad \text{(AD3.8a)}$$

where

$$\lambda_1 = \frac{2L_P}{i_A} + \frac{\mu L_V i_A}{8L_P^2} C_L \quad \text{(AD3.8b)}$$

$$B = -\left(\frac{Y_V}{2} + \frac{2L_P}{i_A} + \frac{2N_r}{i_C}\right) \quad \text{(AD3.8c)}$$

$$C = Y_V\left(\frac{L_P}{i_A} + \frac{N_V}{i_C}\right) + 4\frac{L_P N_r}{i_A i_C} + \frac{\mu N_V}{i_C} \qquad \text{(AD3.8d)}$$

where μ is the non-dimensional density and i_A and i_C are the non-dimensional moment of inertia coefficients - see Notation for definitions.

The frequency of the short period mode is approximately:

$$\omega = \frac{\rho SV}{2M}\left[4\{C + (B + \lambda_1)\lambda_1\} - (B + \lambda_1)^2\right]^{1/2} \quad \text{Hz} \qquad \text{(AD3.8e)}$$

b) Spiral mode. It is not unusual for the spiral mode to be unstable and to take the form of a slow divergence. As a general guide it is likely to be considered satisfactory if the time to double amplitude is at least 20 s. The time to double amplitude is given by:

$$t_2 = \frac{1.386M}{\rho\ SV\ C_L}\left(\frac{L_V N_P - L_P N_V}{N_V L_r - L_V N_r}\right) \quad \text{sec} \qquad \text{(AD3.9)}$$

An unduly large negative value of L_V may result in a more rapid divergence and thus although it must be negative for lateral static stability the dihedral should be chosen to only just meet this condition.

AD3.3.6 Heading change consequent upon engine failure

A common assumption is that subsequent to the sudden failure of the critical engine a period of 2 s elapses before corrective action is taken. During this period the aircraft will yaw and roll. The extent of the rolling motion is a complex function of the lateral aerodynamic derivatives and the vertical offset of the out of balance thrust force relative to the centre of gravity. When corrective action has taken effect a usual requirement is that the total change of heading should not exceed 0.35 radians. Considerable simplification of the analysis of the response of the aircraft is possible if it is assumed that the rolling motion is not significant and this enables a first check to be made of the adequacy of the initial fin/rudder design derived in paragraphs AD3.3.1 and AD3.3.2. It is important to make a subsequent check on the full response of the aircraft. It is suggested that the analysis should be undertaken for a speed of 1.2 times the stalling speed with the aircraft in the take-off configuration.

a) Sideslip angle reached 2 s after engine failure.

$$\beta_2 = \frac{-\mu N_E}{i_C(J^2 + R^2)}\left[1 - e^{-R\tau_2}\left(\cos J\tau_2 + \frac{R}{J}\sin J\tau_2\right)\right]$$

where
$$R = \left(\frac{N_r}{i_C} + \frac{Y_V}{4}\right)$$

and
$$J = \left(\frac{\mu N_V + N_r Y_V}{i_C} - R^2\right)^{1/2}$$

(AD3.10a)

τ_2 is the non-dimensional time equivalent of 2 s

$$\tau_2 = 2\rho / SVM \qquad \text{(AD3.10b)}$$

see Notation for definition of μ, i_A and i_C. R is, in fact, the damping coefficient and J the damped natural frequency in the lateral mode with rolling excluded.

At the end of the 2 s the rate of change of sideslip angle, in terms of non-dimensional time is:

$$\dot{\beta}_2 = \frac{-\mu N_E}{i_C J}\left[e^{-R\tau_2}\sin J\tau_2\right] \qquad \text{(AD3.10c)}$$

b) When 2 s has elapsed the full available corrective rudder is applied, and as a consequence the rate of change of sideslip angle starts to decrease, and eventually becomes zero as the maximum heading change is reached.

Equation (AD3.10c) may be used to give the non-dimensional time τ_0 at which $\dot{\beta}$ becomes zero:

$$e^{-R\tau_0}\sin J\tau_0 = \frac{\dot{\beta}_2 i_C J}{\mu\left(-\zeta N_\zeta + N_E\right)}$$

(AD3.11a)

The right-hand side of Eq (AD3.11a) is known for a given value of the available rudder angle, ζ, and the simplest way of deriving τ_0 is to calculate the value of the left-hand side for various τ_0 until the one which satisfies the equation is found. The value of τ_0 may then be used to find the additional sideslip angle, $\Delta\beta$, which results as the yaw acceleration, $\ddot{\beta}$, is brought to zero:

$$\Delta\beta = -\frac{\mu\left(-\zeta N_\zeta + N_E\right)}{i_c(J^2 + R^2)}\left[1 - e^{-R\tau_0}\left(\cos J\tau_0 + \frac{R}{J}\sin J\tau_0\right)\right] \qquad \text{(AD3.11b)}$$

and the total heading change, β_0, which should be less than 0.35 radians is:

$$\beta_0 = \beta_2 + \Delta\beta \qquad \text{(AD3.11c)}$$

Should β_0 exceed the specified maximum it is necessary to reconsider the rudder effectiveness, N_ζ, and this has a consequence on the required overall value of N_V and hence (S_F/S).

AD3.3.7 Aileron sizing

The aileron size is determined by either the need to maintain wings level in a cross wind or by other roll performance requirements. In the first instance if it is assumed that the aileron span is effectively determined by trailing edge flap requirements the aileron sizing is a matter of determining the aileron chord ratio, $(c_f/c)_A$.

a) Cross-wind case. The cross-wind requirements vary but a typical condition is that the aircraft should be be able to contend with a cross-wind which is 20% of the forward speed. Assuming that this is the case, and that half of the maximum available aileron angle is allowed for control, then:

$$L_\xi \geq 0.4\, L_v \,/\, \xi_{MAX} \qquad \text{(AD3.12)}$$

(A typical value of ξ_{MAX} is 0.28 radians for a simple control.)

b) Roll performance/handling. It is usual for roll performance requirements to be stated in terms of the time taken to roll through a given angle. Often this implies a reversed roll, such as the ability of a civil transport aircraft to roll from a 30° bank angle in one direction to 30° in the other in 7 s when all engines are operating. This is equivalent to being able to achieve 30° bank angle change within 3.5 s of initiating the roll. Longer time is allowed, usually 11 s for the full manoeuvre, when an engine has failed but only part, say 50%, of the aileron movement is then available for the manoeuvre.

The aileron power required to give the specified rolling performance may be expressed as:

$$L_\xi \geq bL_P\Phi_l / \left[V\left\{ t_l + \frac{1}{K_l}(e^{-K_l t_l} - 1) \right\} \right] / \xi_{MAX} \qquad \text{(AD3.13a)}$$

where it has be assumed that ζ_{MAX} of aileron angle is available to achieve a bank angle ϕ_l at a time t_l seconds after initiation of the roll.

b is the aircraft wing span
V is the forward speed

$$K_l = -\rho SVb^2L_P / 2\bar{A} \qquad \text{(AD3.13b)}$$

where $\bar{A}$ is the roll moment of inertia.

c) Critical aileron size requirement. The maximum value of L_ξ derived from the cross-wind case, Eq (AD3.12) or the various roll performance cases, Eq (AD3.13a), is selected to form the basis of the aileron size.

d) Aileron chord ratio. The aileron rolling derivative L_ξ is approximately given by:

$$L_\xi = \frac{-\bar{\eta}}{2} a_{lW} \left\{ \left(\frac{c_f}{c} \right)_A^{0.47} + \frac{0.16}{A} \right\} \left(\Phi_{\xi i} - \Phi_{\xi 0} \right) \qquad \text{(AD3.14a)}$$

where a_{lW} is the lift curve slope of the wing, per radian

$\bar{\eta}$ is the location of the mid-point of the aileron span as a fraction of the wing semi-span

A is wing aspect ratio

$(\Phi_{\xi i}$ *and* $\Phi_{\xi 0})$ are correction factors on the spanwise lift distribution for the inner and outer aileron end locations respectively.

Approximately:

$$\Phi_\xi(\eta) = \left[\left(1 - \frac{0.8}{\beta_M A}\right)(1-\eta)^{1.25} - 1.34\left(12 + A\tan\Lambda_{1/2} - 8\lambda\right)^{2.5}\cos\{\pi(\eta-0.5)\} \times 10^{-4}\right]$$

(AD3.14b)

where

$\beta_M =$ $(1 - M_N^2)^{1/2}$ where M_N is flight Mach number
λ is wing taper ratio
$\Lambda_{1/2}$ is wing sweep along the half chord line
η is the location of one end of the aileron as a fraction of the wing span.

Inner end to give $\Phi_{\xi i}$ and outer end to give $\Phi_{\xi 0}$.

(Given that η_i and η_o are known $\Phi_{\xi i}$ and $\Phi_{\xi 0}$ may be calculated as may $\bar{\eta}$).

Since the value of a_{1W} is also known from the basic wing design, the value of $(c_f/c)_A$ corresponding to the required value of L_ξ may be calculated from Eq (AD3.14a). If $(c_f/c)_A$ is found to be excessively large, say above 0.35, it will result in wing structural layout difficulties. In these circumstances it is necessary to either:

i) Increase the span of the ailerons to the probable detriment of the flap performance or
ii) Consider the use of another device, such as spoilers, to augment the aileron in the critical case.

If cross-wind considerations are critical a further option is to review the wing dihedral in that it has a large impact on the derivative L_V. Any change here should still result in positive lateral stability and the effect on dynamic stability must be ascertained.

Addendum 4

Mass prediction

Notation (for this Addendum only)

(All dimensions are in SI units)

A	Wing aspect ratio
B	Maximum width of fuselage
c_1 to c_5	Coefficients used in Chapter 6 to define mass characteristics
D	Maximum equivalent diameter of fuselage
D_i	Equivalent diameters of fuselage, $i = 0$ to 4, see Figure AD4.1
f_a	Allowable working stress in wing covers [Eq (AD4.1d)]
f_b	Non-dimensional fuselage length parameter; $(L_p - L_n)/D_1$
f_c	Non-dimensional fuselage length parameter; $(L - L_p - L_t)/D_2$
f_n	Non-dimensional fuselage nose length parameter; L_n/D_1
f_t	Non-dimensional fuselage tail length parameter; L_t/D_2
H	Maximum height of fuselage
k_i	Coefficient defining fuselage cross-section parameter, $i = 1$ to 4 [Eq (AD4.3)]
k_5	Coefficient used to define fuselage tail shape [Eq (AD4.4e)]
k_6	Coefficient used to define pressurised fuselage mass [Eq (AD4.5c)]
k_7	Coefficient used to define unpressurised fuselage mass [Eq (AD4.6b)]
k_8 to k_{11}	Coefficients used to define the value of k_7 [Eq (AD4.6b)]
k_{12}	Coefficient used to define vertical fin mass [Eq (AD4.9)]
L	Overall length of fuselage
L_n	Length of fuselage nose, see Figure AD4.1

L_p Distance aft of nose to a particular discontinuity in the cross-section, see Figure AD4.1
L_t Length of tail section of fuselage, see Figure AD4.1
L_T Tail arm; wing quarter mean chord to tailplane quarter mean chord, or
fin when there is no tailplane
m_c Mass of primary wing structural box covers and webs as fraction of
total aircraft mass [Eq (AD4.1b)]
m_r Ratio of primary rib mass to total aircraft mass [Eq (AD4.1c)]
M_b Mass of tail booms [Eq (AD4.7)]
M_f Mass of fuselage [Eqs (AD4.5a) or (AD4.6a)]
M_H Mass of horizontal control/stabiliser [Eq (AD4.8)]
M_{IPS} Mass of ideal primary wing structure box [Eq (AD4.1a)]
M_0 Total mass of aircraft
M_V Mass of vertical control/stabiliser surface [Eq (AD4.9)]
M_W Mass of wing structure
M_{ZW} Zero fuel mass of aircraft [Eqs (AD4.1f) to (AD4.1h)]
n Maximum number of occupants of aircraft
n_l Limit manoeuvre factor
N Ultimate normal manoeuvre factor
$\bar{N}$ Effective ultimate design factor [Eq (AD4.1e)]
N_S Factor in calculation of fuselage nose area [Eqs (AD4.f) to (AD4.1h)]
p Fuselage pressure differential, bar
r Wing inertial relief factor [Eqs (AD4.1f) to (AD4.1h)]
S Wing reference area
S_b Surface area of tail boom
S_f Surface area of fuselage
S_H Reference area of horizontal control/stabiliser [Eq (AD4.8)]
S_V Reference area of vertical control/stabiliser [Eq (AD4.9)]
V_D Structural design speed of aircraft
α Factor defining if air intakes are present in the fuselage [Eq (AD4.6b)]
β Ratio of fuselage width to wing span at wing attachment [Eq (AD4.6b)]
δ Factor defining impact of landing gear on fuselage [Eq (AD4.6b)]
ϵ Factor relating to engines buried in fuselage [Eq (AD4.6b)]
η Factor defining if aircraft is intended for naval operations [Eq (AD4.6b)]
λ Wing taper ratio; tip chord divided by root chord
$\bar{\sigma}$ Ratio of design working tensile stress in a pressurised fuselage
τ Aerofoil thickness to chord ratio at the side of fuselage
ϕ Sweep of wing quarter chord line
ψ Sweep of wing structure [Eqs (AD4.1a) and (AD4.1d]
ω Factor defining a wing strutted to the fuselage [Eq (AD4.6b)]

AD4.1 Introduction

The synthesis of the concept aircraft requires the input of mass data and paragraph 6.4 of Chapter 6 covers this for the initial phase of the design. The data given there have been deliberately kept as simple as is possible and require a knowledge only of those parameters already included in the design procedure. A consequence of this is that there is insufficient detail to enable a complete investigation of the characteristics of a given concept. Once the conceptual design has been established and the analysis procedure outlined in Chapter 9 has been commenced it is desirable to employ more detailed methods of mass prediction. Relevant information is contained in a number of the references given in the Bibliography which forms the Appendix to Chapter 9. The more useful ones are indicated. The information following is complementary to these and may be found to be of some value.

It is necessary to warn of the difficulties of attempting to correlate predicted data with known values of existing aircraft. One obvious reason for this is the improvement of technology with the passage of time. Perhaps less obvious is the considerable variation of the actual content of the mass of a given component. For example is any wing structure which carries through the fuselage defined as part of wing or part of body mass? Are flying control surface actuators part of the control system or part of the power services? Only a detailed mass breakdown can resolve such issues. In practice the best correlation is the total empty mass of the aircraft. Nevertheless the more detailed data that is essential for preliminary design analysis must be based on overall average information.

Mass prediction methods may be derived by various techniques. At their simplest they consist of no more than a direct statement of average actual data, usually referred to the total aircraft mass. Such an approach is often the only feasible one for some components. More useful for design analysis are those methods which, while being based on empirical data, seek to interpret the data in relation to the known primary parameters of the design. Alternatively a simplified theoretical approach may be undertaken but it is invariably necessary to add practically derived corrections. Finally the prediction technique may consist of a detailed approach which, although simplified, follows the design procedure for a given component. To be successful this last method requires a quite detailed definition of the component and ultimately is the mass estimation of the finalised design.

It is convenient to deal with mass prediction data in three groups:

a) Structure, which typically accounts for half, or rather more, of the empty mass. It is dominated by the wing and fuselage contributions but the landing gear is also significant.

b) Powerplant, which is the installed engine component and here includes the fuel system. It usually contributes about one-fifth of the empty mass.

c) Systems and equipment which is taken to include all the other items which go to make up the basic empty mass of the aircraft.

AD4.2 Structure mass

AD4.2.1 Wing

AD4.2.1.1 General

The wing structure is taken to include the primary structure inclusive of the continuity across the fuselage; fixed leading and trailing edges; wing tips and fairings; high lift devices; and control surfaces, but not control operating systems.

The formula given for lifting surface mass in Chapter 6, Eq (6.22a) is primarily based on wing parameters but includes typical allowances for the contribution of the control and stabiliser surfaces. The wing mass is derived by dividing the coefficient c_l of Table 6.7 by the lifting surface factor c_5 of Table 6.10 of Chapter 6. The basic formula is entirely empirical but does include allowance for the effect of all the main geometric factors together with the primary structural design considerations. However, it does not specifically include details of such items as inertial relief, types of high lift devices or any special layout features. Typical values of these effects are covered in the quoted values of the coefficient c_l.

AD4.2.1.2 Theoretical method

To enable an investigation of all the important design parameters it is necessary to use a method, which at least to some extent, is based on a theoretical approach. The following procedure falls into this category. There are three stages of calculation:

a)*The "ideal" mass of the primary structure* is calculated by using equations based on the theoretically required bending strength.

b)*This "ideal" mass is modified by making allowances for departure of the structural concept from the "ideal".*

c)*The mass of the secondary items* such as high lift devices and controls, is estimated using simplified statistical data.

For convenience the procedure is undertaken in terms of the take-off mass of the aircraft, M_0, thus the final result appears as the total wing mass as a fraction of the take-off mass.

The equations are based on the use of aluminium alloy construction, but suggestions are included to indicate the likely effect of the use of carbon-fibre reinforced plastic where this is relevant.

A Ideal primary structure mass

The ideal primary structure mass, M_{IPS} is the sum of two terms, one being that of the structural box covers and spanwise shear webs and the other the ribs required to support them:

$$\left(\frac{M_{IPS}}{M_0}\right) = m_C + m_r \qquad \text{(AD4.1a)}$$

where:

$$m_C = 1920\, A^{1.5} S^{0.5} \bar{N} r (1+\lambda) \sec\phi \sec\varphi / \tau f_a \qquad \text{(AD4.1b)}$$

$$m_r = \frac{3S^{1.25}\tau^{0.5}}{M_0 A^{0.25}}\left[\left(1 - 0.34\lambda + 0.44\lambda^2\right) + 2.2\tau\left(\frac{S}{A}\right)^{0.5}\left(1 - \lambda + 0.72\lambda^2\right)\right] \qquad \text{(AD4.1c)}$$

A, S, λ, and ϕ are, respectively, the usual definitions of aspect ratio, wing area, taper ratio and quarter chord sweep.

τ is the thickness to chord ratio of the aerofoil at the side of the body.

ψ is the sweep of the structure, which in many cases may be taken to be equal to ϕ but it is different in some circumstances, for example a delta wing having a primary structure which is orthogonal to the centreline of the aircraft.

f_a is the allowable working stress of the airframe material. For light alloy:

$$f_a = 1.12\left[\frac{\bar{N} r\, A^{1.75} M_o}{S^{0.75}\tau^{1.5}}(1+\lambda)^{2.5} \sec\varphi \sec\psi\right]^{0.5} \times 10^5\ N/m^2 \qquad \text{(AD4.1d)}$$

$\bar{N}$ is the effective ultimate design factor, which is either about 1.65 times the limit manoeuvre factor, n_l, or where it is greater, the comparable ultimate discrete gust factor. That is $\bar{N}$ is the greater of:

$$1.65n_l$$

$$or \qquad 1.65 + \frac{6.45V_D S}{M_0\left(\frac{2}{A} + sec\varphi\right)} \qquad approximately \qquad \text{(AD4.1e)}$$

r is the factor which allows for inertial relief. The following expressions may be used as relevant:

i) No wing-mounted powerplant or stores:

$$r = 1 - [0.12 + (1-M_{ZW}/M_0)] \qquad \text{(AD4.1f)}$$

ii) Two wing-mounted powerplants:

$$r = 1 - [0.2 + (1-M_{ZW}/M_0)] \qquad \text{(AD4.1g)}$$

iii) Four wing-mounted powerplants:

$$r = 1 - [0.22 + (1 - M_{ZW}/M_0)] \qquad \text{(AD4.1h)}$$

M_{ZW} is the design zero fuel mass, that is the maximum allowable mass when the fuel load is zero.(Maximum payload condition.) If it is not known the term $(1 - M_{ZW}/M_0)$ may be replaced by $(0.1 + 2(\text{Range}) \times 10^{-5})$.
Wing-located stores may be allowed for on the basis that each large store on one side of the aircraft increases the 0.12 factor of Eq (AD4.1f) by 0.02.

B Allowance for departure from the structural ideal and alternative materials

Such considerations as attachments for powerplants, landing gear and stores, wing folding or variable sweep represent penalties to be added to the ideal structure mass ratio given by Eq (AD4.1a). Suggested allowances, also as a ratio of take-off mass, M_0, are given in Table AD4.1A. It will be noted that a braced wing results in a reduction of ideal structure mass, although there is a penalty to the fuselage. Use of carbon fibre reinforced plastics rather than light alloy also results in a reduction of ideal structure mass. The degree of the reduction is dependent upon various design features, especially the number of primary structural joints. In practice the ideal structural mass is not likely to be less than about 60% of that given by Eqs (AD4.1a) and (AD4.1d).

Table AD4.1A Incremental penalties on idealised primary wing structure

Braced attachments for powerplant - two	0.001
- four	0.0015
Attachment for landing gear	0.004
Cutout in primary structure for landing gear	0.01
Attachment of stores	0.004
Scale effect for aircraft; $M_0 < 5700$ kg	
Tapered wing	0.002-0.005
Untapered wing	0.01
Inboard wing fold	0.02
Outboard wing fold	0.005
Variable sweep	$0.03-3.5 \times 10^{-5} M_0^{0.5}$
Braced wing, $M_0 < 5700$ kg (decrement)	-0.022

C Allowance for secondary structure

Table AD4.1B suggests allowance for various items of secondary structure, again as a ratio of aircraft take-off mass, M_0. Use of composite construction results in a reduction of mass relative to light alloy components, the degree of reduction being determined by the extent of use of this material.

Table AD4.1B Incremental allowance for secondary wing structure

Fixed leading and trailing edge with ailerons only	0.02 plus
Trailing edge flaps: Plain or single slotted	0.003 or
Fowler on double slotted	0.006 or
Triple slotted	0.012 and
Leading edge flaps/slots	0.007 plus
Spoilers/air brakes	0.0015 plus
Tips, fairing, etc.	0.002
Reduction for composite moving surfaces (decrement)	-0.005
Additional penalty for aircraft of $M_0 < 140\,000$ kg, up to	0.005

D Total wing mass

The ratio of the total wing mass, M_W to the take-off mass, M_0, is the sum of the values obtained from stages **A** to **C** inclusive.

AD4.2.1.3 Effect of body width on wing structure mass

Both Eqs (6.22a) of Chapter 6 and (AD4.1a) include the mass of the wing within the sides of the fuselage. The mass prediction is based on a typical fuselage width to wing span ratio, β, of 0.1. In some circumstances it is desirable to correct for a different value of β:

i) Method of Chapter 6, Eq (6.22a):
The wing mass derived by using the coefficients c_I and c_5 should be factored by about:

$$0.5\left[1+0.8\left\{\left(1-5\beta^2\right)+\left(1-3\beta\right)\lambda\right\}\right] \qquad \text{(AD4.2a)}$$

(for values of β other than 0.1 and values of λ other than 0.45).

ii) Method of paragraph 4.2.1.2:
The ideal structural mass given by Eq (AD4.1a) should be factored by:

$$1.13[(1-5\beta^2) - 0.0027(1+43\beta)\lambda] \qquad \text{(AD4.2b)}$$

(for values of β other than 0.1 and values of λ other than 0.45).

No change of the other terms is required.

AD4.2.1.4 Wing mass when there is no carry through structure

In some designs the wings are attached to the sides of the body and the fuselage shell is used to give spanwise continuity. In this case the wing mass, M_W, is reduced by the factor $(1 - \beta)$ approximately, but there is a corresponding increase of fuselage structure mass, see paragraph AD4.2.2.5.

AD4.2.2 Fuselage

AD4.2.2.1 General

The contribution of the fuselage to the mass of an aircraft is usually comparable to that of the wing. In general, however, it is more difficult to achieve an accurate prediction due to the wide variety of detail differences which may be present.

The most important parameter is the surface area of the fuselage structure and it is necessary to estimate this as accurately as possible. This is not unduly difficult when the cross-section varies more or less continuously along the length as it does, for example, on transport types. It can be difficult in other cases, for example military combat aircraft.

Apart from surface area an important criterion is whether or not the greater part of the fuselage is pressurised. The simple formulae for fuselage mass given in paragraph 6.4.2.2 of Chapter 6 allows for this, Eq (6.20a) relating to pressurised fuselages and Eq (6.20b) to essentially unpressurised configurations. The surface area is also included in these formulae by virtue of the presence of the length L, breadth B and height H, parameters. The coefficients, c_2, given in Table 6.6 of Chapter 6 are used to cover the effect of major variations in configuration. The following formulae provide a basis for more accurate estimation of fuselage mass but do require a knowledge of more details of the fuselage layout in order to be used.

Other than the effects discussed in paragraph AD4.2.2.5 it is assumed that the mass of wing carry through structure is part of the wing mass, and it is not included in fuselage mass.

AD4.2.2.2 Fuselage structure surface area

A reasonably general representation of a fuselage layout is shown in Figure AD4.1. While this represents the plan view of the fuselage the implied cross-section changes may be related also to the side elevation.

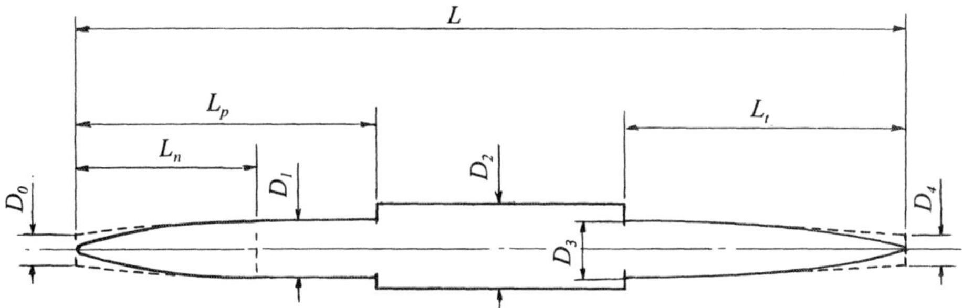

Figure AD4.1 Idealised representation of fuselage dimensions

The lengthwise stations are defined as follows:

- L_n is the distance aft of the nose at which the nose shape effectively blends into a more or less parallel portion.
- L_P is a distance aft of the nose where a cross-section discontinuity can occur, say due to air intakes or a canopy.
- L_t is the distance forward from the aft end of the fuselage over which the cross-section fairs down from a nominally constant value and it can start at a discontinuity.
- L is the overall length (see note * below at Eq (AD4.4c)).

The region between the cross-sections defined by L_p and L_t may be parallel or can vary approximately linearly.

The cross-sections at the various datum points are defined in terms of an equivalent diameter, D_i, where:

$$D_i = k_i\,(B_i + H_i)/2 \qquad \text{(AD4.3)}$$

k_i is a coefficient which depends upon the shape of the cross-section and varies in the range of 1 to 1.27. The lowest value is for a circular/elliptical shape, and the highest for a nominally rectangular one.

B_i and H_i are the local width and height respectively.

D_0 is the equivalent diameter at the nose, usually zero unless a pitot intake is employed, or there is a nose-mounted powerplant.

D_1 is the value where the nose nominally ends, at L_n.

D_2 is the maximum of the values at L_p or $(L - L_t)$ and in practice is the maximum equivalent diameter.

D_3 is the value at L_t.

D_4 is the value at the tail and is often zero.

For convenience the following non-dimensional parameters are introduced:

$$f_n = L_n / D_1$$
$$f_b = (L_p - L_n) / D_1$$
$$f_c = (L - L_t - L_P) / D_2$$
$$f_t = L_t / D_2$$

The structural surface area of the fuselage, S_f, is:

$$S_f = \pi D_1^2 \left[f_b + f_c \left(\frac{D_2}{D_1} \right)^2 + N_S + T_S \right] \quad \text{m}^2 \qquad \text{(AD4.4a)}$$

N_S is a function of the area of the nose:-

For a pointed nose: $N_S = 0.325\,(4f_n^2 + 1)^{0.5}$ (AD4.4b)

For a truncated nose, as with a pitot air intake:

$$N_S = 0.667\, f_n \left(1 + \frac{D_0}{D_1} \right) \qquad \text{(AD4.4c)}$$

*A special case is that of an aircraft where there is a fuselage nose-located powerplant. In this case it is most satisfactory for the fuselage length to be considered as beginning at the bulkhead which denotes the aft end of the powerplant and the nose function, N_S, is then given by Eq (AD4.4c).

T_S is a function of the area of the tail:

For a pointed tail:

$$T_S = 0.322\left(\frac{D_3}{D_1}\right)^2\left(4 f_t^{\,2}+1\right)^{0.5} \qquad \text{(AD4.4d)}$$

For a truncated tail:

$$T_S = k_5\, f_t\left(1+\frac{D_4}{D_3}\right)\left(\frac{D_3}{D_1}\right)^2 \qquad \text{(AD4.4e)}$$

where $k_5 = 0.6$ when D_4 is of the same order as D_3 as when it represents the engine exhaust area.

$k_5 = 0.66$ when D_4 is much less than D_3, as on a general aviation aircraft where the rear fuselage merges into a rudder.

A special case is that of many airliners:

where $D_0 = D_4 = 0$

and $D_1 = D_2 = D_3 = D$, say,

and the surface area becomes, with minor adjustments:

$$S_f = \pi\, D^2\left[\frac{L}{D} - (f_n + f_t) + 0.35\left(4 f_n^2 + 1\right)^{0.5} + 0.34\left(4 f_t^2 + 1\right)^{0.5}\right] \ \text{m}^2 \qquad \text{(AD4.4f)}$$

AD4.2.2.3 Pressurised fuselages

Although the structure of a fuselage is designed by numerous loading cases when the greater part of it is pressurised this is likely to be the dominant consideration. This is especially so for larger cross- sections, such as on wide body airliners. Using the internal pressure as a basis for design results in the fuselage mass being given approximately as:

$$M_f = k_6\left[1+(3.12-0.354\,B)\frac{2\bar{\sigma}}{(1+\bar{\sigma})}\right]\left[\frac{3.56\,p}{\bar{\sigma}^{0.75}}S_f\,B\right] kg \qquad \text{(AD4.5)}$$

where:

S_f is the fuselage surface area given by Eq (AD4.4), m^2

B is maximum fuselage breadth, m

p is the pressure differential under normal operating conditions (as a fraction of atmospheric pressure), bar

$\bar{\sigma}$ is a measure of the corresponding nominal tensile working stress in the fuselage as a fraction of 100 MN/m^2 and may be as much as unity or slightly higher but it is usually somewhat less. In the absence of more accurate information it is suggested that for:

$D_2 = D \leq 2$ m	$\bar{\sigma} = 0.8$
2 m $\leq D \leq$ 6 m	$\bar{\sigma} = 0.8 + 0.05(D-2)$
$D \geq 6$ m	$\bar{\sigma} = 1.0$

k_6 is a coefficient which depends upon the class of aircraft and certain design features. Basically k_6 is unity for a passenger airliner with wing-mounted main landing gear units. The following increments should be used as appropriate:

Two engines mounted on rear fuselage; $k_6 = 0.013$

All main landing gear units mounted on fuselage; $k_6 = 0.05$

Freight aircraft with large, especially ramp, loading door; $k_6 = 0.5(B/7)^2$

Fuselage with no windows and few doors; $k_6 = -0.1$

The equation for fuselage mass estimation given in Chapter 6 at Eq (6.20a) is based on the above equations (AD4.4f) and (AD6.5) with typical values of f_n, f_t and $\bar{\sigma}$. Differences in these parameters and the design features are reflected in the values of c_2 given in Table 6.6 of Chapter 6.

AD4.2.2.4 Basically unpressurised fuselages

It is assumed here that military combat and other types of aircraft where the pressurised volume is small in comparison with the total volume come into the category of unpressurised fuselages. For this class of fuselage it is assumed that the major design criteria may be represented by a simple consideration of the rear fuselage bending together with the design speed, V_D. The fuselage mass is given by:

$$M_f = 0.044 k_7 V_D^{0.74} \left(\frac{L_T}{2D_2} \right)^{0.5} \left[S_f^{0.07} + \frac{0.22 S_f^{0.45} N^{0.32}}{V_D^{0.35}} \right] S_f \quad \text{kg} \qquad \text{(AD4.6a)}$$

V_D is the design speed, see Chapter 7, paragraph 7.2.2, m/s

L_T is the tail arm length, m; this is defined as the nominal distance between the wing and tail quarter mean chord points, or wing and fin for a tailless design

S_f is the fuselage surface area as defined by the relevant parts of Eq (AD4.4), m^2

N is the design ultimate manoeuvre factor

k_7 is a somewhat complex coefficient which depends both on the class of aircraft and various design features

$$k_7 = k_8 + 0.2\epsilon + 0.4\alpha + k_9\delta + 2.7k_{10} + k_{11}(n-1) + 0.1\eta + 0.3\omega \qquad \text{(AD4.6b)}$$

n is the number of seats providing $n \leq 4$, but otherwise zero

ϵ, α, δ, η and ω are unity when the following conditions apply, but otherwise zero

ϵ the engines are buried in fuselage (as opposed to a nose-mounted propeller engine where L is taken aft of engine bulkhead, see paragraph AD4.2.2.2)

α air intakes are in fuselage (as opposed to wing root for buried engines)

δ main landing gear impinges on fuselage, see k_9

η aircraft is designed for naval operations

ω wing is strut braced to the fuselage

k_8 = 1.8 for all military aircraft except small, light types, or

k_8 = 2.0 for twin engine general aviation and feeder aircraft (wing-mounted engines), or

k_8 = 2.0 + 1.5 (20 – S_f) /S_f for light single engine types where $S_f \leq 20$ m^2

k_9 = 0.35 when the main landing gear is wing mounted but retracts into the fuselage, or

k_9 = 0.7 when the main gear is both mounted on and retracts into the fuselage

k_{10} is the ratio of the length of any weapons bay to the overall length of the fuselage

k_{11} = 0.05 for military aircraft, except small light types, or
= 0.2 for light aircraft seating up to 4, or
= 0 for other types of aircraft where $n > 4$

Chapter 6, Eq (6.20b) together with the relevant values of coefficient c_2 from Table 6.6 of Chapter 6 is based on a simplified analysis which uses typical values for the various types of aircraft.

AD4.2.2.5 Mass of wing carry through structure or equivalent

If, for purposes of comparison, it is required to adjust the fuselage mass as given by Eqs (AD4.5) or (AD4.6) to allow for the additional mass of wing carry through structure, reference should be made to paragraph AD4.2.1.4. The mass of the carry through structure is approximately (βM_W), where M_W is either derived from paragraph AD4.2.1.2 (**D**) or from Chapter 6, Eq (6.22a).

When the wing terminates structurally at the sides of the fuselage and the fuselage structure provides the continuity, the fuselage mass derived from Eqs (AD4.6) should be increased by (βM_W), the wing mass being correspondingly reduced, see paragraph AD4. 2.1.4.

AD4.2.2.6. Twin boom layout

Evidence suggests that the total fuselage/boom mass of a twin boom configuration is not greatly different to that of a comparable conventional fuselage. The mass of the basic fuselage can be estimated from Eqs (AD4.5) or (AD4.6a) as appropriate replacing L_T in Eq (AD4.6a) by the length of the fuselage aft of the wing quarter mean chord point. The mass of the booms is of the order of:

$$M_b = 0.25 V_D S_b^{0.75} \tag{AD4.7}$$

where S_b is the total surface area of the boom, m^2.

AD4.2.3 Empennage

AD4.2.3.1 General

Although the total mass of the empennage is often less than 3% of the take-off mass of the aircraft it is of importance in determining the longitudinal position of the centre of gravity. In practice there can be wide variations in the masses of both the horizontal and vertical tail surfaces due to, among other things, differing fuselage layouts and hence tail arms. For convenience of initial mass prediction and synthesis the empennage mass is combined with that of the wing in Chapter 6, paragraph 6.4.3.1 but the contribution of the empennage can be isolated by reference to Table 6.10 of Chapter 6 through the coefficient c_5.

The variation of tail arm mentioned above is reflected in the size of the empennage surfaces and it is found that the two most significant parameters in determining their masses are the surface area and design speed, V_D.

AD4.2.3.2 Horizontal empennage mass

The mass of a tailplane or foreplane may be estimated by:

$$M_H = 0.047\, V_D\, S_H^{\,1.24} \qquad kg \qquad \text{(AD4.8)}$$

where S_H is the plan area of the surface. It should be taken as the area outside the fuselage when the surface consists of two separate parts but otherwise the total plan area should be used.

Typical horizontal tail surface masses lie in the range 0.9 to 1.75% of the overall aircraft mass.

AD4.2.3.3 Vertical empennage mass

The mass of the fin may be estimated as:

$$M_V = 0.065\, k_{12}\, V_D\, S_V^{\,1.15} \qquad kg \qquad \text{(AD4.9)}$$

where S_V is the side area of the fin outside the lines of the fuselage. Care should be taken in evaluating S_V in those cases where the fin/rudder merges into the rear fuselage to ensure that the definition is consistent with the fuselage geometry, see paragraph AD4.2.2.2, Eq (AD4.4e).

k_{12} is dependent upon the vertical location of the horizontal surface relative to the fin. It is unity when the tailplane is not mounted on the fin and 1.5 for a true "T" tail, with appropriate variation between these two extremes.

The mass of the fin usually lies in the range 0.5 to 1.0 % of the all up mass, but can rise to up to 1.2% for a "T" tail configuration.

Table AD4.2 Typical empennage mass fractions

(Relative to aircraft total mass)

Aircraft	Mass range %
Transport and related types with conventional low tail	1.5 to 2.8
Transport and executive aircraft with T tail	2.0 to 3.2
Single-engine light aircraft	2.0 to 3.5
Twin-engine general aviation aircraft	1.9 to 2.4
Interceptors and strike aircraft	1.6 to 3.0
Large bombers	1.5 to 2.4
Tailless configurations	0.6 to 1.6

AD4.2.3.4 Total empennage mass

There is a tendency for total empennage mass predictions to give better correlation with actual values than that of the individual components. This is probably due to differences in the definition of interacting items.

Typical values of the total empennage mass for various configurations are given in Table AD4.2.

The ratio of the vertical to horizontal mass components is normally in the range of 0.5 to 0.85 but fin mass may be greater than the horizontal surface mass for interceptor aircraft which tend to require large vertical surface area.

AD4.2.4 Landing gear

Table AD4.3 gives typical ranges of values of landing gear mass as a fraction of take-off mass for different categories of aircraft. These values are for the complete landing gear. The nosewheel usually contributes about 0.5% of the take-off mass, but it is higher, sometimes greater than 1%, for general aviation and naval aircraft.

The basic mass prediction used in Chapter 6, paragraph 6.4.3.3 effectively assumes a landing gear mass of 4% of the aircraft mass but for convenience includes it within the total systems and equipment item. Reference to Table AD4.3 shows that while 4% is a reasonable mean value there are circumstances where a revised value may be used to give greater accuracy.

More accurate landing gear mass prediction requires the availability of details of the landing gear geometry, dimensions, tyre and brake details and basic loading. Although much of this information arises from the procedures outlined in Addendum 1, there is little merit in using it until a completely detailed mass estimate can be undertaken.

Table AD4.3 Landing gear mass fractions

(Relative to aircraft total mass)

Aircraft	Mass range %
Transport, executive and bomber aircraft with two main landing gear units	3.5 to 4.0
Short field length tactical transports	4.0 to 4.5
Transport aircraft with four main gear units	4.5 to 5.3
Light aircraft with nosewheel layout	5.0 to 6.5
Primary/basic trainers with nosewheel layout	6.0
Light aircraft with tail wheel layout	4.5 to 5.0
Land based military aircraft, other than large bombers	4.0 to 4.5
Naval aircraft	5.0 to 6.0

AD4.2.5 Powerplant related structure

It is usual to include such items as basic nacelle structure and engine pylons as part of the total structure mass. The magnitude of these items is very variable, partly because of the different forms of powerplant possible and partly because of the difficulty of distinguishing between powerplant structure and powerplant installation. For buried engine installations the penalty is small, usually less than 0.5% of the total aircraft mass. Typical values for wing-located engines vary in the range of 1 to 2% of the total mass but it can be as high as 4% for certain multi-engined general aviation types.

Because of the variability it is convenient to include these items within a total powerplant installation allowance and this has been done in Chapter 6, paragraph 6.4.3.2 and Table 6.8, see also the next paragraph.

AD4.3 Powerplant installation

The usual definition of the powerplant installation is that it includes the following items as a minimum:

Accessories needed to enable the engine to functions
Internal pipe work, ducts, baffles, etc.,
Basic engine,
External removable panels and cowlings,
Propeller, where relevant.

In some mass breakdowns it also includes the fuel and oil systems, see paragraph AD4.4.2. For the present purposes the definition is further extended to cover the fixed structure associated with the powerplant installation, as discussed in paragraph AD4.2.5.

Because of the many small items involved it is best to base the mass prediction of the installation on the basic mass of the engine multiplied by an appropriate factor. Typical factors for various powerplants and installations are given in Table 6.8 of Chapter 6. Until the design of the installation is complete there is little point in a more detailed analysis.

AD4.4 Systems, equipment and furnishings

AD4.4.1 General

This group of components is intended to cover all the items not included in the structures group, paragraph AD4.2 and the powerplant group, paragraph AD4.3. In total it can contribute a large part of the overall mass of the aircraft, possibly in excess of 20% and only rarely less than 10%. The group consists of several basic systems and equipment which consist of many small components and which interact with one another. Table 6.9 of Chapter 6 represents an attempt to simplify this complex issue for the purpose of the design synthesis process. The landing gear mass was included for convenience. However, it is usually required to use a more specific approach during the analysis phase of the design. Table AD4.4 provides relevant data for the major contributions to this item. It must be noted that there are many examples which fall outside the ranges of values given. Reference should be made to the following paragraphs when interpreting Table AD4.4.

AD4.4.2 Fuel system

As mentioned in paragraph AD4.3, the fuel system is considered as part of the powerplant installation here and is not included in the items which go to make up the values quoted in Table 6.9 of Chapter 6. However, since the fuel system is sometimes included in the total system mass, relevant data are given in Table AD4.4. Note that these contributions are already covered by Table 6.8 of Chapter 6 and if the powerplant factors quoted there are used no further allowance is required. Military combat aircraft tend to have a high fuel system mass partly because the system is complex and partly because of the provision of tank protection against combat damage.

AD4.4.3 Flying control system

When the control surfaces are operated directly by pilot effort the definition of the flying control system is straightforward. A problem arises when powered actuation is employed since the actuators may be included either as part of the control item or part of the power supplies, often hydraulic. For this reason TableAD4.4 quotes the sum of the flying control and hydraulics systems which may be more useful, especially for military combat aircraft.

AD4.4.4 Power supply systems

AD4.4.4.1 General comments

The power required to operate services on an aircraft is usually provided hydraulically and electrically although pneumatic systems do have some application. It has long been recognised that there are advantages in using only one power supply source, which, because of avionics, has to be electric. Thus electric actuating systems are used on general aviation aircraft and military trainer types. However, the advantages of high pressure hydraulic systems has meant that in the past they have been preferred for situations where large power and rapid response is required. Attempts are continually being made to derive electrically powered systems that can economically meet these requirements, especially using hybrid, electro-hydraulic units. "More" electric aircraft are gaining favour.

Table AD4.4 Percentage contributions of systems and equipment

AIRCRAFT	Fuel System	Fly Cont.	Hyd. Pneu.	FC + Hyd.	Electric System	Instru-ments	Avionics ***	Elec + In + Avionics	Air Cond.	De-ice	Air. Cond. + De-ice	Furnish-ings
Transport - long haul	0.4/0.7	1.0	0.8	1.6/2.4	0.5/2.0	0.2/0.5	0.4/1.5	2.0/3.0	0.6/1.6	0/0.5	1.0/1.7	4.0/6.0
- short haul	1.0/2.0	1.0	0.8	1.6/2.5	1.5/3.0	0.4/0.7	1.5/3.0	2.0/3.0	1.0/1.8	0.7	2.0/2.5	4.0/7.0
- freight	0.4/2.0	1.0	0.8	1.6/2.4	1.0/1.5	0.4/1.5	0.5/1.5	2.0/3.0	0.6	0.4	0.6/1.0	2.0/4.0
Executive	1.0/2.0	1.3/2.0	1.0	2.0/3.0	2.5/5.0	0.4/0.6	1.5/2.8	5.0/8.0	1.4/2.4	0.8	2.0/3.0	6.0/12.0
Single engine gen. aviation	2.5	2.0/3.0	0**	2.5	1.5/4.0	0.5/2.0	0.5/1.0	2.5/5.0	0	0	1.0	1.5/4.0
Multi-eng. G.A. - fixed gear	2.0	1.0/2.0	0.3	1.3/2.0	1.0/2.5	1.0/1.5	1.0/2.5	3.0/5.0	1.0/2.0	0.7	2.0/2.5	3.0/6.0
- retract gear	2.0	1.5/2.0	0**	1.7	2.0/3.0	1.0/1.5	1.0/2.5	4.0/5.0	1.0/2.0	0.7	2.0/3.0	3.0/6.0
Military trainers	1.0/2.0	2.0/3.0	0**	2.0/3.0	4.0/6.0	2.0	3.0/3.5	9.0/11.0	0/3.5	0/0.5	3.0/4.0	2.5/3.5
Strike/interceptors	1.8/3.0	2.0/3.0	2.0/4.0*	4.0/5.0x	1.0/2.0	0.2/1.0	1.0/2.0	3.0/5.0	0.4/1.4	0/0.2	0.6/1.4	0.5/1.0
Bombers	1.0/2.0	0.8/1.5	0.8	1.6/2.4	1.0/3.5	0.2/0.4	0.3/1.0	1.8/3.5	0.8/1.0	0/0.2	1.0	0.5/1.0

***Special installations excluded **assumes all electric actuation *U.S aircraft 1.0 x U.S aircraft 2.0/4.0

AD4.4.4.2 Hydraulic and pneumatic systems
Hydraulic/pneumatic systems typically contribute about 1% of the total mass, or rather less. However, factual data can be confusing when powered flying control systems are used, as discussed in paragraph AD4.4.3.

AD4.4.4.3. Electrical systems
As is to be expected electrical systems are relatively heavy when electrical power is used for primary actuation. A difficulty in analysing the mass of electrical systems is the lack of clear definition of the distinction between electrics, instruments and avionics. It is frequently most satisfactory to use a total value until such time as a detail specification of the relevant components is available. For this reason Table AD4.4 does include guidance in this respect.

AD4.4.4.4 Auxiliary power units and accessory drives
Items not included in Table AD4. 4 include auxiliary power units where fitted and the mass of accessory drive gear boxes. The latter has been conveniently covered in the powerplant installation factor, although really it is part of the power system. Typically an auxiliary power unit contributes about 0.3% of the total mass, but it can be higher in the case of smaller aircraft, such as executive types.

AD4.4.5 Instruments

In addition to the cockpit panel instruments and the supporting devices the instruments item often includes auto-pilot and similar systems. In certain cases the instrument mass, being relatively small, is combined with electrics or avionics.

AD4.4.6 Avionics (electronic) systems

This item can be very variable, especially for light aircraft and military combat types, depending upon the level of equipment fitted. In the case of civil transport aircraft intended to operate on a world-wide basis a more or less standard avionics fit is required. It is likely to have a mass in the region of 1000 to 1500 kg for a long range aircraft, but, perhaps, half of this for locally operated short haul types.

Special military avionics fits, such as for anti-submarine work, are in reality a major part of the payload and can account for of the order of 10% of the total mass.

AD4.4.7 Environmental control systems

AD4.4.7.1 General

Air conditioning and pressurisation, where fitted, are effectively part of one system. It is usual for gas turbine powered aircraft to use engine compressor offtakes to provide the required air. Hot air, and sometimes pneumatic, de-icing systems use the same air source. Because of this it is often convenient to treat both air conditioning and de-icing systems as one system under the general description of environmental control. Table AD4.4 gives an indication of the possible total environmental system mass as well as the individual contributions.

AD4.4.7.2 Air conditioning, pressurisation and oxygen

Although Table AD4.4 gives air conditioning mass as a percentage of total aircraft mass it is to be expected that in practice it is more directly a function of the number of occupants for which the aircraft has been designed.

The approximate mass of the air conditioning system is given by the following equations, where n is the maximum number of occupants for which the aircraft has been designed.

i) Pressurised airliners and executive aircraft; subsonic:

$$(4n + 120)\ \text{kg} \qquad \text{(AD4.10a)}$$

ii) Short haul pressurised feeder line aircraft:

$$(4n + 60)\ \text{kg} \qquad \text{(AD4.10b)}$$

iii) Unpressurised feeder line aircraft:

$$(2n + 20)\ \text{kg} \qquad \text{(AD4.10c)}$$

iv) Light general aviation aircraft, allow 1 kg for each occupant

v) Military trainers, allow 30 kg for each occupant

vi) Military combat aircraft, inclusive of oxygen provision, 150 kg per occupant is typical

vii) Supersonic long range cruise aircraft:

$$(12n + 400)\ \text{kg} \qquad \text{(AD4.10d)}$$

AD4.4.7.3 De-icing

De-icing provision varies considerably and gives the highest penalty on smaller passenger carrying aircraft. As mentioned in paragraph AD4.4.7.1 hot air de-icing is usual for aircraft powered by gas turbines, but there are exceptions, for example where the penalty of long hot air ducts is greater than alternative de-icing techniques. Electric thermal de-icing introduces significant power requirements but forms of electrical impulse de-icing may become practical and could result in reduction in de-icing system mass.

AD4.4.8 Furnishings

The decision of the standard of furnishings rests with the operator but it is possible to give some indication of the usual mass values for this component. TableAD4.4 gives furnishing mass as a percentage of take off-mass for various classes of aircraft but, as is the case with air conditioning, probably a better parameter is the number of occupants.

Table AD4.5 Furnishings mass

(For each occupant)

Aircraft	Furnishings mass Per occupant - kg
Long range passenger	40
Short haul passenger	25
Feeder line types	15
Executive aircraft	up to 100
Light general aviation	6 to 10
Military trainers	30 to 65*
Combat aircraft	at least 100*

* Assumes use of ejection seats

Table AD4.5 provides this comparison. The mass contribution for passenger aircraft is seen to be significant. Furnishings includes seats, trim, false bulkheads, galleys, etc. and possibly toilets and certain components of air conditioning systems.

AD4.4.9 Armament and crew protection

Most military combat aircraft carry some level of fixed armament, such as guns or bomb carriers. Some combat aircraft also incorporate armour protection for the crew. While there can be considerable variation here the allowance is typically of the order of 1% of the total mass, but not infrequently it is as high as 2.5%.

AD4.4.10 Miscellaneous items

Almost invariably there are a number of items which do not clearly fall in any one of the categories so far discussed. Among these may be mentioned the external paint. However, it should be possible to allocate most items to one of the given categories and the provision for miscellaneous components is likely to be less than 0.5% of the total.

AD4.5 Total aircraft mass

AD4.5.1 Empty mass

In the present context the basic empty mass of the aircraft is considered to be the sum of all those components covered in paragraphs AD4.2 to AD4.4 inclusive. This definition is not the only accepted one, for example the furnishings and other operator specified items may not be included here, but placed with the operational items discussed in paragraph AD4.5.2.

AD4.5.2 Operating empty mass (OEM)

The operating empty mass is defined here as the basic empty mass plus those items necessary for the aircraft to be operational. Thus these additional operational items include crew, crew equipment, food, water and safety equipment. Typical allowances for this are given in Chapter 6, paragraph 6.4.2.3.

AD4.5.3 Disposable mass

The disposable mass is the sum of the payload and fuel appropriate to a given mission. It is possible to exchange allowances between these two items within the volumetric restrictions of payload and fuel capacity.

AD4.5.4 All up mass and take-off mass (AUM and TOM)

The all up mass is simply the sum of the operating empty mass and the disposable load. For most purposes this is the same as the take-off mass. On some large aircraft allowance is made for fuel used up to take-off, so take-off mass may be slightly less than the total, or ramp, value.

Addendum 5

Examples of the synthesis procedure

AD5.1 Introduction

Chapter 3 outlines the close interrelation between the flight regime and the class of powerplant. From this it follows that the derivation of the spreadsheets used for the first stage of the synthesis procedure is primarily dependent upon the powerplant model. The following main categories can be established:

a) Piston engine/propeller. This is applicable to relatively low speed flight in which compressibility effects are small, if not negligible.

b) Turbopropeller. Used for moderate subsonic flight speeds, the performance of turbopropeller equipped aircraft is affected to some degree by compressibility effects, especially under cruise conditions.

c) High bypass ratio fan. Applicable for flight up to high subsonic speeds where compressibility effects are of major significance. This class of aircraft is typified by transport and business jets.

d) Low bypass ratio engine - subsonic. Compressibility effects are of significance for this application, which includes certain military combat and trainer types, and may involve operation at very high altitude.

e) Low bypass ratio engine - supersonic. It is probable that reheat will be used to augment transonic and supersonic engine performance in this application.

Category (c) above has already been adequately covered by the airliner used to demonstrate the typical application of the synthesis process in Chapter 8. Whilst it may be argued that category (d) is really a special case of category (e) where the flight regime is limited to subsonic speeds, the complexity of the latter is such as to make the former worthy of individual consideration. The examples which follow cover categories (a), (b), (d) and (e) and appropriate spreadsheets have been formulated so that they may be readily adapted to various types of aircraft which use the given classes of powerplant.

AD5.2 Piston engined, two seat, aerobatic trainer

AD5.2.1 Introduction

The early biplane primary training aircraft were replaced by monoplanes during the World War 2 period. The tandem seating arrangement was, however, retained. Certain of these two seat monoplanes were developed into competition aerobatic aircraft but were themselves replaced by more compact, higher performance, single seat aircraft. At the same time primary trainers developed into side-by-side seat configurations as this gave simpler instruments and controls and easier communication between the instructor and pupil. More advanced military turboprop basic trainers have retained the tandem seat arrangement since this is more representative of combat aircraft.

Side-by-side seat primary trainers are not really suitable for advanced aerobatic training due to the displacement of the pilot off the centreline of the aircraft. Usually they possess only limited inverted flight capability. Hence the example chosen to demonstrate the application of piston engine characteristics is a tandem seat trainer. Unlimited inverted flight capability requires the use of a special engine oil system as well as basic aerodynamic considerations. It may be regarded as an updated version of the earlier monoplane trainers.

AD5.2.2 Specification

The relatively simple specification for this aircraft follows:

a) Two seat tandem arrangement. The rear pilot should be located over the region of the wing trailing edge and the front pilot near to the centre of gravity. The rear cockpit is the primary pilot station.

b) Unsupercharged piston engine driving a variable pitch propeller. (Past experience suggests that the power required is likely to be in the region of 250 KW.)

c) Fixed, tricycle, landing gear. (Fixed for simplicity and tricycle rather than tailwheel layout to facilitate ground operations, especially landing.)

d) Standard aerobatic limit normal manoeuvre accelerations of +6g and –3g.

e) Unlimited upright and inverted manoeuvres, including spinning.

f) Aileron induced roll rate of about 180°/ s at maximum speed.

g) Maximum speed in level flight 72 m/s at cruise altitude.

h) Cruise speed 55 m/s at 915 m altitude (3000 ft) (nominally 200 km/h)

i) Mean rate of climb to cruise altitude at least 10 m/s.

j) Manoeuvre speed, V_A, 70 m/s and design speed, V_D, of 120 m/s.

k) Maximum stall speed in landing configuration 25 m/s, and landing approach at $1.25V_{STALL}$.

l) Range 930 km.

m) Take-off length, factored, 440 m.

n) Landing length, factored, 600 m.

o) Ceiling, at least 4000 m.

AD5.2.3 Development of the configuration

Aircraft in this category almost invariably have a low wing. The reasons for this include:

a) Good upwards view, which is important in aerobatic manoeuvres.

b) Ease of access to the cockpits and egress in emergency.

c) Minimal blanking of the tail by the wing in high nose down attitude, as in a spin.

d) Ease of mounting the main landing gear.

e) Some degree of protection in ground accidents, including less probability of ground loops and overturning.

Large, powerful, ailerons are necessary for roll control. This suggests that the wing should not be too highly tapered to ensure adequate aileron chord at the tip, a taper ratio of 0.4 being typical. Further a small wing quarter chord sweep of, say, about 8.5°, enables the aileron hinge to be conveniently more or less unswept. However, the large ailerons do result in reduced wing span being available for the trailing edge flaps. This is not serious as the high manoeuvre requirement results in a relatively low wing loading. The main purpose of the flaps is to reduce aircraft nose up attitude during the approach to landing.

The need for both upright and inverted manoeuvre capability suggests that the aerofoil section should be close to a symmetric form although a small amount of camber should be acceptable. For the same reason the wing setting on the body should be small to avoid unduly high body incidence in inverted flight, but a consequence of this is a high fuselage attitude in low speed upright flight unless flaps are extended.

As it is highly desirable to pass the primary wing structure below the cockpit floor it is helpful to place a limit of, say 0.15, on the aerofoil thickness to chord ratio.

Spin recovery suggests that the fin should be located longitudinally such that it is not blanketed by the horizontal tail in the spinning attitude. Ideally this is achieved by locating the fin just forward of the horizontal tail but the need for powerful yaw control usually results in it being located aft of the tailplane with the latter mounted high on the fuselage.

The fuselage layout follows directly from the stated requirements and configuration. The two basic modules are the powerplant and the crew/cockpit. These are shown in Figure AD5.1 which is the preliminary layout of the fuselage. Points worthy of comment are:

i) The somewhat raised rear cockpit to facilitate the view of the rear pilot.
ii) Provision for an "aerobatic" fuel tank just behind the engine bulkhead. This would include a device to ensure continuity of fuel feed in all aircraft attitudes.
iii) Location of nose landing gear support structure from the bottom of the engine mounting.
iv) Access to the cockpits by separate sliding canopies located on common rails. This is a preferred, if heavy, solution. The alternatives of side opening or separate canopies hinged at their rear end cannot be opened in flight and must be jettisoned for emergency egress. Sliding canopies may be opened for ventilation and direct vision in adverse weather conditions.

It is intended that the greater part of the fuel will be carried in tanks located in the inboard wing leading edges.

It is visualised that significant use would be made of reinforced plastics in the construction of the airframe. Most likely the fuselage would consist largely of glass reinforced materials with local carbon fibre reinforcement. On the other hand it may be expected that carbon-fibre reinforced material will form the basis of the lifting surface construction, glass fibre reinforcement possibly being adequate for the control surfaces.

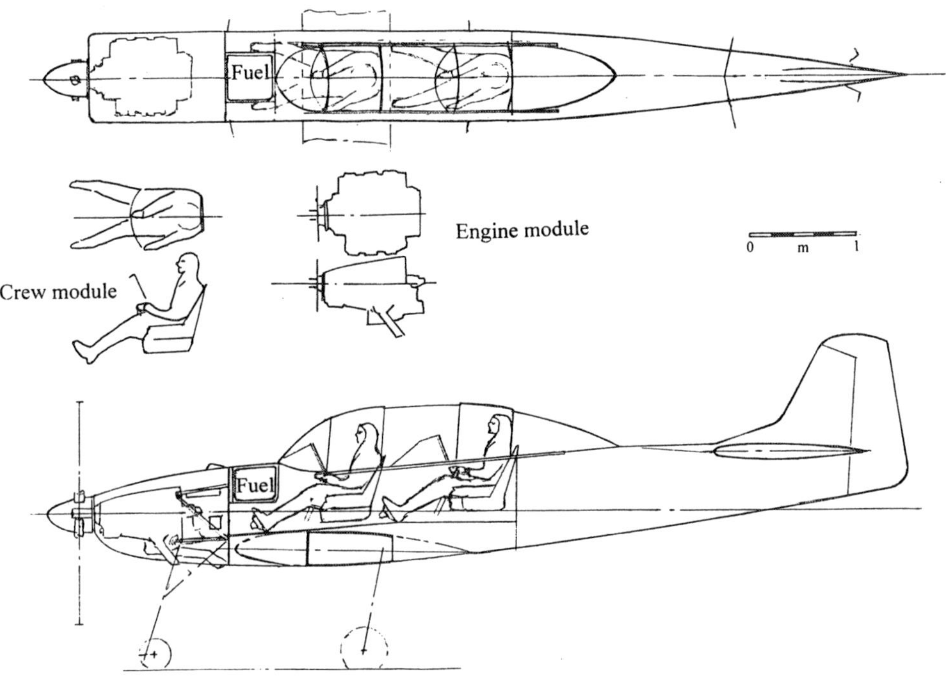

Figure AD5.1 Two seat aerobatic trainer - fuselage layout

AD5.2.4 Spreadsheet formulation (SPREADSHEET AD5 .1)

AD5.2.4.1 Initial inputs and assumptions

The basic parameters selected for analysis are aspect ratio, A, and thickness to chord ratio, t/c. Although not relevant in this example the number of engines, N_E, is provided for since this is necessary in conversion between thrust and power for piston engines. The Requirements are derived from the specification.

Assumed values

M1/Mo Ratio of start of climb mass to take-off values, taken to be 0.99 (Ch.7, para 7.3.6)

S^-0.1 Wing area parameter, initially taken to be 0.76 for trainers (Ch .6, Figure 6.1), but subject to subsequent correction

Rw	Wetted area factor taken to be 4 for a single propeller engine aircraft (Ch. 6, Table 6.3)
Type Fac.	Assumed to be 2.25 (Ch.6, Table 6.4)
TE flap	Single slotted flap assumed, to be used only for approach and landing
SS alpha SS gamma	Second segment climb coefficients, not relevant for the single engine case [See Ch.7, Eq (7.4a)]
Flap Fac	Flap drag factor, single slotted flap here (Ch.6, para 6.3.2.2)
a Cruise	Speed of sound at cruise altitude
(T/Mg)ass	Assumed static thrust to weight ratio for initial calculation (Ch.3, Table 6.3B suggests 0.35 to 0.55 for specialist propeller aerobatic aircraft) (note that the value is corrected during analysis).
Tan gam d	Tangent of descent angle in approach to land [Ch.7, Eq (7.6)]
Mu G	Mean stopping deceleration [Ch.7, Eq (7.6)]
Sigma ceil	Relative density at ceiling (4000m assumed here)
Cl Fac Z	The factor relating climb speed to datum value (Ch.7, para 7.6.2.3) (value can be corrected subsequently).

AD5.2.4.2 Initial calculations

Ml/Mo	Ratio of landing to take-off mass (Ch.7, Table 7.1)
Mcr/Mo	Mass ratio at start of cruise (derived from M1/Mo as given by factor in Ch.7, Table 7.2).
(Clm)o ae	Basic aerofoil maximum lift coefficient (Ch.6, para 6.2.4.2)
Del flp (TO)	Lift coefficient increment due to flap at take-off setting, zero here.
Del flp (L)	Lift coefficient increment due to flap at landing setting (Ch.6, Table 6.1 suggests 1.0 for single slotted flap, but reduced here to allow for large span ailerons)
(Cl use)o	Usable cruise lift coefficient (Ch.6, para 6.2.4.6)
(Cdz)o	Zero lift drag coefficient at zero Mach number [Ch.6, Eq (6.13a), no laminar flow]
Landg (L)	Landing distance parameter [Ch.7, Eq (6.6a), no reverse thrust]
Man (Cl)o	Basic aerofoil manoeuvre limit lift coefficient, used in conjunction with manoeuvre speed V_A. Assumed to be the same as (Clm)o ae here
(Clm)o TO	Maximum basic aerofoil/flap take-off lift coefficient [(Clm)o ae plus del flp (TO)]
(Cl)us o	Basic unstick lift coefficient, no sweep [Ch.6, Eq (6.4)]
(Cl max)o	Basic maximum aerofoil/flap landing lift coefficient [(Clm) o ae plus del flp (L)]
(Cl)a o	Basic approach lift coefficient, no sweep, [Ch.6, Eq (6.3) based on approach at 1.25 V_{STALL}]

AD5.2.4.3 Preliminary calculations

cos delta Delta deg	Details of wing quarter chord sweepback, derived from the assumed value of 8.5 degrees approximately
Wave Dr F	Wave drag factor [term in Ch.6, Eq (6.13a) raised to power of 20]
(Cdz)cr	Zero lift drag coefficient in cruise [Ch.6, Eq (6.13a)]
(Cd)co	Corrected climb out zero lift drag coefficient [Ch.6, Eq (6.15)]
(Kv)o	Induced drag factor at zero Mach number [Ch.6, Eq (6.14a)]
(Kv)cr	Induced drag factor at cruise Mach number [Ch.6, Eq (6.14a)]
Cl max (L)	Landing maximum lift coefficient, [(Cl max)o corrected for sweep]
Cl us	Unstick lift coefficient [(Cl)us o corrected for sweep]
Cl a	Approach lift coefficient [(Cl)a o corrected for sweep]
Cl use	Usable lift coefficient in cruise [(Cl use)o corrected for sweep]
Va calc	Maximum approach speed consistent with the required landing distance [Ch.7, Eq (6.6b)].
Va	The lower of Va calc and, in this case, $1.25V_{STALL}$
(Mg/S)o ld	Take-off equivalent wing loading (derived from Va, Cl a and (Ml/Mo)
(Mg/S)o gt	Take-off equivalent wing loading to meet gust sensitivity criteria [Ch.5, Eq (5.9b) (not applicable to this case)
Cor L length	Corrected value of landing length if Va is limited by stalling conditions [Ch.6, Eq (6.6a)] (does not apply here)
Man (Mg/S)	Take-off equivalent wing loading to the manoeuvre condition (uses Man Cl and Mcr/Mo, with manoeuvre speed Va)
Man Cl	Available manoeuvre lift coefficient [Man (Cl)o corrected for sweep]

AD5.2.4.4 First stage analysis

The range of take-off wing loading, (Mg/S)o, used for investigation is selected on the basis of past experience and to include (Mg/S)o ld (Ch.5, Table 5.3).

Take-off:

1st app	The first approximation of the thrust to weight ratio required to meet the take-off requirement is calculated using the already assumed value (T/Mg)ass [Ch.7, Eq (7.1b), right-hand side]
(T/Mg)o	The second evaluation uses the first approximation (if acceptable convergence is not obtained the procedure must be repeated)

Acc. Stop:

(T/Mg)o	Thrust to weight ratio needed to meet acceleration-stop requirement in required take-off distance [Ch.7, Eq 7.3a)]

Sec Seg Climb:
Although the second segment climb requirement does not apply to single engined aircraft, as has already been recognised by setting the coefficients to zero, it is included here for completeness.

Tau co	The climb out thrust factor is determined on the assumption that the climb out speed is 1.1 times the unstick speed [derived from Ch.7, Eq (7.4c), without using the terms in (Mg/S) and T_{CO}]
(Cd)co/Cl	The ratio of the equivalent zero lift drag coefficient to lift coefficient in the climb out flap condition
(T/Mg)o	The equivalent take-off thrust to weight ratio [uses Ch.7, Eq (7.4a) in conjunction with Ch.7, Eq (7.4c)]
Y ceil	The altitude factor needed to evaluate constant EAS climb conditions up to the required ceiling [uses σ_2 relevant to ceiling altitude, 4000 m assumed - sigma ceil, with Ch.7, Eq (7.18c)]
Y cr	The altitude factor similar to above but relevant to climb up to cruise altitude of 915 m (uses σ_2 for cruise case)

Mean R/C to Cruise Altitude:

Fact Qv	Factor in climb prediction [Ch.7, Eq (7.15b)]
f (drag)	The climb drag factor [Ch.7, Eq (7.15d)]
1st app	The first approximation of the required thrust to weight ratio (T/Mg) to give the specified mean rate of climb, assuming the factor X to be unity [Ch.7, Eq (7.18e)]
X	Corrected value of X derived by using 1st approximation of (T/Mg)
(T/Mg)1	Corrected value of (T/Mg) using revised value of X [Ch.7, Eq (7.18e)]
Tau Cl	The rate of climb condition thrust factor [uses Ch.7, Eq (7.11e) with appropriate climb speed and the relative density at start of climb]
(T/Mg)o	The equivalent take-off thrust to weight ratio (derived from the same procedure as for the second segment climb)

Landing-rev thrust:

(T/Mg)o	Selected as the maximum value derived from all the other conditions
Landg L	Landing distance parameter [Ch.7, Eq (7.6a), with reverse thrust]
L length	Reverse thrust landing length follows from previous term

End of climb:

Fact Qv	[Ch.7, Eq (7.15b)]
(T/Mg)1	The thrust to weight ratio at the start of climb needed to give a residual rate of climb of 0.5 m/s at the ceiling [Ch.7, Eq (7.19b)]

Tau ceil	The ceiling thrust factor [uses Ch.3, Eq (3.11e) with appropriate values]
(T/Mg)o	Equivalent take-off thrust to weight ratio [derived from the same procedure as the second segment climb]

Cruise:

Cl	The lift coefficient appropriate to start of cruise conditions
Cd	The total drag coefficient (zero lift plus induced) at start of cruise [uses (Cdz)cr, (Kv)cr and Cl]
L/D	Cruise lift to drag ratio (Cl/Cd)
(T/Mg)cr	Cruise thrust to weight ratio (inverse of L/D)
(T/Mg)o	The equivalent take-off thrust to weight ratio [derived from Ch.3, Eq (3.11g) with other relevant parameters]

Max. speed:
The maximum speed analysis follows the same procedure as that used for cruise, with the specified speed and altitude.

Summary:
All the first stage analysis calculations are summarised, together with the manoeuvre, landing and, where relevant, gust sensitivity wing loadings previously derived. This summary is used to produce the diagram of (T/Mg)o vs (Mg/S)o which illustrates the interaction of the various requirements. Inspection of the diagram suggests that the design condition for further examination is that given by the intersection of the climb and landing lines. The appropriate value of (Mg/S)o of 733.713 N/m^2 is included in all the performance calculations.

Results:
The values of the various parameters relevant to (Mg/S)o of 733.713 N/m^2 are given. Included is an evaluation of the Structural Parameter, SP [using Ch.5, Eq (5.8a)] and the equivalent take-off power to weight ratio, (P/Mg)o [uses Ch.3, Eq (3.11a)].

AD5.2.4.5 Second stage assumptions and input data

Climb path:

Climb EAS	The assumed climb speed factor of Z equal to one is equivalent to a speed of 1.458Qv, or about 50 m/s. Subsequent revision may be necessary if the wing loading changes significantly on optimisation
Cl EAS H2 Cl EAS sig	The height, H2, at end of climb and corresponding relative density are as specified. Climb Mach number follows from climb EAS
Cl Mn	Climb Mach number

Assumed:

Lambda	Wing taper ratio is set at 0.4 as discussed previously (Ch.5, para 5.3.3)
(P/Mg)eng	Bare engine power to weight ratio [Ch.6, Eq (6.26f)]
Op It Fac	Nominal value of 3 kg/crew chosen for operational items
AppFuel/Mo	Fuel used in descent and approach as ratio of take-off mass (Ch.7, para 7.4.4)
V bar Vv bar	Horizontal and vertical tail volume coefficients (selected from Ch 8, Table 8.1A)

Input data:

Fus L Fus B Fus H	Fuselage length (aft of engine bulkhead), *L*, maximum width, *B*, and maximum height, *H*, derived from initial fuselage layout (Figure AD5.1)
c1	Lifting surface mass coefficient, the value for a propeller driven military trainer of composite construction has been chosen (Ch.6, Table 6.7)
c2	Fuselage mass coefficient, allowing for composites (Ch.6, Table 6.6)
c3 c4 c5	Mass coefficients for powerplant installation, c3, systems, c4, and lifting surface ratio, c5, (from Ch.6, Tables 6.8 to 6.10 respectively)
Payload	Two crew members, assumed to be 86 kg each
N bar	Effective ultimate normal acceleration factor [taken as 1.65 times the limit value, Ch.6, Eq (6.22a)]
N eng	Single engine
Overall L	Overall length of fuselage, including spinner from layout (Figure AD5.1)
Del lw	Centre of gravity of wing mass relative to overall centre of gravity (c.g.) (assumed to be 0.1 wing mean chord aft of overall c.g.)
l Fus	Location of fuselage structure centre of gravity aft of nose (assumed from layout)
l Tail	Location of tail unit centre of gravity aft of nose from layout (Figure AD5.1)
l PP	Location of powerplant centre of gravity aft of nose from layout
l SYS	Location of system centre of gravity aft of nose (assumed from layout)
l PAY	Location of payload centre of gravity aft of nose from layout
l OP IT	Location of operational items (assumed to be as payload)
Del lwg fue	Wing fuel assumed to be in wing root leading edge (taken as 0.33 m forward of overall centre of gravity as estimate derived from initial layout, Figure AD5.1, but subject to checking subsequently)
l fus fuel	Location of fuselage fuel from layout (Figure AD5.1)
l nose gr	Location of nose landing gear (assumed from layout, Figure AD5.1)

Del l mn gr Location of main landing gear (assumed to be on the wing and positioned 0.2 wing mean chord aft of the overall centre of gravity but subject to correction subsequently when landing gear is defined in detail, see Addendum 1)

AD5.2.4.6 Second stage calculations

(S/Mo)^0.45 Power of wing loading used in lifting surface mass calculation
p bar Cabin pressure differential, not relevant in this case

Cruise:
Reqd (T/Mg) Reciprocal of cruise lift to drag ratio
Av (T/Mg) Available cruise thrust to weight ratio [derived from estimated static power to weight ratio using Ch.3, Eq (3.11f)]
Av/Reqd Ratio of previous two values
(c) des Nominal datum specific fuel consumption (Ch.3, para 3.6.3.4)
(c) od Off design specific fuel consumption [assumes that power required is proportional to thrust required at a given set of conditions, Ch.3, Eq (3.15a)]

Climb:
Z Constant EAS climb speed correction factor [Ch.7, Eq (7.15c)]
X1 Climb correction factor for propeller engines [Ch.7, Eq (7.18a)]
Fact Qv Climb prediction factor [Ch.7, Eq (7.15b)]
(T/Mg) Thrust to weight ratio available for climb [Ch.7, Eq (7.11d)]
(Vv) EAS Mean rate of climb in EAS climb up to cruise altitude [Ch.7, Eq (7.15a)] (slight discrepancy from specified value is a consequence of the slightly higher assumed climb speed)
Dist EAS Ground distance covered in constant EAS climb [Ch.7, Eq (7.21) with cosine term assumed to be one]
Wf/(Mg)o Weight ratio of fuel used in constant EAS climb [Ch.7, Eq (7.20a)]

Descent:
Desc Dist Distance covered in descent and landing [Ch.7, Eq (7.56)]

Masses:
Mc1/Mo Start of cruise mass ratio, allows for new estimate of fuel used in climb
M fus Mass of fuselage structure [uses value of c2 in Ch.6, Eq (6.20b)]
c1 bar Factor in prediction of lifting surface mass [Ch.6, Eq (6.24)]
Mpp/Mo Mass ratio of powerplant installation [uses value of c3, calculated static power to weight ratio and assumed (P/Mg)eng]
Msys/Mo System mass ratio (coefficient c4)

M op it	Mass of operational items (product of Op It Fac and number of crew)
M fixed	Fixed mass (sum of fuselage, payload and operational items)
Net range	Range to be covered in cruise (specified range less ground distance covered in climb and descent)
Log 10	Logarithm to base 10 of mass ratio of start to end of cruise [uses Ch.7, Eq (7.53b)]
Mc1/Mc2	Cruise mass ratio (antilog of previous value)
Mc2/Mo	End of cruise mass as ratio of take-off value (uses Mc1/Mo and Mc1/Mc2)
Mf/Mo	Total mass of fuel as ratio of take-off mass [Mo less (Mc2 plus approach fuel)]
Kappa Mo	Sum of terms directly proportional to mass, as fraction of mass (fuel, systems and installed powerplant)
x bar 0.25/ root chord	Location of 0.25 aerodynamic mean chord on wing centreline chord [uses wing geometry as subsequently defined and Ch.8, Eq (8.7b)]

AD5.2.4.7 Second stage analysis and calculated take-off mass

The second stage analysis consists of two parts. The first uses the calculated values to deduce the actual take-off mass of the aircraft, and the second then uses this value to calculate the wing position on the body required to bring the overall centre of gravity to the 0.25 aerodynamic mean chord point. Thence, using the assumed volume coefficients, the horizontal and vertical tail sizes are calculated.

(Mo)est1	This is a first estimate of take-off mass derived by assuming, arbitrarily, that the lifting surface mass is 12% of the total
(Mo)est2	Initially this is the same value as (Mo)est1, entered directly as a number
Kappa*Mo	Mass of terms directly proportional to take-off mass [product of Kappa Mo and (Mo)est2]
Mlift surf	Mass of lifting surfaces [uses c1bar with (Mo)est2)]
(Mo)calc	Initially this is the sum of (Mfixed), (Kappa*Mo) and (Mlift surf) to give the total mass. This is the target cell used for optimisation by invoking SOLVER©. The value is minimised, in this case by changing the aspect ratio, thickness to chord ratio, wing loading and (Mo)est2 subject to the following constraints:- Thickness to chord ratio to be less than, or equal to, 0.15. Wing loading to be less than, or equal, to the equivalent landing and manoeuvre values (Mo)calc to be equal to (Mo)est2
error	The difference between (Mo)calc and (Mo)est2, as a check on the optimisation.

l CG	Location of 0.25 aerodynamic mean chord point aft of nose of aircraft derived from longitudinal balance of all the mass items
l WG APX	Location of leading edge of wing centreline chord aft of nose [derived from l CG and x bar (0.25)/root chord]
l TL ARM	Distance from centre of gravity to nominal centres of pressure of vertical and horizontal tails (derived from l CG and l tail)
S Hor Tail S Vert Tail	Areas of horizontal and vertical tail surfaces (derived from wing area, wing mean chord, l TL ARM and volume coefficients)

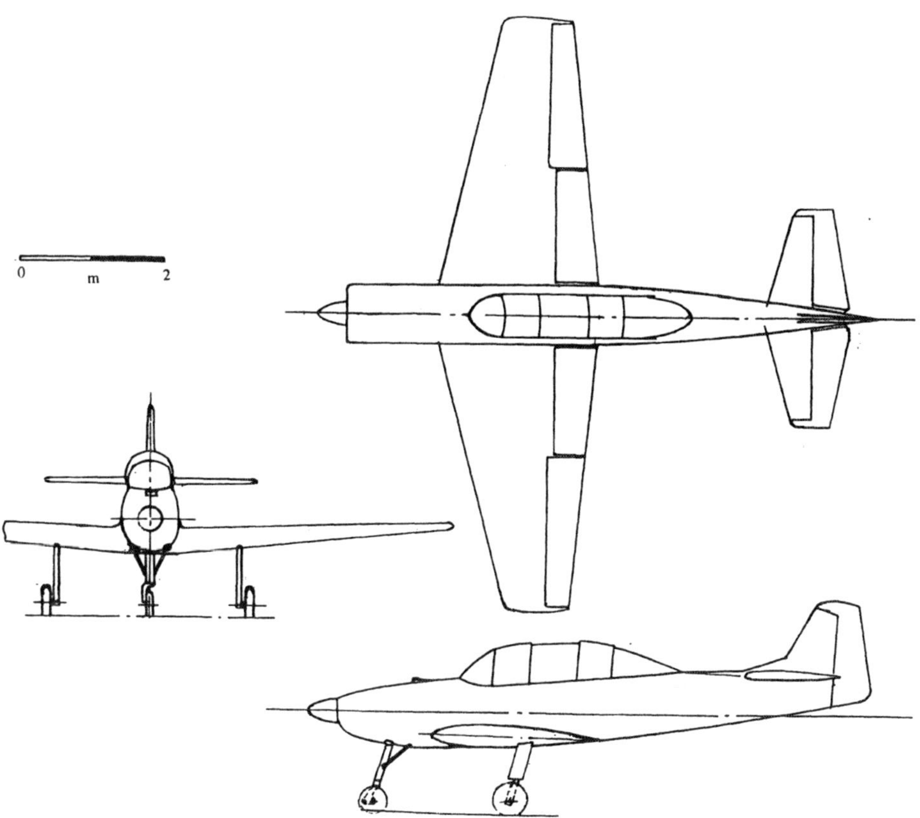

Figure AD5.2 Two seat aerobatic trainer - general arrangement

AD5.2.4.8 Summary of final results

The summary of final results is directly derived from the previous calculations.

Individual masses, which are all used in the centre of gravity calculation, are obtained as follows:

i) Wing mass is isolated from the tail mass by using coefficient c_5 in conjunction with the total lifting surface mass.
ii) Landing gear in this case is assumed to account for 5% of the 16% allocated to systems, with the nose gear being 1% of the total aircraft mass (see Addendum 4)
iii) Fuselage fuel mass of 30 kg is derived by consideration of both the volume available and a reasonable quantity for aerobatics.
iv) Wing area, *S*, follows from (Mg/S)o and the wing geometry from aspect ratio, *A*, taper ratio, lambda, and sweepback (Del 0.25).
iv) Engine power follows from (P/Mg)o and the propeller diameter from Ch.3, Eq (3.9k).

AD5.2.5 Preliminary general arrangement

The geometric information derived from the optimisation procedure has been used in conjunction with the fuselage layout to produce a first general arrangement drawing of the aircraft. Use has been made of the typical horizontal and vertical tail characteristics given in Chapter 8, Table 8.1B. Figure AD5.2 shows the layout of the aircraft, which can form the basis for a more detailed analysis. This must include a re-evaluation of lift and drag characteristics, more specific consideration of longitudinal and lateral control and stability, revised mass evaluation and review of landing gear layout. A consequence of this analysis may be a need to revisit the optimisation process with revised assumptions.

AD5.3 Twin turboprop feeder line /commuter aircraft (30 seats)

AD5.3.1 Introduction

The market for small and medium sized feeder line aircraft has traditionally been met by twin turbopropeller engined designs. There is now a trend towards the use of aircraft equipped with fan engines for the role but it remains to be seen whether this development will prove to be economically justified. While the use of fan engines confers a degree of flexibility in the layout, since large diameter propellers do not have to be considered, there is no doubt that they possess a higher specific fuel consumption. The argument for the fan engines includes greater attractiveness to passengers, relative simplicity due to the elimination of the propeller reduction gearboxes and potentially higher cruise speeds. The issue is whether the lower maintenance costs and possibly higher passenger load factors can compensate for the greater fuel costs.

In spite of the current trends a twin turbopropeller feeder line aircraft has been selected to illustrate the design synthesis procedure for the turbopropeller class of powerplant.

AD5.3.2 Specification

The specification chosen for the aircraft follows.

a) Maximum payload of 30 passengers at an equivalent mass of 93.4 kg per person to be carried over 2000 km still air range with only landing reserve fuel.

b) Take-off field length not to exceed 1500 m in standard atmosphere conditions.

c) Landing field length not to exceed 1250 m.

d) Approach speed (at 1.3 times stalling speed) not to exceed 55 m/s.

e) Normal operating altitude 6 km, but up to a maximum of 8 km.

f) Normal cruise speed 139 m/s true airspeed.

g) Maximum level speed 154 m/s true airspeed ($M_N = 0.5$ at 8 km altitude).

h) Structural design speed, V_D, 160 m/s equivalent air speed.

i) Ultimate normal acceleration factor, N, of 6 (subject to possible revision when mass of aircraft has been established).

j) Total crew of 3, inclusive of a steward.

The following supplementary requirements are introduced to enable the procedure to be used for other classes of aircraft, such as turboprop trainers. They are not, in reality, really relevant to this example.

i) Mean rate of climb up to initial cruise altitude, 6.5 m/s
ii) Sustained manoeuvre of 4 deg/s at 120 m/s and sea level
iii) Instantaneous manoeuvre of 4 *g* at 125 m/s and sea level (limit case) (125 m/s equivalent airspeed is presumed to be the manoeuvre speed, V_A).

AD5.3.3 Configuration

There are really only two major considerations to be resolved in this example.

The first, which has a bearing on the second, is the vertical location of the wing on the fuselage. The issues involved are comprehensively covered in Chapter 4, paragraph 4.4.1.

Some designs in this category of aircraft use a low wing while others have the wing located across the top of the fuselage. From this it may be deduced that there is no clear best solution. The altitude of operation in the present case implies the need for pressurisation, and hence a nominally circular cross-section fuselage, and this is of relevance here. A low wing passing below the cabin floor enables a compact cross-section to be derived although it is unlikely for there to be suitable underfloor baggage volume on this size of aircraft. The low wing also facilitates landing gear mounting and stowage, albeit at some penalty due to cutouts in primary structure. Against this the propeller ground clearance dictates that the powerplants must be located across the top of the wing surface which is aerodynamically inefficient. A high wing overcomes this difficulty but also introduces a possible problem with location of the landing gear. With a large aircraft the consequence is a fuselage-located main landing which inevitably implies a narrow track and fuselage mass penalty. However, for a relatively small aircraft as is the situation here it is possible to locate the gear on the wing and retract it behind the powerplant providing some displacement of the engine exhaust is accepted. The high wing also influences the fuselage cross-section as, unless it is totally above the cross-section, it effectively displaces the cabin floor down resulting in a larger diameter to maintain a given floor width. This does, however, mean that there may be volume for baggage below the floor and the low height of the floor above the ground line can afford easy access for passengers without the need for airport facilities. Other issues to be considered include buoyancy of the aircraft if it is forced to alight on water and general crashworthiness.

The second issue is the cabin layout, especially the number of seats there are across the width. It is clear that to provide a total of 30 seats there could be 10 rows of 3 abreast or 8 at 4 abreast. Reference to Chapter 4, paragraph 4.5 suggests that the former might give a better balance of overall fuselage length and diameter. However, it is also necessary to consider a probable requirement to stretch the aircraft, up to say 40 seats, in which case the 3 abreast layout is likely to result in an unduly long fuselage. To resolve the issue it is necessary to investigate the possible cross-sections in more detail. To do this the following assumptions are made:

i) Seat pitch 0.78 m.
ii) One galley to serve snacks, etc.
iii) One cross aisle only in the actual cabin area.
iv) One toilet.
v) Cabin floor width to be no less than 0.4 m less than maximum cabin width.
vi) Aisle height to be as near to 2 m as feasible but 1.83 m acceptable.
viii) Structure and trim add 0.2 m to cabin width to obtain external dimensions.

From Chapter 4, paragraphs 4.5.3.3 and 4.5.3.4 the cabin width and length may be estimated for both 3 and 4 abreast seating. Figure AD5.3 shows the cross-sections assuming that the fuselage is circular. In the 3 abreast seating case an aisle height of 1.9 m can be achieved by having a step up to the seating. However, the aisle effectively occupies all the fuselage height and a high wing would have to be passed totally over the fuselage. Clearly in this situation it is likely that a low wing will enable a better wing/fuselage junction to be designed. There is no room for baggage below the floor. Should a high wing be required with 3 abreast seating it is really necessary to increase the diameter. Reference to other aircraft in this category suggests that the centre wing structure is likely to be about 0.35 m deep of which , say, 0.25 m might be within the fuselage depth to give a good wing/fuselage fairing. To achieve this with the 3 abreast seating layout it is necessary to increase the diameter of the fuselage, as shown in Figure AD5.3. When the same requirement is applied to the 4 abreast configuration only a small step is required in the floor to achieve 1.9 m aisle height and the high wing can be accommodated within the cross-section determined solely by the seating/aisle width. There is room below the floor for baggage and this is much improved if a small increase in diameter is accepted to enable the floor to be level across its whole width. Fuselage mass is an important consideration and it is possible to compare the four cross-sections and corresponding cabin lengths by using Eq (6.20a) of Chapter 6. Chapter 4, Table 4.3 suggests that for this class of aircraft the cabin occupies about half the total fuselage length.

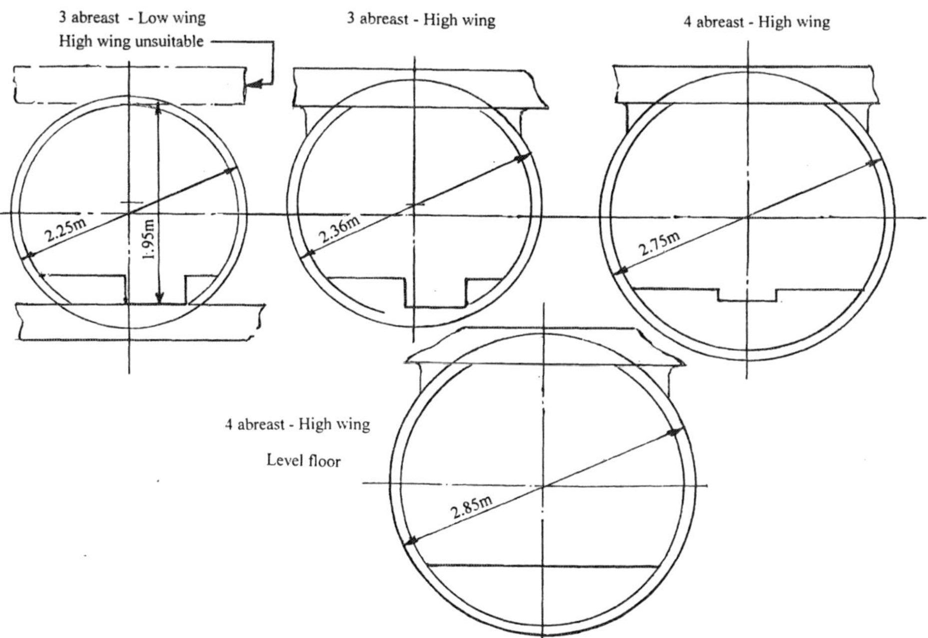

Figure AD5.3 Commuter aircraft - fuselage cross-sections

Table AD5.1 Commuter aircraft - fuselage cross-section comparison

	CASE	CABIN WIDTH	EXTERNAL DIAMETER	FUSELAGE WIDTH	LENGTH / DIAMETER	MASS RATIO	NOTES
				30 SEATS			
1	3 ABREAST	2.05	2.25	20.76	9.32	1.0	LOW WING
2	3 ABREAST	2.16	2.36	20.76	8.84	1.07	HIGH WING
3	4 ABREAST	2.55	2.75	17.64	6.41	1.02	HIGH WING
4	4 ABREAST	2.65	2.85	17.64	6.18	1.07	HIGH WING LEVEL FLOOR
				STRETCHED TO 40 SEATS			
1A	3 ABREAST	2.05	2.25	23.9	10.62	1.18	
2A	3 ABREAST	2.16	2.36	23.9	10.13	1.26	
3A	4 ABREAST	2.55	2.75	19.2	6.98	1.14	
4A	4 ABREAST	2.65	2.85	19.2	6.74	1.19	

Using the basic low wing 3 abreast seating case as the datum, Table AD5.1 summarizes the results for a 30 seat layout. The 3 abreast low wing layout is inevitably lighter, but if a high wing is selected for other reasons the 4 abreast arrangement is at least as good. The lower part of Table AD5.1 shows the results when the fuselage is stretched to enable 40 passengers to be carried, where it has been assumed that the length increase is solely due to the additional rows of seats. The advantage of 4 abreast seating becomes apparent although it should be noted that the overall length to diameter ratio in the 30 seat design case is somewhat low. It may be concluded that a 30 seat requirement is a marginal case, below which a low wing is clearly preferable when associated with a pressurised fuselage. The case for a high wing combined with 4 abreast seating is a strong one when more than 30 seats are required.

As a consequence of the above considerations, especially with respect to stretch potential, the chosen configuration will be based on a high wing and 4 abreast seating layout with a level floor. This gives a basic fuselage diameter of 2.85 m and length of 17.64 m for the 30 seat case, subject to refinement as the design proceeds. The conventional nose landing gear layout will be adopted. The vertical position of the horizontal tail may need investigation although it should be acceptable to locate it at the top of the rear fuselage rather than on the fin.

AD5.3.4 Spreadsheet formulation (SPREADSHEET AD5.2)

AD5.3.4.1 Initial inputs and assumptions

The parameters selected for optimisation are aspect ratio and thickness to chord ratio, together with provision to vary the number of engines. The Requirements are derived from the specification.

Assumed values:

M1/Mo	Ratio of start of climb mass to take-off value, taken to be 0.99 (Ch.7, para 7.3.6) (subject to subsequent correction)
S^-0.1	Wing area parameter (Ch.6, Figure 6.1), subject to subsequent change
Rw	Wetted area factor taken to be 5.5 (Ch.6, Table 6.3)
Type Fac	Taken to be 1.4 (Ch.6, Table 6.4)
TE flap	Single slotted flap assumed, no leading edge devices
SS alpha SS gamma	Second segment climb coefficients for twin-engine aircraft [Ch.7, Eq (7.4a)]
Flap Fac	Flap drag factor, relevant to single slotted flap (Ch.6, para 6.3.2.2)
a Cr (Initl)	Speed of sound at initial cruise altitude (6 km)
(T/Mg)ass	Assumed static thrust to weight ratio for initial calculation (Ch.3, Table 6.3B suggests 0.3 to 0.35, assume 0.32 - it is corrected during analysis)
Tan gam d	Tangent of descent angle in approach to land [Ch.7, Eq (7.6)] (5° descent angle assumed)
Mu G	Mean stopping deceleration, 0.4 assumed here [Ch.7, Eq (7.6)]
TO ke	Take-off distance correction factor, 0.12 taken assuming that engines are not thrust limited [Ch.7, Eq (7.1a)]
z	Number of blades on propeller, assumed to be 5 here
nDp	Product of propeller diameter and rotational speed, taken to be 75 m/s here (Ch.3, para 3.6.2.3)
a sustn	Speed of sound at sustained turn altitude (sea level here)
Delta deg	Assumed wing quarter chord sweepback, degrees (taken as zero here)

AD5.3.4.2 Initial calculations

Ml/Mo	Ratio of landing to take-off mass (Ch.7, Table 7.1 for short haul)
Mcr/Mo	Ratio of start of cruise mass to take-off mass (derived from M1/Mo as given by factor in Ch.7, Table 7.2)
(Clm)o ae	Basic aerofoil maximum lift coefficient (Ch.6, para 6.2.4.2)
Del flp (TO)	Lift coefficient increment due to flap at take-off (Ch.6, Table 6.1)
Del flp (L)	Lift coefficient increment due to flap at landing (Ch.6, Table 6.1)
(Cl use)o	Usable cruise lift coefficient at zero Mach number [Ch.6, Eq (6.13a)]
(Cdz)o	Incompressible zero lift drag coefficient [Ch.6, Eq.(6.13a)]

Landg (L)	Landing distance parameter [Ch 7, Eq (7.6a) - no reverse thrust]
Man(Cl)o	Basic aerofoil manoeuvre lift coefficient, used for instantaneous manoeuvre requirement at speed, V_A, in this case [same as (Clm)o ae]
Cr Mn (Initl)	Initial cruise Mach number, ratio of true cruise speed and local speed of sound
Mn Max	Mach number at maximum required speed at initial cruise altitude, as above
Mmo/Mo	Ratio of mass at maximum speed design condition to take-off mass. Here it is arbitrarily assumed to be 0.94 based on use of fuel after start of cruise value Mcr/Mo

AD5.3.4.3 Preliminary calculations

cos delta	Cosine of previously assumed quarter chord sweepback
Wave Dr F	Wave drag factor [term in Ch.6, Eq (6.13a) raised to power of 20]
(Cdz)cr	Zero lift drag coefficient in cruise [Ch.6, Eq (6.13a) - no laminar flow]
(Cd)co	Corrected climb out zero lift drag coefficient [Ch.6, Eq (6.15) with an additional factor of 1.1 to allow for the slipstream effect]
(Kv)o	Induced drag factor at zero Mach number [Ch.6, Eq (6.14a) - the engines are assumed to be located below the wing]
(Kv)cr	Induced drag factor in cruise [Ch.6, Eq (6.14a)]
Ins Man Cl	Lift coefficient available for instantaneous manoeuvre [Man(Cl)o corrected for sweepback]
(Cdz)sustn	Zero lift drag coefficient at sustained turn condition [Ch.6, Eq (6.13a)]
(Kv) sustn	Induced drag factor at sustained turn condition [Ch.6, Eq (6.14a)]
(Beta)sustn	Product of zero lift drag coefficient and induced drag factor at sustained turn condition (previous two values)
Sustn n	Normal acceleration in sustained turn [Ch.7, Eq (7.39b)]
Cl max	Maximum lift coefficient, sum of (Clm)o ae and del flp (L), corrected for sweepback [Ch.6, Eq (6.2)]
Cl us	Unstick lift coefficient [0.8 of sum of (Clm)o ae and del flp (TO), corrected for sweepback and with 1.15 factor to allow for slipstream, [Ch.6, Eq (6.4)]
Cla	Approach lift coefficient at 1.3 times the stall speed, corrected for sweepback and with 1.1 allowance for slipstream [Ch.6, Eq (3)]
Cl use	Usable lift coefficient in cruise, (Cl use)o corrected for sweepback
Va calc	Maximum approach speed consistent with the required landing distance [Ch.7, Eq (6.6b)]
Va	The lower of Va calc and Va as specified, in this case Va specified.
(Mg/S)o ld	Take-off equivalent wing loading derived from Va, Cl a and (Ml/Mo) [Ch. 7, Eq (7.6e) without reverse thrust]

(Mg/S)o gt Take-off equivalent wing loading to meet gust sensitivity requirement [Ch.5, Eq (5.9b)]

Cor L length Corrected value of landing length when Va is limited by specified value [Ch.6, Eq (6.6a)]

Ins (Mg/S)o Take-off equivalent wing loading determined by instantaneous manoeuvre requirement (uses Inst Man Cl with Mcr/Mo at speed V_a)

AD5.3.4.4 First stage analysis

The range of take-off wing loading, (Mg/S)o, has been selected by reference to Chapter 5, Table 5.3 and checked to ensure that it covers the already derived value of (Mg/S)o ld.

Take-off:

1st app The already assumed value of (T/Mg)ass is used to make a first estimate to meet the take-off requirement [Ch.7, Eq (7.1b), where the assumed value is used only on the right-hand side]

(T/Mg)o The second evaluation uses in the right-hand side of the equation the previous approximate value. (If acceptable convergence is not achieved the calculation should be repeated.)

Acc. Stop:

(T/Mg)o This is the take-off thrust to weight ratio required to meet the engine failed acceleration-stop requirement [Ch.7, Eq (7.3a)]

Sec Seg Climb:

Tau co The climb out thrust factor is determined on the assumption that the climb out speed is 1.1 times the unstick speed in hot and high conditions. The value is for one engine [Ch.7, Eqs (7.4d) and (7.10c)]

(Cd)co/Clu The ratio of equivalent zero lift drag coefficient to unstick lift coefficient at the climb out flap condition.

(T/Mg)o The equivalent take-off thrust to weight ratio [uses Ch.7, Eq (7.4a) in conjunction with Ch.7, Eq (7.4d)]

Climb:

Y The altitude factor appropriate to end of the constant equivalent airspeed climb [Ch.7, Eq (7.18d) - Qv value from Mean R/C condition below]

Z Speed correction factor, climb here assumed to be at 1.14 times minimum drag speed [Ch 7, Eq (7.15c)]

Veas Climb speed, equivalent airspeed [Ch 7, Eq (7.15c)]

Mean R/C to Cruise Altitude:

Fact Qv Factor in climb prediction [Ch.7, Eq (7.15b)]

f (drag)	The climb drag factor [Ch.7, Eq (7.15d)]
1st app	First approximation of the required thrust to weight ratio at start of climb, (T/Mg), to give the specified mean rate of climb, assuming the X factor to be unity [Ch.7, Eq (7.18e)]
(T/Mg)1	Corrected value of (T/Mg)1 using revised value of X1 from next column.
X1	Corrected value of X1 derived by using first approximation to (T/Mg) [Ch.7, Eq (7.18a)]
Tau Cl	Rate of climb condition thrust factor [uses Ch.7, Eq (7.18g) but does not include the terms involving (*Mg/S*) and thrust, T_l]
(T/Mg)o	The equivalent take-off thrust to weight ratio [uses the remaining terms of Ch.7, Eq (7.18g) together with tau climb]

Landing-rev thrust:

(T/Mg)o	Selected as maximum value derived from all other conditions
Landg L	Landing distance parameter [Ch.7, Eq (7.6a) with reverse thrust]
L length	Reverse thrust landing length follows from the previous term

Sustained Turn:

D bar	Zero lift drag to weight ratio [Ch.7, Eq (7.13e)]
(T/Mg)sustn	Thrust to weight ratio required to maintain sustained turn at the given condition [Ch.7, Eq (7.40a), n is the sustained turn rate value]
(T/Mg)o	Equivalent take-off thrust to weight ratio [derived from previous value by use of Ch.7, Eq (7.55a)] (advance ratio assumed to be greater than one)

End of climb:

Fact Qv	Factor in climb prediction [Ch.7, Eq (7.15b)]
(T/Mg)1	Thrust to weight ratio required at start of constant equivalent airspeed climb to achieve a residual rate of climb of 1.5 m/s at the cruise altitude, assuming same climb conditions as for rate of climb analysis [uses Ch.7, Eq (7.19b)]
(T/Mg)o	Equivalent take-off thrust to weight ratio [derived from previous value by use of Ch.7, Eq (7.18g)] (advance ratio assumed to be greater than one)

Cruise:

Cl	The lift coefficient appropriate to start of cruise conditions
Cd	The total drag coefficient (zero lift plus induced) at start of cruise [uses (Cdz)cr, (Kv)cr and Cl]
L/D	Cruise lift to drag ratio (Cl/Cd)
(T/Mg)cr	Cruise thrust to weight ratio (inverse of L/D)

(T/Mg)o The equivalent take-off thrust to weight ratio [derived from Ch.7, Eq (7.55a)]

Max speed:
The maximum speed analysis follows the same procedure as that used for cruise, with the specified speed, altitude and mass ratio Mmo/Mo.

Summary:
All the first stage analysis calculations are summarised, together with the instantaneous manoeuvre, landing and gust sensitivity wing loading previously derived. The summary is used to produce the diagram of (T/Mg)o vs (Mg/S)o which illustrates the interaction of the various requirements. Inspection of the diagram suggests that the design condition for further examination is that given by the intersection of the second segment climb, the maximum speed and the landing lines. The appropriate value of (Mg/S)o of 3187.2121 N/m^2 is included in all the performance calculations.

Results:
The values of the various parameters relevant to (Mg/S)o of 3187.121 N/m^2 are given. Included is an evaluation of the Structural Parameter, SP [using Ch.5, Eq (5.8a)] and the equivalent take-off power to weight ratio (P/Mg)o [uses Ch.3, Eq (3.10a) in conjunction with other relevant values of wing loading, etc.].

AD5.3.4.5 Second stage assumptions and input data

Climb path:

Climb EAS The assumed climb speed factor of nominally one is equivalent to a climb speed of 1.458Qv, or about 90 m/s. Subsequent revision to this value may become necessary if the wing loading changes significantly on optimisation

ClEAS H2
ClEAS sig
Cl Mn
The height, H2, at the end of climb and the corresponding relative density as specified. Climb Mach number, where relevant, follows from Climb EAS

Assumed:

Lambda Wing taper ratio chosen to be 0.5 initially by comparison with comparable designs

(P/Mg)eng Bare engine power to weight ratio [Ch.6, Eq (6.26c) - advanced design of moderate size]

Op It Fac Operational items factor [Ch.6, para 6.4.2.3 and Eq (6.21a)]

AppFuel/Mo Fuel used in descent and approach as ratio of take-off mass (Ch 7, para 7.4.4)

V bar Vv bar	Horizontal and vertical tail volume coefficient (Selected from Ch 8, Table 8.1A for turboprop transports)
Input data:	
Fus L Fus B Fus H	Fuselage overall length, *L*, maximum width, *B*, and maximum height, *H*, as derived from consideration of fuselage configuration in para AD5.3.3
c1	Lifting surface mass coefficient, value for short range propeller transport [Ch.6, Eq(6.23c)]
c2	Fuselage mass coefficient, 4 abreast transport (Ch.6, Table 6.6)
c3	Powerplant installation mass factor (Ch.6, Table 6.8)
c4	Systems mass factor, taken to be 0.20 here (Ch.6, Table 6.7)
c5	Lifting surface mass ratio (Ch.6, Table 6.10)
N bar	Effective normal acceleration factor [Ch.6, Eq.(6.22a)]
Prop eta	Propeller efficiency in cruise, assumed initially to be 0.85 as a typical value, but subject to subsequent correction when actual installed power is known [Ch.3, Eq (3.9e)]
del l w	Centre of gravity of wing mass relative to overall centre of gravity aircraft (c.g.) (assumed to be 0.1 wing mean chord aft of c.g.)
l Fus	Location of fuselage structure centre of gravity aft of nose (assumed to be 0.45 of Fus L)
l Tail	Location of tail unit centre of gravity aft of nose (taken as 0.9 of FusL)
l PP	Location of powerplant centre of gravity relative to aircraft centre of gravity [assumed to be 0.85 of the mean wing chord divided by the cosine of the wing quarter chord sweep forward of the aircraft centre of gravity, Ch.8, para 8.10.3(f)]
l SYS	Location of systems mass (excluding landing gear) aft of fuselage nose (assumed to be 0.45 of Fus L)
l PAY	Location of payload mass aft of fuselage nose (assumed to be 0.45 of Fus L)
l OP IT	Location of operational items centre of gravity aft of fuselage nose (assumed to be 0.45 Fus L)
Del l Fuel	Location of centre of gravity of fuel relative to aircraft centre of gravity (assumed to be 0.1 wing mean chord aft of c.g.)

AD5.3.4.6 Second stage calculations

(S/Mo)^0.45	Power of wing loading used in lifting surface mass calculation
p bar	Cabin pressure differential [taken as (0.74 - Fin Cr sigma to power of 1.235)]

Cruise:

Reqd (T/Mg)	Reciprocal of cruise lift to drag ratio
Av (T/Mg)	Available cruise thrust to weight ratio [derived from estimated static thrust to weight ratio (T/Mg)o by inversion of Ch.7, Eq (7.55a)]
Av/Reqd	Ratio of two previous values
c(des)	Design datum specific fuel consumption [Ch.3, Eq (3.14a) with the available static power, Po, assumed to be 1000 kW; it could be subsequently corrected but the effect is small]
c(od)	Off design specific fuel consumption [in the absence of any other data assumed to be the same as c(des)]

Climb:

Z	Constant equivalent airspeed climb speed correction factor [Ch.7, Eq (7.15e) corrected for slightly revised climb speed]
X1	Climb correction factor [Ch.7, Eq (7.18a)]
Qv	Factor in climb prediction [Ch.7, Eq (7.15b)]
(Vv) EAS	Mean rate of climb in constant equivalent airspeed climb up to initial cruise altitude [Ch.7, Eq (7.15a)]
Dist EAS	Ground distance covered in constant equivalent airspeed climb [Ch.7, Eq (7.21) with the cosine term assumed to be unity]

Descent:

Desc Dist	Distance covered in descent and landing [Ch.7, Eq (7.56)]

Climb EAS:

(c)o	Specific fuel consumption in climb condition [Ch.3, Eq (3.14a) with the initial climb Mach number and assumed engine power of 1000 kW]
Wf/(Mg)o	Climb fuel usage as ratio of take-off mass [Ch.7, Eq (7.20b)]

Masses:

Mc1/Mo	Start of cruise mass as ratio of take-off mass [M1/Mo-Wf/(Mg)o]
M fus	Mass of fuselage structure [Ch.6, Eq (6.20a)]
c1 bar	Factor in prediction of lifting surface mass [Ch.6, Eq (6.24)]
Mpp/Mo	Mass ratio of powerplant installation [uses value of c3, calculated static power/weight ratio and the assumed (P/Mg)eng]
M/sys/Mo	Systems mass ratio (coefficient c4)
Mop it	Mass of operational items (product of the Op It Fac and number of passengers plus 85 kg allowance for each member of the crew)
M fixed	Fixed mass (sum of the fuselage, payload and operational items)

Net Range	Range to be covered in cruise (Specified range less ground distance covered in climb and descent)
Log 10	Logarithm to base 10 of mass ratio of start to end of cruise [Ch.7, Eq (7.53a)]
Mc1/Mc2	Cruise mass ratio (antilog of previous value) (uses Mc1/Mo and Mc1/Mc2)
Mc2/Mo	Ratio of end of cruise to take-off mass
Mf/Mo	Total mass of fuel as ratio of take-off mass [Mo less (Mc2 plus approach fuel)]
Kappa Mo	Sum of terms directly proportional to mass, as fraction of mass (fuel, systems and installed powerplant)
x bar (0.25)/ root chord	Location of 0.25 aerodynamic mean chord on wing centreline chord [uses wing geometry as subsequently defined and Ch.8, Eq (8.7b)]

AD5.3.4.7 Second stage analysis and calculated take off mass

The first part of the second stage analysis uses the calculated values to deduce the actual take-off mass of the aircraft. The second part then uses this value to make a first estimate of the wing position along the body which brings the centre of gravity to the 0.25 aerodynamic mean chord position. Use of the assumed volume coefficients then enables the horizontal and vertical tail sizes to be calculated.

(Mo)est1	The first estimate of take-off mass derived by assuming, arbitrarily, that the lifting surface mass is 12% of the total
(Mo)est2	Initially this is the same as (Mo)est1, entered directly as a number
Kappa *Mo	Mass of terms directly proportional to take off mass [Product of Kappa Mo and (Mo)est2]
M lift surf	Mass of lifting surfaces [uses c1bar with (Mo)est2]
(Mo)calc	Initially this is the sum of (M fixed), (Kappa*Mo) and (M lift surf) to give the total mass. This is the target cell used for optimisation by invoking SOLVER©. The value is minimised, in this case by changing the aspect ratio, thickness to chord ratio, wing loading and (Mo)est2, subject to the following constraints: Structural Parameter, SP, to be no more than 17 Thickness to chord ratio to be less than, or equal, to 0.15 (Mo)calc to be equal to (Mo)est2 Wing loading to be less than, or equal to, the equivalent landing and manoeuvre values but greater than, or equal to, the gust sensitivity value
error	The difference between (Mo)calc and (Mo)est2 as a check on the optimisation.

l CG	Location of 0.25 aerodynamic mean chord point aft of the fuselage nose, derived from longitudinal balance of all the masses [the nose landing gear is assumed to be at 10% of fuselage length back from the nose and main landing gear at 1.1 times (l CG) back from the nose]
l WG APX	Location of leading edge of wing centreline chord aft of the nose [derived from l CG and x bar (0.25)/root chord]
l TL ARM	Distance between the centre of gravity and nominal centres of pressure of vertical and horizontal tails (derived from l CG and l Tail)
S Hor Tail S Vert Tail	Areas of horizontal and vertical tail surfaces (derived from wing area, wing mean chord, l TL ARM and assumed volume coefficients)

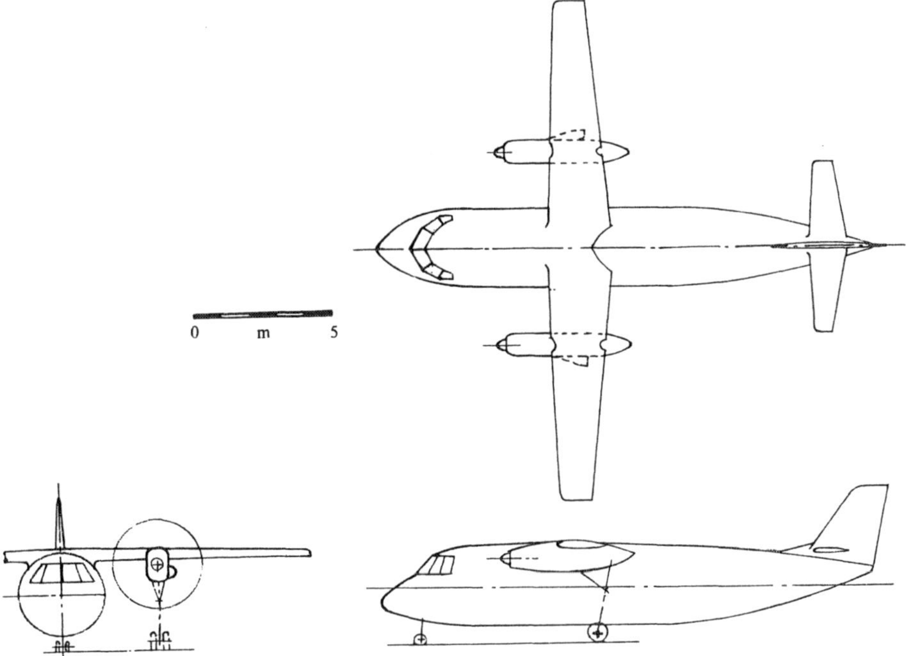

Figure AD5.4 Commuter aircraft - general arrangement

AD5.3.4.8 Summary of final results

The summary of final results is directly derived from the previous calculations.

Individual masses, which are all used in the centre of gravity calculations, are obtained as follows:

i) Wing mass is isolated from the tail mass by using coefficient, C_5, in conjunction with the total lifting surface mass.
ii) Landing gear is assumed to account for 4.6% of the 20% allocated to systems, with the nose gear being 0.15% of the total landing gear value.
iii) Fuel is all assumed to be located in the wing.
iv) Wing area, S, follows from (Mg/S)o and wing geometry from aspect ratio, A, taper ratio, lambda, and sweepback, (Del 0.25).
v) Engine power follows from (P/Mg)eng and propeller diameter from Ch.3, Eq (3.91).

AD5.3.5 Preliminary general arrangement

The geometric information derived from the optimisation procedure has been used in conjunction with the cabin/fuselage geometry derived from Table AD3.1 to produce a first general arrangement drawing of the aircraft. Use has been made of the typical horizontal and vertical tail characteristics given in Chapter 8, Table 8.1B. Figure AD5.4 shows the layout of the aircraft, which can form the basis for a more detailed analysis. This must include a review of all the assumptions made together with a more detailed mass and control/stability investigation. A consequence of this analysis may be a need to re-visit the optimisation process with improved assumptions.

AD5.4 Uninhabited, high altitude, long range reconnaissance aircraft (low bypass ratio engine)

AD5.4.1 Introduction

There exists a requirement for a reconnaissance aircraft which is capable of a large enough radius of operation to enable it to be positioned over a target area virtually anywhere in the world after launch from European or Pan-American bases. If the aircraft flies at moderate subsonic speed and allowance is made for, say, 3 h on station the resulting endurance is of the order of 24 h. To minimise the vulnerability of the aircraft to ground based missiles and manned interceptors it should possess low observability characteristics and operate at altitudes in excess of 24 km. The time and altitude of the flight suggest strongly that an uninhabited design solution is preferable. The aircraft should be capable of operation from existing airfields with a minimum of special support equipment and be able to fly autonomously over the cruise and on-station flight phases.

The demanding requirements imply design considerations which are, in many respects, quite different from those of more conventional subsonic aircraft which fly at lower altitudes and over shorter distances. Typical design values may not apply in this case.

AD5.4.2 Powerplant

The type of powerplant to be utilised is a major consideration. While the flight speed is not too critical it clearly should be chosen so that the fuel consumption is minimised and the cruise phase of the operation is not too extended. This suggests cruising at a Mach number of 0.6 or somewhat higher. This speed is appropriate to propeller propulsion or possibly a bypass turbojet. In spite of the high altitude a rocket engine is totally ruled out because of the long endurance. Further consideration of this issue is justified.

a) Propeller propulsion. Reference to Chapter 3, paragraph 3.6.2.5, Eq.(3.9a) shows that for a given propeller thrust coefficient and propeller tip speed, nD_p, the thrust is proportional to the air density and square of the propeller diameter, Dp. At altitudes above 22 km the air density is at least an order of magnitude less than that of typical propeller cruise conditions, and hence, everything else being equal, the diameter must be three to four times as large. Alternatively one propeller could be replaced by several to give the required disc area. Whichever approach is taken there are complications in the layout of the aircraft and probable mass penalties. Another consideration is the provision of power for driving the propeller. Several possibilities exist:

i) Electrically driven by solar power, but this requires large surface area for the solar cells and consequent susceptibility of the aircraft to turbulence, especially at lower altitudes.
ii) A piston engine, which would need several stages of supercharging in order to attain a sufficiently high intake manifold pressure. Superchargers designed for such a purpose are bulky and heavy and rely upon gas turbine technology.
iii) Shaft turbine engine of special design to handle the low air density at the intake.

A detailed study would be required to ascertain which of the latter two possibilities is preferable, but in either case an expensive, specially designed, powerplant system is required.

b) Bypass turbojet. Some aircraft powered by low bypass ratio turbojet engines operate up to altitudes approaching 20 km. While the air density at about 24 km altitude is half that at 20 km it is not unreasonable to presume that it is possible to adapt an engine of this type to maintain combustion at these higher altitudes. This being the case the use of such a powerplant greatly simplifies the layout of the aircraft since it is much more compact than propeller propulsion. Some aircraft in this category, such as the Teledyne - Ryan Global Hawk, employ a powerplant of this type. While a bypass ratio of the order of five would confer a relatively low specific fuel consumption it is by no means obvious that such an engine could be adapted to operate at altitudes in excess of 24 km. On the other hand an engine having a bypass ratio of, say, rather less than unity might possibly be amenable to modification to achieve the desired high altitudes. Such an engine, possibly adapted from a supersonic military design, would have the disadvantage of a relatively high specific fuel consumption.

On balance it is concluded that the most promising powerplant is the relatively low bypass ratio turbojet and the use of this type will be assumed.

AD5.4.3 Specification

The specification used as a basis for this design is as follows:

a) Still air radius of operation 8900 km (4800 nm)

b) 3 hours on station at extreme radius of operation

c) Altitude over target at least 24.4 km (nominally 80,000 ft)

d) Cruise speed at least $M_N = 0.65$

e) Field length for take-off 1000m and for landing 800 m

f) Payload to include:

- i) means of observing ground targets by day and night in all weather conditions and automatic identification and prioritisation of static and moving targets
- ii) presentation of real time image display to remote operators
- iii) payload mass allowance to be 600 kg at a density of 400 kg/m^2

g) Ground operations to use existing facilities as far as is possible

h) Navigation in cruise and over target to be autonomous

i) Low observability characteristics are required, but not to the undue detriment of performance

AD5.4.4 Configuration development

The requirement for 3 h on station over the target is equivalent to a flight of about 2100 km at a Mach number of 0.65 and 24.4 km altitude. Thus the total still air range is approximately 20,000km and will take about 28.5 h to complete.

There is no doubt that flight at 24.4 km altitude will imply a relatively low wing loading.

This has two major effects:

i) There is unlikely to be a need for high lift devices to meet the field requirements.

ii) The aircraft will be sensitive to air turbulence and although the ride comfort criterion of Chapter 5, paragraph 5.5.3 does not apply here some form of automatic ride control is likely to be needed.

The very long range/endurance requirement demands a high lift to drag ratio to restrict fuel consumption to an acceptable level. Thus both zero lift and induced drag must be made as low as possible and the wing loading chosen to ensure near to maximum lift to drag cruise conditions. The wing Reynolds number, defined as the product of the true flight speed and mean chord divided by the kinematic viscosity, will be relatively low in cruise conditions. This raises the possibility of some measure of natural laminar flow over the wing and, maybe, the nose region of the fuselage. Providing the surface is not contaminated during take-off and initial climb a laminar flow over the forward 45% of the chord is plausible. An aerofoil section of the NASA LS (1) 0.4 series or MS (1) 0.3 series mentioned in Chapter 5, paragraph 5.2.2.1 is likely to be suitable and should enable a high cruise lift coefficient to be achieved. A high aspect ratio is indicated to reduce induced drag to a minimum although there is likely to be an implied maximum due to practical wing span limitations. Overall the aircraft may be assumed to have characteristics similar to those of a high performance sailplane with an inevitably high wing mass and structural design parameter.

Low observability requirements suggest that the powerplant should be located within the fuselage preferably with the air intake positioned on the top of the fuselage above a high mounted wing. A high wing also has the advantage of giving the maximum wing tip ground clearance in the fully loaded, static, condition. Some degree of sweepback is beneficial, both for stealth reasons and to bring the centre of gravity aft to facilitate engine location but it should not be so high as to compromise cruise efficiency. A quarter chord sweep of about 10° is likely to be acceptable, but this must be reconsidered subsequently.

Tail stability and control surfaces are preferable to canard ones and since the vertical and horizontal surface should not be perpendicular to one another for stealth reasons this could be one of the rare situations where a "butterfly" tail layout is preferable.

Operation from existing airfields without the need for special equipment suggests the use of a fairly conventional landing gear layout. However, there is a likely to be a need for outrigger units to support the high aspect ratio wings and it may be convenient to use a single, central, main leg which retracts into the fuselage.

It is clear that in this example the wing configuration and location will greatly influence the layout of the fuselage. Therefore while it is necessary to complete a first layout of the fuselage in order to undertake the whole conceptual design procedure, it is possible that a revised fuselage may have to be contemplated subsequently, and the process repeated.

An initial layout of the fuselage is shown in Figure AD5.5 and this is sufficient to give a reasonably good indication of the overall dimensions. However, it may need to be refined when the wing and tail geometry are determined.

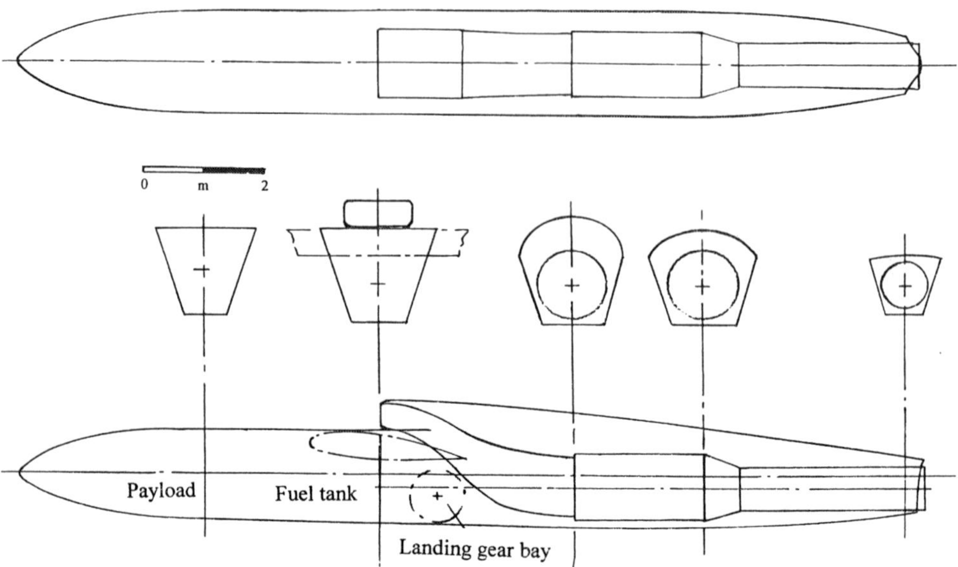

Figure AD5.5 Uninhabited aircraft - fuselage layout

AD5.4.5 Powerplant model

The demanding high altitude and long endurance requirements of the specification imply the need for an accurate model of the powerplant characteristics. In this case the following procedure was adopted:

i) A datum powerplant type was selected. This was the General Electric F118-GE-101 engine as used in the later versions of the Lockheed U2 high altitude reconnaissance aircraft. This was assumed to have an overall pressure ratio of 34, a bypass ratio of 0.76 and a mass flow of 134 kg/s.

ii) These assumed values were used in the Cranfield University "Turbomatch" program to produce a performance deck (see Palmer, J. R. - "The TURBOMATCH Scheme for Gas Turbine Performance Calculations", S.M.E. Cranfield Institute of Technology, October 1983). The performance deck covered operating conditions from sea level static up to a Mach number of 0.7 and 24.5 km altitude.

iii) The low bypass ratio engine thrust model of Chapter 3, paragraph 3.6.2.2 and Table 3.2 was then compared with the performance deck. It was found that to obtain close agreement at altitudes above 9 km it was necessary to use a value of 0.85 for the altitude lapse factor, s, rather than the 0.8 suggested in Table 3.2. This did result in some discrepancies at lower altitudes but this is considered to be acceptable as these conditions are not critical phases of the flight.

iv) The specific fuel consumption model of Chapter 3, paragraph 3.6.3.1 and Eq (3.12a) was also compared with the derived performance deck. It was found that a good match was obtained over all the altitude range by selecting c^l to be 0.9 and by increasing the 0.28 factor of the M_N term to 0.4. This gave an exact match at a Mach number of 0.65 at high altitudes and was otherwise slightly conservative.

AD5.4.6 Spreadsheet formulation (SPREADSHEET AD5.3)

AD5.4.6.1 Initial inputs and assumptions

The parameters selected for primary optimisation are aspect ratio and thickness to chord ratio, but provision is also made for the adjustment of bypass ratio subject to the representation of the model described in the previous paragraph. An aspect ratio of 35 and thickness to chord ratio of 0.21 were selected as the initial values in this case by reference to high performance sailplane characteristics.

The Requirements were derived directly from the specification of paragraph AD5.4.3 together with the interpretation of paragraph AD5.4.4. For completeness certain additions have been made:

Va max	Approach speed to be 30 m/s to ensure adequate control in landing
Vd	The structural design speed to be 180 m/s EAS. This enables a Mach number of 0.65 to be flown at altitudes above 6 km with appropriate margins.
N	Ultimate normal acceleration factor of 3.375 (based on a limit value of 2.25 in this case, but subsequent analysis could indicate a more severe condition due to discreet gust conditions)
N bar	Effective ultimate design factor of 3.7125 (taken to be 1.65 times the limit manoeuvre factor, see Ch.6, para 6.4.3.1)
SusTrn Rate	Sustained turn rate of 1 deg/s at cruise altitude
SusTrn V	Sustained turn velocity of 194 m/s which is $M_N = 0.65$ at 24.5 km altitude
SusTrn Sig	Relative density of 0.03488 at 24.5 km altitude

The assumptions made are:

M1/Mo	Ratio of start of climb mass to take-off value, taken to be 0.99 (Ch.7, para 7.3.6)

S^-0.1	Wing area parameter initially taken to be 0.60, but subject to subsequent correction (Ch.6, Figure 6.1, large combat aircraft value)
Rw	Wetted area factor taken to be 3 as for a sailplane (Ch.6, Table 6.3)
Type Fac	Taken to be 1.1 for this aircraft (Ch.6, Table 6.4)
TE flap	No flaps fitted
SS alpha SS gamma	Second segment climb out factors, not relevant here and set to zero
Flap Fac	0.133 for no flap (Ch.6, para 6.3.2.2)
a Cruise	Speed of sound at nominal cruise altitude (24.5 km)
Cl EAS sig	Relative density at altitude where constant equivalent airspeed climb is assumed to end (6 km)
PP Fac -s	Altitude dependency of powerplant thrust, 0.85 here [see para AD5.4.5, also Ch. 3, para 3.6.2.2 and Eq (3.9b)]
Lam Chrd	degree of laminar flow predicted, taken as 0.45 [Ch.6, para 6.3.2.1, Eq (6.13a)]
Gust Sen	Gust sensitivity switch, 0 for off and 1 for on [see Preliminary calculations, (Mg/S)o gt]
a sus tn	Speed of sound at sustained turn condition (cruise altitude in this case)

AD5.4.6.2 Initial calculations

Ml/Mo	Landing mass ratio (derived from Ch.7, Table 7.1)
Mcr1/Mo	Mass ratio at start of first stage of cruise, in the case where cruise commences below 11 km altitude (uses Ch.7, Table 7.2)
(Clm)o ae	Basic aerofoil maximum lift coefficient, assumed to be 1.5 here [as Ch.6, para 6.2.4.2 but could be somewhat higher for the section chosen, see (Cl use)o]
Del flp (TO)	Take-off high lift device, lift increment (zero here as no flap)
Del flp (L)	Landing high lift device lift increment (zero here)
(Cl use)o	Allowable lift coefficient in cruise, taken to be 0.8 [Ch.6, para 6.2.4.6 suggests 0.65, but increased here because of selection of MS (1) 0.3 series aerofoil section, see paragraph AD5.4.4]
(Cdz)o	Zero lift drag coefficient at zero Mach number [Ch.6, Eq (6.13a) with laminar flow]
Mcr2/Mo	Mass ratio at start of cruise in the case where the cruise commences above 11 km altitude (derived from Ch.7, Table 7.2 with Mcr1/Mo)
Mcl1/Mo	Mass ratio at the start of the first phase of a constant Mach number climb (derived from Ch.7, Table 7.2 with M1/Mo)

AD5.4.6.3 Preliminary calculations

cos delta	Cosine of quarter chord sweep, derived from aerofoil data and critical Mach number, (Mn)crit
Delta deg	Quarter chord sweep
Wave Dr F	Wave drag factor [term in Ch.6, Eq (6.13a) raised to power of 20]
(Cdz)cr	Zero lift drag coefficient at cruise Mach number [Ch.6, Eq (6.13a)]
(Cdz)co	Corrected climb out zero lift drag coefficient [Ch.6, Eq (6.15)]
(Kv)o	Induced drag factor at zero Mach number [Ch.6, Eq (6.14a)]
(Kv)cr	Induced drag factor at cruise Mach number [Ch.6, Eq (6.14a)]
Tau(ClMn1)	Thrust ratio at start of constant Mach number climb (Ch.3, para 3.6.2.2)
Tau(ClMn2)	Thrust ratio at start of constant Mach number climb when cruise is above 11 km altitude (Ch.3, para 3.6.2.2, 11km conditions)
(Cdz)sustn (Kv)sustn	Drag values for sustained turn conditions [Ch.6, Eqs (6.13a) and (6.14a)]
(Beta) sustn	Product of previous two values
Sustn n	Normal acceleration factor in sustained turn [Ch.7, Eq (7.39b)]
Cl max	Maximum lift coefficient [sum of (Clm)o ae and del flp (L) corrected for sweepback, Ch.6, Eq (6.2)]
Cl us	Unstick lift coefficient [sum of (Clm)o ae and del flp (L) factored by 0.8 and corrected for sweepback, Ch.6, Eq (6.4)]
Cl a	Approach lift coefficient [Cl max factored by 0.6, Ch.6, Eq (6.3)]
Cl use	Usable lift coefficient in cruise [(Cl use)o corrected for sweepback]
Va calc	Maximum approach speed consistent with required landing distance [Ch. 7, Eq (7.6b)]
Va	The lower of Va calc and Va as specified, in this case Va as specified
(Mg/S)o ld	Take-off equivalent wing loading derived from Va and Ml/Mo [Ch.7, Eq[(7.6e), no reverse thrust]
(Mg/S)ogt	Take-off equivalent wing loading to meet gust sensitivity requirement [Ch. 5, Eq(5.9b)] (Gust Sen switch used to equate to zero here)
Cor L legth	Corrected value of landing length when Va is limited by specified value [Ch.6, Eq (6.6a) no reverse thrust]

AD5.4.6.4 First stage analysis

The range of take-off wing loading, (Mg/S)o, has been selected somewhat arbitrarily on the basis of the usable cruise lift coefficient, cruise speed and air density at 24.5 km altitude. Although it was not found to be necessary in this case it is possible for a situation to arise where the range would need to be extended to cover the eventual optimum value.

Take-off:

1st app	First estimate of required take-off thrust to weight ratio [uses Ch.7, Eq (7.1b) with (T/Mg)o set to zero in right-hand side]
(T/Mg)o	Second evaluation of take-off thrust to weight ratio [uses value of (T/Mg)o from first approximation in right-hand side of equation; no further corrections considered to be needed here]

Acc. Stop:

(T/Mg)o	Take-off thrust to weight ratio required to meet engine failed acceleration-stop requirement [Ch.7, Eq (7.3a)]

Sec Seg Climb:

Tau co	Not relevant to this case, the second segment climb condition, but
(Cd)co/Clu	the values are based on a speed of 1.1 times the unstick speed to give tau co
(T/Mg)o	[uses Ch.7, Eq(7.4a)]

Sustained turn:

D bar	Zero lift drag to weight ratio in sustained turn case [Ch.7, Eq (7.13e)]
(T/Mg)sustn	Thrust to weight ratio required in sustained turn [Ch.7, Eq (7.40a)]
(T/Mg)o	Equivalent take-off thrust to weight ratio (allows for mass reduction and altitude effect on powerplant)
Cl case	Lift coefficient in sustained turn, to check acceptability of value

Start of cruise:

Fac Qm	Factor in climb prediction [Ch.7, Eq (7.22b)]
(T/Mg)o-1	Equivalent take-off thrust to weight ratio required to achieve a residual rate of climb of 1.5 m/s at a cruise altitude of 11 km [Ch.7, Eq (7.23b); included to cover the case when the climb ends at or below 11 km]
(T/Mg)o-2	Equivalent take-off thrust to weight ratio required to achieve a residual rate of climb of 0.5 m/s at a cruise altitude above 11 km [Ch.7, Eq (7.27b) is assumed to apply to this condition although strictly it is for altitudes up to 20 km, not the 24.5 km here; this implies that the speed of sound is constant from 11 km up to 24.5 km which is about 1% in error]
Cl case	Lift coefficient at start of 24.5 km altitude cruise condition
Cd	Total drag coefficient at start of 24.5 km altitude cruise
L/D	Cruise lift to drag ratio (Cl case/Cd)
Mn*L/D	Product of cruise Mach number and lift to drag ratio, for reference

Landing-rev thrust:

(T/Mg)o	Selected as maximum value derived from all other conditions
L length	Reverse thrust landing length [uses derived (T/Mg)o in Ch.7, Eq (7.6e)]

Summary:
All the first stage analysis calculations are summarised together with the landing and gust sensitivity wing loadings previously derived. The summary is used to produce the diagram of (T/Mg)o vs (Mg/S)o. Inspection of the diagram shows, as could be anticipated, that the critical performance condition is start of cruise at 24.5 km altitude and that a wing loading somewhat less than the landing equivalent value may be the optimum. While the actual value can only be determined by the optimisation process, a take-off wing loading of about 700 N/m^2 appears to be likely. (Subsequent optimisation gives a value of 681.66 N/m^2.)

Results:
The values of the various parameters relevant to (Mg/S)o of 681.66 N/m^2 are given. Included is an evaluation of the Structural Parameter, SP [using Ch.5, Eq (5.8a)].

AD5.4.6.5 Second stage assumptions and input data

Climb path:

Climb EAS	The relatively high installed thrust enables the initial climb to be made at the structural limit cruise speed, V_C, taken here to be 150m/s EAS
Cl EAS H2 Cl EAS sig	Constant EAS climb assumed to end at 6km altitude, with the relative density value at that altitude
Cl Mn	Remainder of climb assumed to be at constant Mach number of 0.65, the limiting critical value
Sig cr 1	First stage of constant Mn climb ends at 11 km altitude with relative air density value given for that altitude
Sig cr 2	Second stage of constant Mn climb ends at 24.5km altitude with appropriate relative air density (Set to zero should cruise start below 11 km altitude)
End Cl 1	Altitude at the end of first stage of constant Mn climb, 11 km as above

Assumed:

Lambda	Wing taper ratio chosen to be 0.35 by comparison with high performance sailplanes
(T/Mg) eng	Bare engine thrust to weight ratio [Ch.6, Eq (6.26a) taken to be 6 in this case]
Op It Fac	Operational items factor assumed to be zero in this case
AppFuel/Mo	Fuel used in descent and approach as ratio of take-off mass (Ch.7, para 7.4.4)
V bar Vv bar	Horizontal and vertical tail volume coefficients (selected from Ch.8, Table 8.1A as for sailplanes)

Input data:

Fus L Fus B Fus H	Fuselage overall length, *L*, maximum width, *B*, and maximum height, *H*, as derived from the initial fuselage layout, Figure AD5.5
c1	Lifting surface mass coefficient, value for long range subsonic aircraft assumed to be 0.0007 (Ch.6, Table 6.7, this requires further investigation when the actual take-off mass is determined. Allows for use of composites)
c2	Fuselage mass coefficient, value for land based combat aircraft with allowance for use of composites (Ch.6, Table 6.6 reduced to 0.03 due to absence of manned cockpit requirement)
c3	Powerplant installation factor (Ch.6, Table 6.8)
c4	Systems mass factor, taken as general military value (Ch.6, Table 6.9 reduced to 0.11 as payload already includes avionics items)
c5	Lifting surface mass ratio (Ch.6, Table 6.10)
Del lw	Centre of gravity of wing mass relative to overall centre of gravity of aircraft (c.g.) (assumed to be 0.1 wing mean chord aft of aircraft c.g.)
l Fus	Location of fuselage structure centre of gravity aft of nose (estimated from fuselage layout)
l Tail	Location of tail unit centre of gravity aft of nose (assumed to be 0.9 of Fus L)
l PP	Location of powerplant centre of gravity aft of nose (estimated from fuselage layout)
l SYS	Location of systems mass, excluding landing gear, aft of nose (estimated from fuselage layout)
l PAY	Location of payload mass aft of nose (estimated from fuselage layout)
l OP IT	Location of operational items mass aft of nose (estimated from fuselage layout, although in this case the allocated mass is zero)
Del l Fuel	Location of centre of gravity of fuel mass relative to aircraft centre of gravity (all fuel is assumed to be located in the wing and at 0.1 wing mean chord aft of the aircraft centre of gravity)

AD5.4.6.6 Second stage calculations

(S/Mo)^–0.45	Power of wing loading used in lifting surface mass calculation.
p bar	Differential pressure appropriate to cruise altitude relative to datum 24.5 km (not actually used here as the aircraft is uninhabited)

Cruise:

Reqd (T/Mg)	Reciprocal of cruise lift to drag ratio gives thrust to weight ratio at start of cruise

Av(T/Mg)	Available cruise thrust to weight ratio [derived from estimated static thrust to weight ratio (T/Mg)o in conjunction with Ch.3, Eq (3.7a) and Table 3.2]
Av/Reqd	Ratio of two previous values
c(des)	Design datum specific fuel consumption [Ch.3, Eq (3.12a) modified as stated in para AD5.4.5]
c(od)	Off design specific fuel consumption [Ch.3, Eq (3.12b)]

Constant EAS climb:

Z	Speed factor in climb [(Ch.7, Eq (7.15c)]
Tau Mn 1	Thrust factor appropriate to Mach number at the start of constant equivalent airspeed climb and sea level [Ch.7, Eq (7.7a) and Table 7.2]
(Vv) EAS	Average climb rate in constant equivalent airspeed climb [Ch.7, Eq (7.15a) with Eqs (7.15d), (7.16a) and (7.16b)]
Dist EAS	Ground distance covered in constant equivalent airspeed climb [Ch.7, Eq (7.21) with the cosine term assumed to be unity]

Constant Mach number climb to 11 km altitude:-

Qm	Factor in climb prediction [Ch.7, Eq (7.22b) allowing for fuel used in constant equivalent airspeed climb and take-off]
Tau 1	Thrust factor at Mach number of 0.65 and start of constant Mn climb altitude [Ch.3, Eq (7.7a) and Table 7.2]
(Vv) Mn 1	Average climb rate in constant Mn climb up to 11 km altitude [Ch.7, Eq (7.22a)]
Dist Mn 1	Ground distance covered in constant Mn climb up to 11 km altitude [Ch. 7, Eq (7.25) with the cosine term assumed to be unity]

Constant Mach number climb from 11km up to cruise altitude:-

Tau 2	Thrust factor at Mach number of 0.65 and 11km altitude [Ch.3, Eq (7.7a) and Table 7.2]
(Vv) Mn 2	Average climb rate in constant Mn climb from 11 km to cruise altitude [Ch.7, Eq (7.26a) with powerplant factor, *s*, taken as unity above 11 km, with allowance made for reduced rate of climb near ceiling, factor X2]
X2	Allowance for reduced rate of climb near ceiling [Ch.7, Eq (7.26b)]
(Dist Mn2)	Calculated ground distance covered in climb above 11km altitude. [Ch.7, Eq (7.25) with the cosine term assumed to be zero]
DistMn2	Actual ground distance covered in climb above 11 km altitude. From the previous cell, or zero when the cruise starts below 11 km altitude

Descent:

Desc Dist	Ground distance covered in descent to land [Ch.7, Eq (7.56)]

Cl EAS:

(c)o	Specific fuel consumption at start of constant equivalent airspeed climb [Ch.3, Eq (3.12a)]
Wf/(Mg)o	Ratio of fuel used in constant equivalent airspeed climb to take-off weight [Ch.7, Eq (7.17)]

Cl Mn:

(c)1	Specific fuel consumption at start of const. Mn climb [Ch.3, Eq (3.12a)]
Wf/(Mg)o	Ratio of fuel used in constant Mn climb to 11 km to take-off weight [Ch.7, Eq (7.24)]

Cl Mn 2:

(Wf/(Mg)o)	Ratio of fuel used in constant Mn climb from 11 km to cruise altitude to take-off mass [Ch.7, Eq (7.28)] (See also Cl Mn 2: Wf/(Mg)o at far end of this row)

Masses:

Mc1/Mo	Ratio of start of cruise mass to take off mass (Mo less fuel used in take off, constant equivalent airspeed, and two stages of constant Mn climbs)
M fus	Fuselage mass [Ch.6, Eq (6.20b)]
c1 bar	Factor in prediction of lifting surface mass [Ch.6, Eq (6.24b)]
Mpp/Mo	Mass ratio of powerplant installation [uses value of c3, calculated thrust to weight ratio and assumed (T/Mg)eng]
Msys/Mo	Systems mass ratio (coefficient c4)
Mop it	Mass of operational items (Zero in this case)
M fixed	Fixed mass (Sum of fuselage mass, payload and operational items mass)
Net range	Range to be covered in cruise (specified range less ground distance covered in climbs and descent)
Log 10	Logarithm to base 10 of mass ratio of start to end of cruise [Ch.7, Eq (7.50b)]
Mc1/Mc2	Cruise mass ratio (antilog of previous value)
Mc2/Mo	Ratio of end of cruise to take off mass
Mf/Mo	Total mass of fuel as ratio of take-off mass [Mo less (Mc2 plus descent fuel)]
Kappa Mo	Sum of terms directly proportional to mass, as ratio of mass (fuel, systems and installed powerplant masses)
x bar (0.25)/ root chord	Location of 0.25 aerodynamic mean chord on wing centreline chord. [uses geometry as subsequently defined and Ch.8, Eq (77b)]
Cl Mn 2 Wf/(Mg)o	Selection of fuel weight ratio in second phase of constant Mn climb from (Wf/(Mg)o) or equated to zero when cruise starts below 11 km altitude

AD5.4.6.7 Second stage analysis and calculated take-off mass
The first part of the second stage analysis uses the calculated values to deduce the actual take off mass of the aircraft. The second part then uses this value to make a first estimate of the wing position along the body which brings the centre of gravity to the 0.25 aerodynamic mean chord position. Use of the assumed volume coefficients then enables the equivalent tail sizes to be calculated, based on the horizontal and vertical values.

(Mo)est1	A first estimate of the take-off mass derived by assuming that the lifting surface mass is 12% of the total.
(Mo)est2	Initially this has the same value as (Mo)est1, and is entered directly as a number
Kappa*Mo	Mass of the terms directly proportional to M_0 [uses c1bar with (Mo)est2]
M lift surf	Mass of lifting surfaces [Uses c1 bar with (Mo) est2]
(Mo) calc	Initially this is the sum of (M fixed), (Kappa*Mo) and (M lift surf) to give the total mass. This is the target cell used for optimisation by invoking SOLVER©. The value is minimised in this case by changing aspect ratio, thickness to chord ratio, wing loading and (Mo)est2, subject to the following constraints: (Mo)calc to be equal to (Mo)est2 Wing loading to be less than or equal to the equivalent landing and manoeuvre values (when applicable the wing loading to be greater or equal to the gust sensitivity value) Thickness to chord ratio to be equal to or less than 0.21 Cruise lift coefficient to be equal to or less than 0.788 Structural parameter, SP, to be equal or less than 40 (aspect ratio is also constrained at a later stage, see paragraph AD5.4.7)
error	The difference between (Mo) calc and (Mo)est2 as a check on the optimisation
l CG	Location of 0.25 aerodynamic mean chord point aft of the fuselage nose, derived from the longitudinal balance of all the mass items (the nose landing gear and main landing gear positions being derived from the fuselage layout)
l WG APX	Location of leading edge of wing centreline chord aft of the nose [derived from l CG and x bar (0.25)/root chord]
l TL ARM	Distance between the centre (Derived from l CG and l Tail)
S Hor Tail S Vert Tail	Equivalent areas of horizontal and vertical tail surfaces (Derived from wing area, mean chord, l TL ARM and assumed volume coefficients)

A5.4.6.8 Summary of final results

The summary of the final results is directly derived from the previous calculations.

Individual masses, as used for the centre of gravity calculations, are obtained as follows:

i) Wing mass is isolated from the tail mass by using coefficient c_5 in conjunction with the total lifting surface mass.
ii) Landing gear is assumed to account for 4% of the 11% allocated to systems with the nose gear being 0.15 of the total landing gear value.
iii) Initially all the fuel has been assumed to be located within the wing.
iv) Wing area, S, follows from (Mg/S)o and the wing geometry follows from the aspect ratio, A, taper ratio, lambda, and sweepback (Del 0.25).
v) Sea level static thrust follows from (T/Mg)o and (Mg)o.

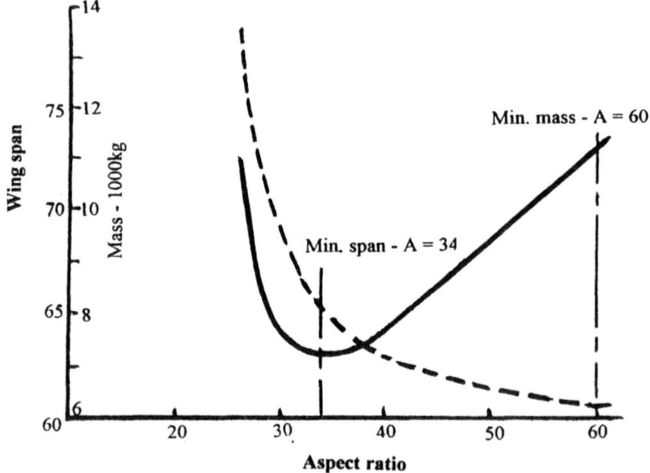

Figure AD5.6 Uninhabited aircraft - effect of aspect ratio constraint

AD5.4.7 Use of results to examine wing span

As was anticipated the optimum solution in this case coincides with a very high aspect ratio and correspondingly large wing span. An aspect ratio of almost 60 is required with a wing span of 73.2 m and with a take-off mass of 6210 kg. The structural parameter has a value approaching 40 and even if it is possible to construct a wing of this configuration with adequate strength it can be predicted that its flexibility would make ground operations virtually impossible. The effect of constraining the aspect ratio is illustrated in Figure AD5.6. As the aspect ratio is reduced towards about 34 the wing span also decreases and so does the structural parameter. However, total mass, of course, does increase. Further reduction of aspect ratio below 34 results in increased wing span, due to the increased wing area required to meet the cruise lift coefficient limit condition as the mass increases. It is not possible to use an aspect ratio below 26 which gives a wing span of about 72 m. In passing it is worth noting that the whole range of possible aspect ratios from 26 up to 60 have thrust requirements which can be met by the size of powerplant used for the fuselage layout.

While other solutions may be acceptable, and possibly preferable, it is concluded that for the present purposes the best solution is the one which gives the minimum wing span. The aspect ratio of 34 is within high performance sailplane experience and the indicated structural parameter of 27.4 is not unreasonable. The take-off mass and thrust values are 8164 kg and 42930 N respectively. Even here the wing span is large at 63.2 m and the operational feasibility of the concept demands very careful consideration.

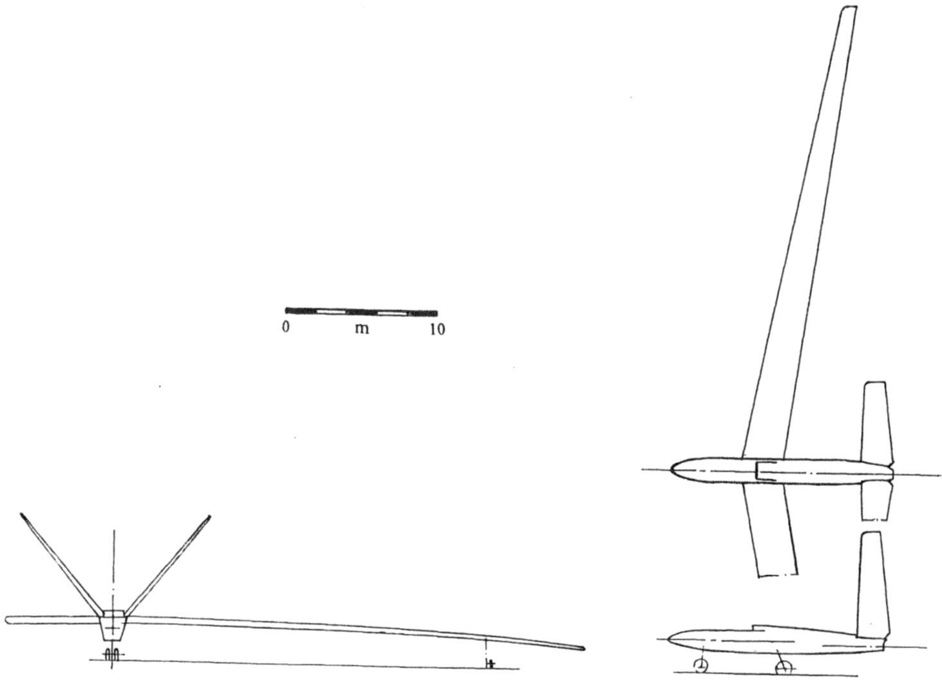

Figure AD5.7 Uninhabited aircraft - general arrangement

AD5.4.8 Preliminary general arrangement

The data for the case having an aspect ratio of 34 are shown in Spreadsheet AD5.3, and the geometric information can be used in conjunction with Figure AD5.5, the fuselage layout, to produce a general arrangement drawing of the aircraft. This is shown in Figure AD5.7 where use has been made of typical tail geometry from Chapter 8, Table 8.1B and the equivalent horizontal and vertical tail areas to derive a butterfly configuration. More detailed analysis may now be undertaken. This must include an initial design of the wing structure to assess the spanwise bending flexibility as well as a review of all the assumptions made to derive the concept. In many respects this design is only marginally achievable and detailed analysis might possibly indicate that the required performance, especially range and cruise altitude, may have to be reduced somewhat to yield a practical design.

AD5.5 Short take-off/vertical landing supersonic combat aircraft

AD5.5.1 Introduction

There have been many proposals for a supersonic successor to the British Aerospace Harrier. A key issue in the design of such an aircraft is the means of deriving and applying the vertical thrust required for low speed operations. Methods investigated include:

a) Independent direct lift engines. The weight and volume penalties of this concept rules it out for an advanced combat aircraft where the conventional performance requires an installed thrust to weight ratio of the order of one in any case.

b) Tilting main propulsion engines. The usual configuration suggested is for two engines to be located out along the wing. This can introduce layout constraints but the main objection is the difficulty of roll control in the event of the failure of one engine.

c) Development of the Rolls-Royce Pegasus engine used in the Harrier by adding fuel in the forward, bypass, flow to augment thrust. This so-called "plenum chamber burning" modification introduces problems of ground erosion and ingestion of hot gases into the air intakes. Apart from these problems the propulsive system is large in frontal cross-section and aircraft layout is restricted by the need to locate the engine at the centre of gravity of the aircraft.

d) Retention of the pair of rear nozzles of a Pegasus-type engine, but splitting the exhaust flow between them and a further deflected exhaust located towards the rear of the aircraft. This arrangement requires the basic powerplant to be well forward in the fuselage so that much of the volume is occupied by it and the exhaust systems. The likelihood of hot gas ingestion is still present even though the nozzles are relatively further aft than with plenum chamber burning. Nevertheless this concept was adopted for the Boeing X-32s design for the JSF (Joint Strike Fighter) competition. The hot gas ingestion problem requires the introduction of an air screen derived from fan flow between the intakes and the two side hot gas nozzles. One advantage of this concept is that the modifications needed to a conventional engine are relatively small.

e) Movement of the main powerplant aft, retaining a single, rear deflecting nozzle. There are various ways of obtaining vertical thrust balance about the longitudinal centre of gravity:

i) Provision of a forward remote lift nozzle supplied by air bled from the main engine bypass flow. In one version of this arrangement additional fuel is added at the front nozzle to augment its effect, leading to the RALS concept (Remote Augmented Lift System). However, this also introduces hot air ingestion difficulties. As an alternative the augmentation may be dispensed with but this implies a much larger bleed air requirement and the strong possibility of a need for a variable cycle engine.

ii) The introduction of an extra, forward, fan to the engine with Pegasus type nozzles deflecting its flow downwards for vertical flight. This so-called 'tandem fan' configuration also results in a long powerplant.

iii) The provision of one or two special vertical lift engines in the forward region of the fuselage, as on the YAK 141 aircraft. Since this part of the system is used only for vertical flight it represents a substantial design penalty.

iv) As an alternative to dedicated lift engines it is possible to incorporate a vertical lift fan driven from the main engine. Compressed air drive is advantageous in giving layout flexibility but is inefficient in propulsion terms. An alternative is a direct mechanical drive to the fan through a 90° gearbox. The Lockheed X-35s JSF design used such a system where it is claimed that the presence of the front fan flow effectively prevents the possibility of hot gas ingestion.

Whichever concept is employed it is necessary to provide means for pitch, roll and yaw control of the aircraft during vertical flight. This requires the introduction of control nozzles fed by air bled from the main powerplant. Roll control nozzles are inevitable for the accepted situation where a single propulsion system is located within the fuselage. However, pitch and yaw control may possibly be provided from the main vertical lift nozzles by such means as nozzle deflection and variation of the fore and aft thrust split.

In summary, it may be concluded that all feasible vertical lift systems introduce some penalties, the relative severity of which is critically dependent upon the layout of the aircraft. Vertical lift requirements inevitably reduce the flexibility in the layout of the design due to the need to balance the thrust components about the centre of gravity, regardless of the actual concept employed.

AD5.5.2 Configuration background and powerplant

Over a period of time the College of Aeronautics, Cranfield University, has investigated the application of several vertical lift systems in various project studies. The coverage has included plenum chamber burning and remote augmented and un-augmented lift systems in conjunction with a deflecting rear nozzle. The S-95 project originally developed by R. Hewson and M. Rosa, used this latter concept and forms a starting point for the present example (see S-95 project specification, DAeT 9500, prepared by Fielding, Howe, Rosa, Smith, Young and Jones and reported by Smith. H, in College of Aeronautics Aerogram, Vol. 8 No 4, June 1997). The aircraft configuration was based on a close-coupled canard layout associated with a low wing and it incorporated low observability features.

The powerplant used for the S-95 project was a "selected bleed variable cycle" concept designed in the School of Mechanical Engineering, Cranfield University (see ASME91-GT-388, June 1991, by Do Nascimento and Pilides). This powerplant was intended to operate with a bypass ratio of one at Mach numbers below about 0.9, but with a bypass ratio of 0.4 at higher Mach number. Provision was made for afterburning which could be used in either mode. In the high bypass ratio mode a valve located at the end of the low pressure compressor could be opened to supply a large quantity of air to the un-augmented front nozzle system. Change of engine cycle was achieved by closing this low pressure valve and opening a high pressure valve located between the intermediate and high pressure compressor stages. This high pressure bleed was equivalent to the bypass airflow and was subsequently mixed with the core flow ahead of the afterburning stage. An important feature of this engine concept was that the low pressure, high bypass ratio, mode was only unavailable at the front nozzle system and therefore its application to forward thrust was dependent upon the deflection capability of the front nozzles. In the case of the S-95 design study the front nozzle deflection was limited to small change from the vertical to provide control force components. Hence the high bypass ratio mode could only be used for nominally vertical flight and, possibly, during subsonic manoeuvres. For conventional flight at both subsonic and supersonic speeds the engine operation was in the lower bypass ratio mode. This difficulty could be overcome by arranging for the deflection of the front nozzle system to an aft, horizontal mode. The consequent penalty on the layout of the aircraft would have to be balanced against the better potential performance in subsonic cruise. However, the use of the lower bypass ratio in all conventional flight is comparable with that of an aircraft using a constant cycle powerplant matched for a compromise between subsonic and supersonic flight.

For the purposes of the present example the powerplant system is assumed to be the same as that described for the S-95 study, thereby providing a basis for the design and facilitating comparison. For the purpose of powerplant modelling the engine is assumed to be of conventional design with no cycle variation. It will be further assumed that there is a provision for sufficient low pressure bleed to enable a longitudinal thrust split of 40 - 60% between the front and rear nozzles, respectively, to be achieved.

AD5.5.3 Specification

The specification is conveniently presented in three parts, namely point performance, mission requirements and supplementary data.

a) *Point performance:-*

i) Level flight: $M_N = 1.1$ at sea level
$M_N = 1.2$ at 11 km altitude (without afterburning)
$M_N = 1.6$ at 11 km altitude

ii) Sustained turn rates: 16 deg/s at $M_N = 0.5$ and sea level
11 deg/s at $M_N = 0.9$ and sea level
12 deg/s at $M_N = 0.9$ and 6 km altitude
8 deg/s at $M_N = 1.2$ and 6 km altitude

iii) Instantaneous turn rates: 20 deg/s at $M_N = 0.55$ and 3 km altitude (without the use of thrust deflection)
25 deg/s at $M_N = 0.5$ and sea level

iv) Specific Excess Power: 150m/s at $M_N = 1.4$ and 9 km altitude
300m/s at $M_N = 0.9$ and sea level
190 m/s at $M_N = 0.9$ and 6 km altitude

v) Field performance: Short take-off roll of 125 m; dry concrete at Inernational Standard Atmosphere (ISA)+15°C zero wind (full internal weapons and fuel)
Vertical landing at appropriate mission landing mass at ISA+15°C and in head winds of up to 10 m/s

b) Mission requirements:

i) Air superiority - *Mission A*
Weapon load: 2 x short range air to air missiles
4 x medium range air to air missiles
(any combination of 4 missiles carried internally)
1 x 20 mm M61 A1 Vulcan gun with 400 rounds
Warm up and taxi (5 min at ground idle) and short take-off
Accelerate and climb to $M_N = 0.8$ and 11 km altitude
Cruise for 436km at $M_N = 0.8$ and 11 km altitude
Accelerate to $M_N = 1.5$ at 9 km altitude
Supersonic combat at $M_N = 1.5$ and 9 km altitude: one 360° sustained 3g turn
Descend to sea level and $M_N = 0.9$

Subsonic combat at $M_N = 0.9$ and sea level: two 360° sustained 8g turns (all missiles and ammunition expended by end of subsonic combat)
Climb to 11 km altitude and $M_N = 0.8$
Cruise for 436 km back to base
Descend to sea level and hover for 1 min out of ground effect
Vertical landing with at least 5% internal fuel remaining

ii) Offensive support - *Mission B*
Weapon load: 4 x air to ground missiles (e.g Maverick)
2 x short range air to air missiles
(any combination of 4 missiles carried internally)
1 x 20 mm M61 A1 Vulcan gun with 400 rounds
Warm up and taxi (5 min at ground idle) and short take-off
Accelerate to $M_N = 0.6$ at 80 m altitude (nominally sea level)
Cruise at $M_N = 0.6$ at 80 m altitude for 200 km
Accelerate to $M_N = 0.8$ at 80 m altitude
Cruise at $M_N = 0.8$ at 80 m altitude for 100 km
Fly at $M_N = 0.9$ at 80 m altitude for 2 minutes
Subsonic combat at $M_N = 0.9$ and sea level: four 360° sustained 8g turns. (all missiles and ammunition expended by end of combat).
Cruise at $M_N = 0.8$ at 80m altitude for 100km
Cruise at $M_N = 0.6$ at 80m altitude for 200km
Hover for 1 min out of ground effect
Vertical landing with at least 5% internal fuel remaining

c) Supplementary data

i) Avionics fit
Integrated communications, navigation and identification
Internal navigation and global positioning system
Controls and displays as relevant
Radio altimeter
Digital computers
Airborne radar
Radar warning system
Radar jamming system
Weapons management system

ii) Structural design
Design speed, V_D, 386 m/s EAS
Design maximum Mach number, M_D, 1.8
Limit normal manoeuvre factors at combat mass: +9 and −3.
Ultimate factor 1.5
Design vertical landing velocity 4.45 m/s
(Runway CBR of 8 for 100 passes at design landing mass - see paragraph AD1.3.3)

(Bird impact of 1.8 kg at 260 m/s and sea level)

iii) General
Chaff and flare dispensers required
Low observability characteristics required
Consideration of vulnerability, reliability and maintainability
Life cycle costs to be as low as feasible

(Note that not all of these requirements are relevant to the conceptual design phase)

AD5.5.4 Configuration development

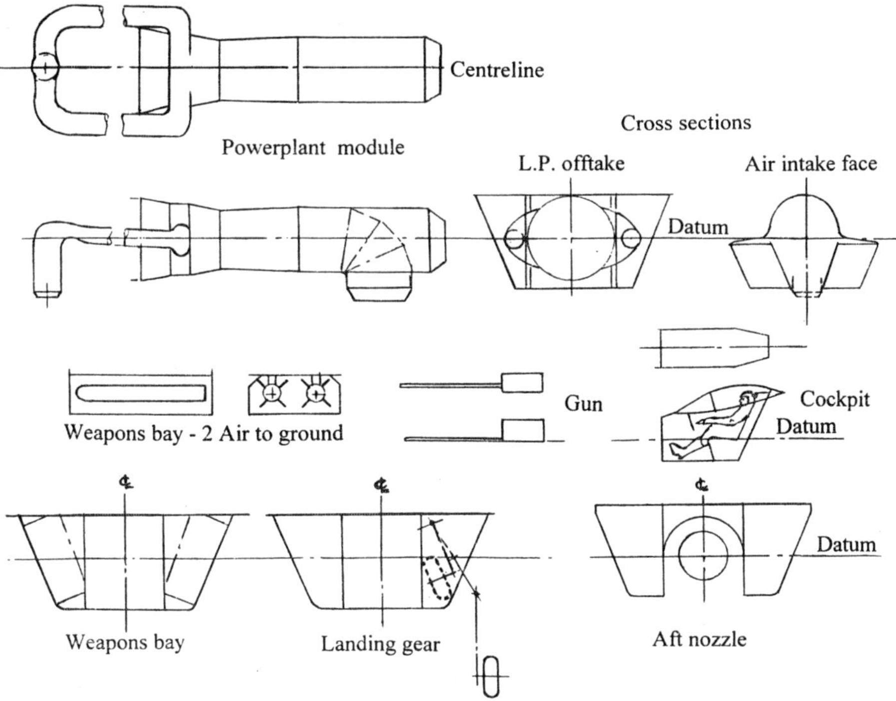

Figure AD5.8 STOVL combat aircraft - fuselage modules

The S-95 project study referred to in paragraph AD5.5.2 was based on a close-coupled canard configuration with a low wing. The wing was blended into a rather wide body with the weapons being carried on semi-submerged pallets located either side of the powerplant installation. The air intake system was split to pass either side of the forward nozzle, the pair of inlets being positioned just behind and below the cockpit region. To avoid acoustic damage the fuselage behind the main, rear, propulsive nozzle was split laterally to take the form of a rudimentary twin boom configuration. Each boom carried an outwards inclined fin. A wing quarter chord sweep of 20° associated with an aspect ratio of 3 resulted in a diamond-shaped planform, which was reproduced on the canard surface. Both leading and trailing edge flaps were found to be necessary to meet the demanding manoeuvre requirements in spite of the advantage of the close-coupled canard layout in this respect Although careful attention was given to detail stealth considerations the association of the canard with semi-submerged weapons resulted in unfavourable characteristics from this aspect. A further difficulty became apparent in that the structural layout of the low wing was seriously compromised by the need to gain access for engine removal.

As a consequence of the perceived disadvantages of the configuration selected for the S-95 study the present concept is based on the use of a high wing which allows the engine to be removed by lowering it through access doors in the fuselage underbelly. As it is difficult to effectively combine a close-coupled canard with a high wing a conventional tailed configuration will be adopted. Provision will be made for the carriage of any four missiles in two weapons bays located in the sides of the fuselage, alongside the powerplant. The penalty on the stealth characteristics of the external carriage of two further missiles will be accepted on the basis that an operational compromise is acceptable. Two further side fuselage bays positioned behind the weapons bays will be used to accommodate the retracted main landing gear units. The nose inlet and rear fuselage layout of the S-95 will be retained, albeit being subject to detail change. Likewise the wing quarter chord sweep of nominally 20° will be kept, at least initially.

While the fuselage layout of a supersonic aircraft must be considered together with the lifting surface configuration in order to determine an acceptable longitudinal area distribution, it is possible to investigate an initial layout by consideration of the cross-section shapes needed to accommodate the various items located within the fuselage. As suggested in Chapter 4 this is best done by utilising appropriate combinations of modules such as powerplant, crew, weapons, etc. Figure AD5.8 shows the relevant modules for this design together with cross-sections developed from them. These form the basis of the layout of the fuselage as shown in Figure AD5.9, from which the overall dimensions may be established. The lifting surfaces shown in Figure AD5.9 are those derived at the conclusion of the initial design synthesis, see paragraph AD5.5.6. They are included for completeness since the fuselage layout cannot be finalised without them, but a notional arrangement has to suffice for the commencement of the synthesis process. The effective length to diameter ratio needed to evaluate the wave drag contribution of the fuselage is derived by finding the equivalent diameter of the area defined by the maximum fuselage cross-section less the engine intake face area, see Chapter 6, paragraph 6.3.3.4.

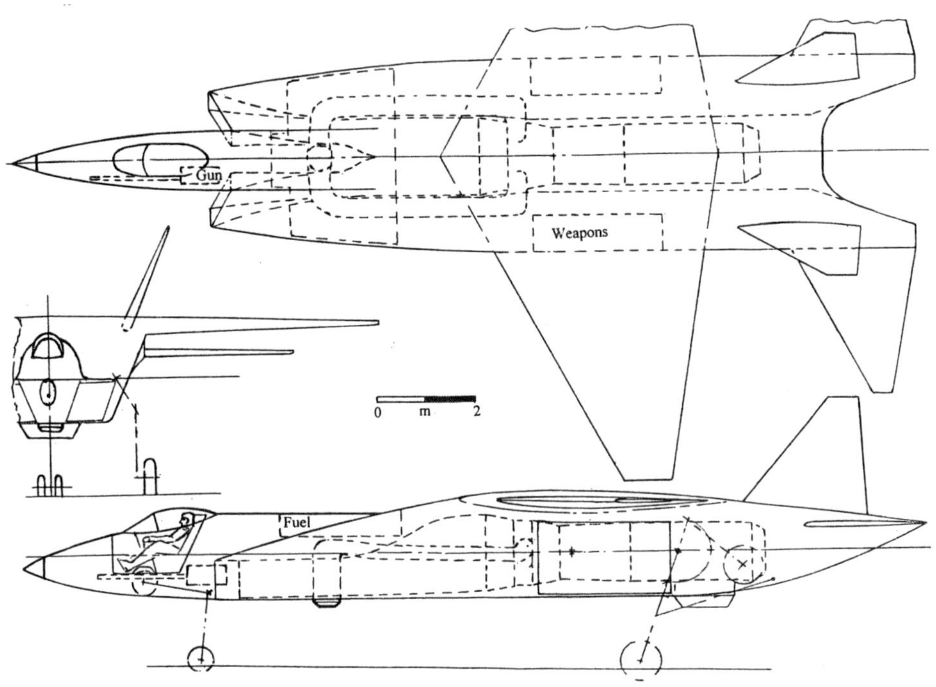

Figure AD5.9 STOVL combat aircraft - fuselage and lifting surface layout

AD5.5.5 Spreadsheet formulation (SPREADSHEET AD5.4)

AD5.5.5.1 Initial inputs and assumptions

The parameters selected for optimisation are wing aspect ratio, thickness to chord ratio, quarter chord sweep, engine bypass ratio and number of engines. However, in this case the latter two parameters have been not been varied. The Requirements have been derived directly from the specification given in paragraph AD5.5. Engine bypass ratio has been taken as 0.4, see paragraph AD5.5.2.

Assumed values are:

M1/Mo	Ratio of climb to take-off mass, taken to be 0.99 as standard (Ch.7, para 7.3.6)
S^-0.1	Wing area parameter (Ch.6, Figure 6.1 gives 0.665 for combat aircraft as an initially assumed value, to be subsequently corrected)
Rw	Wetted area factor (Ch.6, Table 6.3 suggests 4 to 5, assume 5 here)
Type Fac.	Assume 1.2 (Ch.6, Table 6.4)
Ko	Area distribution factor (Ch.6, para 6.3.3 suggests 1.7 to 2.0 for a "reasonable" combat aircraft area distribution, assume 1.9 in this case)
Kf	Fuselage shape factor (Ch.6, para 6.3.3, assume 1.0 as a conservative value when associated with Ko =1.9)
Kw	Wing shape factor in induced drag term (Ch.6, para 6.3.3 gives 0.2 when conical camber is not used)
S/l^2	Aircraft shape parameter (Ch.6, Table 6.5 suggests values in the range 0.14 to 0.2 for conventional tail with stealth features, assume 0.14 as an initial value, to be subsequently corrected when the actual geometry becomes defined)
Flap Fac	1.0 for single slotted flaps which are assumed here (Ch.6, para 6.3.2.2)
Cl EAS sig	Air density at end of constant equivalent airspeed climb, assume that this climb phase ends at 7.5 km for the first estimation of the fuel usage
Cl Mn sig	Air density at end of constant Mn climb, assume that this climb constant is up to 11 km altitude
Trans sig	Altitude for transonic acceleration is 9 km as given in the specification (Mission A - Air Superiority)
SS alpha SS gamma	Second segment climb parameters, not applicable to this example
a Cr1	Speed of sound for cruise 1 condition, sea level case
a Cr 2	Speed of sound for cruise 2 condition, cruise at 9 km altitude as given in specifications (Mission A - Air superiority), speed of sound
Wing	Assumed wing configuration is variable camber leading edge flaps (Ch.6, para 6.2.5, variable geometry leading edge)
Del L	Lift coefficient increment at low speed (Ch.6 Table 6.2 gives 0.4 for the chosen configuration)
Del H	Lift coefficient increment at high speed (Ch.6, Table 6.2 gives 0.8 for the chosen configuration)
(T/Mg)ass	For the first approximation for use in calculations assume (T/Mg)o is 0.8 (actual value is not critical to final output).
PPFac s	Engine data for low bypass ratio suggests a value of 0.8 (See Ch.3, para 3.6.2.2)
F tau	Engine reheat thrust factor (Ch.3, para 3.6.2.3 data for the engine model indicates 1.13)

c'	Datum specific fuel consumption factor (Ch.3, para 3.6.3.2, 0.96 assumed here)

AD5.5.5.2 Initial calculations

M-ClMn/Mo	Ratio of start of constant Mn climb to take-off mass (Ch.7, Table 7.2 gives the fuel used in constant equivalent airspeed climb, hence this ratio may be derived)
Mcr1/Mo	Ratio of start of cruise 1 to take-off mass, cruise is at sea level, so the ratio is M1/Mo
MTrns/Mo	Ratio of transonic acceleration to take-off mass (this is derived from the mass used in the constant equivalent airspeed and constant Mn climbs Ch.7, Table 7.2)
Mcr2/Mo	Ratio of start of cruise 2 to take-off mass (derived from mass at start of transonic acceleration by assuming fuel used in the transonic acceleration is $0.01M_0$, Ch.7, para 7.7.4.5)
Ml/Mo	Landing mass ratio (derived from Ch.7, Table 7.1)
l/d body	Effective overall length to diameter ratio of body, value of 8 is derived from the initial fuselage layout (see Figure AD5.9)

AD5.5.5.3 Preliminary calculations

The large number of point performance requirements leads to an extensive set of preliminary calculations.

The first row contains general details:

(Cdz)o	Zero lift drag coefficient in incompressible flow [Ch.6 Eq (6.13a)]
cos delta	Cosine of wing quarter chord sweep
(Cl)L	Maximum lift coefficient at low speed [Ch.6, Eq (6.8)]
(Cl)H o	Maximum lift coefficient in high speed conditions [Ch.6, Eq (6.7)]
(Cl)us	Unstick lift coefficient [taken to be 0.75 times (Cl)L]
(Cl)a	Approach lift coefficient [approach assumed to be a 1.2 times stall speed and hence value is 0.7 times (Cl)L]
(Cl)Instn 1 (Cl)Instn 2	Lift coefficients at instantaneous turn conditions 1 and 2 [product of (Cl)H o and (1-0.25 respective Mn), Ch.6, Eq (6.7)]
Va calc	Calculated approach speed to meet the landing length requirement [Ch.7, Eq (7.6b)]
Va	The lower of Va calc and Va as specified, in this case the latter value.
WDF cr1	Wave drag factor corresponding to the first (subsonic) cruise case [the term in brackets to power of 20 in Ch.6, Eq (6.13a)]
(Cdw) bar	Coefficient of the volume wave drag coefficient [Ch.6, Eq (6.17b)]

The second row of the preliminary calculations gives the values of the zero lift drag coefficient for the numerous flight conditions. These can be identified with the specification point performance requirements except for (Cdz)9 and (Cdz)10sb which are the values for Mach numbers of 0.9 and 1.0 using the subsonic equation, and (Cdz)10sp (Cdz)11, (Cdz)12, (Cdz)15 and (Cdz)16 which are for Mach numbers of 1.0, 1.1, 1.2, 1.5 and 1.6 respectively based on the supersonic equation [Ch.6, Eqs (6.13a) and (6.17a) respectively]

The third row gives the values of the induced drag factor K_V, for the same conditions as those covered for the zero lift coefficients [Ch.6, Eqs (6.14a) and (6.18) respectively].

The fourth and fifth rows give the thrust parameters (τ). D for the dry and W for the wet (afterburning) engine conditions respectively [Ch.3, Eqs (3.7a), (3.7b) and (3.8) and Table 3.2]. The altitudes are appropriate to the specification cases.

The sixth row give the values of Beta, the product of (Cdz) and (Kv), for the various conditions outlined above. Also in this row are:

(Cdz/Cl)co	The ratio of (Cdz) and (Cl) in the climb out condition with flaps extended
(Cdw/Cdz)	The ratio of the volume wave drag and basic zero lift drag coefficients used in the Cr2 supersonic cruise case

The seventh row of the preliminary calculations gives the normal acceleration factors associated with the required instantaneous and sustained turn rates. In the case of the instantaneous turns the calculated values are compared with the structural design limit of $9g$ and the lower value taken [Ch.7 (Eq 7.39b)]. The other terms in the seventh row are:

(Mg/S)o ld	The take-off equivalent wing loading for a conventional landing as defined by approach speed Va [Ch.7, Eq (7.6a) without reverse thrust]
(Mg/S)o gt	Take-off wing loading equivalent to gust sensitivity case (Not used on this example)
(Mg/S)oIT1 (Mg/S)oIT2	Take-off equivalent wing loadings of the instantaneous turn cases based on the available lift coefficients
(T/Mg)oVL	Take-off equivalent thrust to weight ratio to enable the vertical landing to be undertaken [the ratio of landing to take-off mass factored by 1.15 to allow for control, Ch.7, para 7.4.5]

The eighth row gives values of the parameters used for calculation of the supersonic climb and transonic acceleration.

Supersonic climb:

Mn'	Mach number/height change factor [Ch 7, Eq (7.34f)]
Mn bar	Mach number change factor [Ch 7, Eq (7.34h)]
tau bar	Thrust ratio correction factor [Ch 7, Eq (7.34g)]
del	Zero lift drag correction factor [Ch 7, Eq (7.34j)]
Kv bar	Induced drag correction factor [Ch 7, Eq (7.34l)]
A bar B bar C bar	Terms to correct thrust, zero lift drag and induced drag respectively [Ch.7, Eqs (7.34c), (7.34d) and (7.34e)]

Transonic acceleration:

T a bar	Thrust correction factors [Ch 7, Eqs (7.44a) and (7.44b)]
T b bar	

AD5.5.5.4 First stage analysis

The range of take-off wing loading (Mg/S)o has been selected by reference to Chapter 5, Table 5.3 and extended at both low and high values because of the special characteristics of this requirement. The already derived value of the equivalent landing case is covered as are the instantaneous manoeuvre values.

Conventional take-off:

1st app	The previously assumed value, (T/Mg)ass, is used to make a first estimate of the required (T/Mg)o [Ch.7, Eq (7.1b)]
(T/Mg)o	The second calculation of (T/Mg)o uses the first approximation (a further iteration could be needed if convergence is not satisfactory)

Acc stop:

(T/Mg)o	The thrust to weight ratio needed to meet the engine failed condition in take-off within the specific conventional distance [Ch. 7, Eq (7.3b)]

Sec Seg Climb:

(Tau)co Dry	Thrust factor at climb out speed [Ch.3, Eq (3.7a) and Table 3.2 no afterburning]
(T/Mg)o D	The equivalent take off thrust to weight ratios[(Uses Ch.7, Eq (7.4a), no afterburning, but zero here since condition is not relevant]
(Tau)co W	Thrust factor at climb out speed [Ch.3, Eq (3.7a) and Table 3.2, with afterburning]
(T/Mg)o W	The equivalent take off thrust to weight ratios[(Uses Ch.7, Eq (7.4a), with afterburning, but zero here since condition is not relevant]

Deflected Thrust Take-Off:

(T/Mg)	This performance condition is best analysed by using assumed values of (T/Mg) to derive the equivalent wing loadings
Theta	Optimum thrust deflection angle [Ch.7, Eq (7.5d))]
(Mg/S)o	Wing loading corresponding to assumed (T/Mg) values to enable take-off to be achieved in the specific ground roll [Ch.7, Eq (7.5c)]
(T/Mg)o	Corrected value of (T/Mg) allowing for control requirements [the assumed value of (T/Mg) divided by 0.92, see Ch.7, para 7.3.5]

Specific Excess Power (SEP):

The required (T/M)o values are calculated for four conditions, subsonic 1, 2 and 3 and supersonic, ss:

D bar — Drag factor relevant to flight condition [Ch.7, Eq (7.13e), using case data]

(T/Mg)o D, (T/Mg)o W — Required dry and wet thrust to weight ratios to give the specified SEP [Ch.7, Eq (7.13f), using case data]

Sustained Turn:

The required (T/Mg)o values calculated for four conditions, subsonic 1, 2 and 3 and supersonic, ss:

D bar — Drag factor relevant to flight condition [Ch. 7, Eq (7.13e), using case data]

(T/Mg)o D, (T/Mg)o W — Required dry and wet thrust to weight ratios to give the specified sustained turn rate [Ch.7, Eq (7.40a), using case data]

Maximum Mach number at low level:

D bar, (T/Mg)o D, (T/Mg)o W — Drag factor and thrust to weight ratios required to achieve specified low level Mach number (derived from same equation as sustained turns with unit normal acceleration)

Maximum Mach number at high level:

D bar, (T/Mg)o D, (T/Mg)o W — Drag factor and thrust to weight ratios required to achieve specified high level Mach number at assumed combat mass (derived from same equations as above)

Start of Subsonic Cruise - Cr1 at Mn = 0.9 at sea level:

Fac Qm — Factor in climb prediction at constant Mach number [Ch.7, Eq (7.22b)]

(T/Mg)o D, (T/Mg)o W — Dry and wet thrust to weight ratios needed to achieve a residual rate of climb of 1.5 m/s at the high level cruise altitude of 9 km at a Mach number of 0.9 [Ch.7, Eq (7.23b)]

Cl, Cd — Case lift and drag coefficients and lift to drag ratio appropriate to start of climb condition at sea level (for Mn = 0.9 cruise at sea level)

L/D — Ratio of two previous terms

Start of Supersonic Cruise - Cr 2 (end of supersonic climb at 17 km altitude):

Fac Qvm — Factor in prediction of supersonic climb [Ch.7, Eq (7.34b)]

(T/Mg)o D, (T/Mg)o W — Dry and wet thrust to weight ratios required to achieve a residual rate of climb of 1.5 m/s at supersonic cruise altitude of 17 km [Ch.7, Eq (7.36b)]

Transonic Acceleration at 10 km altitude:

Da bar, Db bar — Drag factors for acceleration from Mn = 0.9 to 1.0, and Mn = 1.0 to 1.2 [Ch.7, Eqs (7.44a) and (7.44b) respectively]

E bar } Factors in evaluation of transonic acceleration [Ch.7, Eq (7.45c)]
F bar }
(T/Mg)o W Afterburning thrust to weight ratio required to achieve the transonic acceleration in the specified time [Ch.7, Eq (7.45c)]

Supersonic Cruise 2 at Mn = 1.5 at 9 km altitude:
Case Cl Case lift and drag coefficients and lift to drag ratio at start of Mn = 1.5
Cd supersonic cruise
L/D Ratio of the two previous terms

Summary:
The first summary table states the dry thrust to weight ratios as a function of wing loading for the relevant performance conditions. Also included are the wing loading values previously derived for the conventional landing, gust sensitivity (zero in this case as it is not relevant), the instantaneous turns and deflected thrust take-off. The summary values have been used to produce the left hand of the two diagrams showing the variation of thrust to weight ratio with wing loading. The second summary table gives the corresponding values for afterburning thrust to weight ratios, where this engine condition is relevant. The right hand of the two diagrams represents these values. It can be seen that the first instantaneous turn case places an upper limit on the wing loading of 4435 N/m^2 which is, of course, independent of thrust to weight ratio. The supersonic sustained turn requirement determines the required thrust to weight ratio for both engine conditions, although the supersonic specific excess power case is almost as critical. Interestingly the deflected thrust take-off and vertical landing conditions, evaluated only for dry engine conditions, are well below the required value if the dry thrust necessary to meet the manoeuvre cases is installed. However, when reheat is used to meet the critical manoeuvre and SEP cases the matching is much better. The required dry equivalent thrust to weight ratio, corresponding with the wing loading of 4435 N/m^2 is 1.05 for the optimised condition shown.
Results:
The results for the given parametric case are shown in the box below the Summary.

AD5.5.5.5 Effect of change of aspect ratio and thickness to chord ratio on the required thrust to weight ratio.
It is of interest to investigate how the aspect ratio and thickness to chord ratio influence the required thrust to weight ratio. This has been undertaken for the ranges $2 \leq A \leq 5$ and $0.04 \leq t/c \leq 0.07$. (Note that the optimised value of t/c as given in Spreadsheet AD5.4 is outside this range at 0.09014.) The data required for this analysis are shown below the Results box and are illustrated in the carpet plot located immediately below the two diagrams of (T/Mg)o variation with (Mg/S)o. It can be seen that increase of thickness to chord ratio results in a relatively small increase of required thrust to weight ratio while

increase of aspect ratio is accompanied by a significant reduction of required thrust to weight ratio. (Because it is necessary to input the particular result for a given combination of parameters into the data list, the results presented here become invalid if any of the stage one input data or calculations are changed.)

AD5.5.5.6 Second stage analysis; input data and assumptions

Requirements for Missions:

The specification given in paragraph AD5.5.3 defines two specific missions, *Mission A* - Air superiority and *Mission B* - Offensive support. The performance conditions appropriate to these are listed in the Requirements for Mission Box.

Assumptions:

lambda	Assumed taper ratio of 0.15 as being typical for this class of aircraft (Ch. 5, Table 5.2)
(T/Mg)eng	Basic thrust to weight ratio of engine, from given engine data (assumed to be 6.5, see Ch.6, para 6.4.3.2)
Crew	One crew member only
Op It Fac	100 kg assumed (Ch.6, para 6.4.2.3)
App Fuel/Mo	Ratio of approach fuel to take-off mass, value of 0.01 assumed (Ch.7, para 7.4.4)
V bar	Horizontal tail volume coefficient, 0.4 assumed (see Ch.8, Table 8.1A)
Vv bar	Vertical tail volume coefficient, 0.065 assumed (see Ch.8, Table 8.1A)

Input data:

Fus L	Fuselage length determined from layout, Figure AD5.9.
Fus B	Fuselage maximum width determined from layout
Fus H	Fuselage height determined from layout
c1	Lifting surface mass coefficient from Ch.6, Table 6.7 (low value of 0.0062 assumed to allow for use of composites)
c2	Fuselage mass coefficient from Ch.6, Table 6.6 (low value of 0.036 assumed, again to allow for composite construction)
c3	Powerplant installation mass coefficient (Ch.6, Table 6.8, value of 1.5 chosen assumes fixed geometry intakes but does allow for front air duct and nozzle)
c4	Systems mass coefficient (Ch.6, Table 6.9, value of 0.19 allows for landing gear and advanced control systems)
c5	Lifting surface factor (Ch.6, Table 6.10, typical value of 1.24 is assumed)

Location of component mass centres, based on the layout Figure AD5.9:

Del lW	Location of wing structure mass assumed to be 0.1 mean aerodynamic chord aft of 0.25 MAC [nominal centre of gravity of aircraft (c.g.)]
l Fus	Fuselage centre of gravity position aft of nose, derived from layout
l Hor Tl	Horizontal tail position aft of nose, derived from layout assumptions
l Vert Tl	Vertical tail position aft of nose, derived from layout assumptions
Del l PP	Powerplant centre of gravity aft of aircraft centre of gravity, derived from powerplant data and assumed balance of nozzle forces in vertical flight about aircraft centre of gravity
l SYS	Assumed location of systems mass aft of nose, again based on layout
Del l PAY	Location of payload mass aft of aircraft centre of gravity, derived from layout
Del l wg fue	Location of centre of wing fuel mass aft of centre of gravity of aircraft, assumed to be the same as the wing structure mass
l fus fu	Location of fuselage fuel mass aft of aircraft nose, derived from layout
Del l mngr	Location of main undercarriage mass aft of aircraft centre of gravity, derived from layout
l nose gr	Location of nose undercarriage mass aft of nose, derived from layout

AD5.5.5.7 Second stage calculations

The second stage calculations are undertaken for each mission separately.

Mission A

(S/Mo)^−0.45	Power of wing loading used in lifting surface mass calculations, initially assumed to be 0.66 (Ch.6, Figure 6.1)
p bar	Cabin pressure differential (taken to be 0.272 bar, but not actually needed for subsequent analysis in this case)

Initial Climb A1- Mn = 0.8 to 11 km altitude:

Tau Mn 1	Thrust factor at Mach number 0.8 and sea level [Ch.3, Eq (3.7) and Table 3.2)]
(Cd)zClA1	Zero lift drag coefficient at Mn = 0.8 [Ch.6, Eq (6.12a)]
(Kv)ClA1	Induced drag factor at Mn = 0.8 [Ch.6, Eq (6.13a)]
(Qm)ClA1	Factor in constant Mach number climb [Ch.7, Eq (7.22b)]
(Vv)Mn1	Mean rate of climb from sea level to 11 km altitude at Mn = 0.8 [Ch.7, Eq (7.22a)]
DistClA1	Distance covered in climb [Ch.7, Eq (7.25) including cosine term]
(c)CrA1/A3	Specific fuel consumption (dry) at start of Mn = 0.8 climb [Ch.3, Eq (3.12a)]

Cruise A1- Mach number 0.8 at 11 km altitude:

Dist Cr A1	Required cruise distance (specified radius of operation less distance covered in climb to 11 km)
Cr A1 Cl	Lift coefficient at start of Mn = 0.8 cruise at 11 km altitude
Reqd(T/Mg)	Required thrust to weight ratio in Mn = 0.8 cruise at 11 km (the reciprocal of the lift to drag ratio obtained from the required lift coefficient and the previously calculated drag terms at Mn = 0.8)
Av(T/Mg)	Available thrust to weight ratio derived from static thrust to weight ratio with allowance for the fuel used in climb
Av/Reqd	Ratio of available to required thrust to weight ratio
(c)des.	Design specific fuel consumption at Mn = 0.8 and 11 km [Ch.3, Eq (3.12a)]
(c)od	Off design specific fuel consumption [Ch.3, Eq (3.12b)]

Transonic Acceleration at 9 km altitude, Mach number 0.9 to 1.2 (wet):-

(c)0.9	Specific fuel consumption (wet) at Mn = 0.9 and 9 km altitude [Ch.3, Eq 3.(13)]
(c)1.2	Specific fuel consumption (wet) at Mn = 1.2 and 9 km
(Wf/Mg)o	Ratio of fuel used in transonic acceleration to take-off weight [Ch.7, Eq (7.46a)]

Supersonic Acceleration/Climb A2 - to Mn = 1.5 at 9.001 km altitude (wet):
(Note supersonic acceleration/climb must have a nominal change in altitude to avoid a singularity in the equations, hence the small increment in altitude shown here.)

Mn'ssCl	Factor in supersonic climb prediction [Ch.7, Eq (7.34f)]
(Vv)ssCl	Mean rate of climb in supersonic climb [Ch.7, Eq (7.34a)]
Dist ss Cl	Ground distance covered in supersonic climb [Ch.7, Eq (7.38)]

Supersonic Cruise A2 - Mn = 1.5 and nominally 9 km altitude (wet):

Reqd(T/Mg)	Required thrust to weight ratio in cruise (reciprocal of previously determined lift to drag ratio)
Av(T/Mg)	Available thrust to weight ratio derived from static thrust and relevant thrust factor (wet), with allowance for fuel used in climbs and transonic acceleration
Av/Req	Ratio of available to required thrust to weight ratios
(c)des	Design specific fuel consumption (wet) at Mn = 1.5 and 9 km altitude [Ch. 3, Eq (3.13)]
(c)od	Off design specific fuel consumption [Ch.3, Eq (3.12b)]

Combat at 9 km altitude and Mn = 1.5 (wet):

D bar	Drag factor at Mn = 1.5 and 9 km altitude based on assumed combat mass ratio (Mcr2/Mo) [Ch.7, Eq (7.13e)]

Reqd(T/Mg) Required thrust to weight ratio in combat manoeuvre [Ch.7, Eq (7.40a)]
Av(T/Mg) Available thrust to weight ratio, based on recalculated combat mass
Av/Req Ratio of available to required thrust to weight ratios
(c)des Design specific fuel consumption (wet) [Ch.3, Eq (3.13)]
(c)od Off design specific fuel consumption [Ch.3, Eq (3.12b)]
Dist (ss) Distance aircraft travels in combat manoeuvre of one 360° sustained turn at 3*g*
(Wf/Mg)o Ratio of fuel used to take-off weight during supersonic climb/acceleration and combat

Combat at Mn = 0.9 and sea level (dry):
(Mg/S)sub Wing loading at subsonic combat conditions (allows for all fuel used previously during the flight)
D bar Drag factor at Mn = 0.9 and sea level [Ch.7, Eq (7.13a)]
Reqd(T/Mg) Required thrust to weight ratio in combat manoeuvre [Ch.7, Eq (7.40a)]
Av(T/Mg) Available thrust to weight ratio at Mn = 0.9 and sea level and wing loading condition appropriate to subsonic combat, (Mg/S)sub
Av/Req Ratio of available to required thrust to weight ratios
(c)des Design specific fuel consumption at Mn = 0.9 and sea level (dry) [Ch.3 Eq (3.12a)]
(c)od Off design specific fuel consumption [Ch.3, Eq (3.12b)]
Dist (sb) Distance aircraft travels in combat manoeuvre of two 360° sustained turns at 8*g*
(Wf/Mg)o Ratio of fuel used to take-off weight in subsonic combat turns.

Return Climb A3 - Mn = 0.8 (dry):
(Mg/S) ClA3 Wing loading at start of return climb, allows for all fuel used previously
(Qm) ClA3 Factor in climb prediction using case data [Ch.7, Eq (7.22b)]
(Vv) MnA3 Mean rate of climb in return climb [Ch.7, Eq (7.22a)]
DistCl A3 Ground distance covered in return climb [Ch.7, Eq (7.25)]

Return Cruise A3 - Mn = 0.8 at 11 km:
DistCrA3 Distance to be covered in return cruise (specified value less climb distance)
(Mg/S)strt Wing loading at start of return cruise (allows for all fuel used previously)
(Cl)CrA3 Lift coefficient at start of return cruise (derived from Cruise A1 value with allowance for reduced wing loading)
Reqd(T/Mg) Required thrust to weight at start of return cruise (reciprocal of lift to drag ratio)
Av(T/Mg) Available thrust to weight at start of return cruise

Av/Reqd	Ratio of available to required thrust to weight ratios (Derived from Cruise A1 case with allowance for wing loading difference)
(c)od	Off design specific fuel consumption (dry) [Based on previous value of (c)des and Ch.3, Eq (3.12b)]

Masses:

M fus	Mass of fuselage structure [Ch.6, Eq (6.20b)]
c1 bar	Factor in prediction of lifting surface mass [Ch.6, Eq (6.24)]
Mpp/Mo	Mass ratio of powerplant installation [uses value of c3 associated with the static thrust to weight ratio and the assumed (T/Mg)eng]
Msys/Mo	Systems mass ratio (coefficient c4)
M/op it	Mass of operational items (product of the Op It Fac and number of crew)
M fixed	Fixed mass (sum of fuselage structure, payload and operational items) (Note this is the initial payload and so it reduces when weapons are used)

Fuel usage:

T.O	Assumed fuel used in take-off as ratio of take-off weight (0.01)

Cl Al:-

(Wf/Mg)o	Ratio of fuel used to take-off weight in Mn = 0.8 climb [Ch.7, Eq (7.24)]

Cr A1:

log (M1/M2)	Logarithm of mass ratio over Mn = 0.8 cruise [Ch.7, Eq (7.50b)]
M1/M2	Antilog of previous value
M2/Mo	Ratio of mass at end of Mn = 0.8 cruise to take-off value (allows for fuel used in initial climb)

Cr A2:

log(M1/M2)	Logarithm of mass ratio over supersonic cruise[Ch.7, Eq (7.50b)] (It is zero in this case since there is no supersonic cruise requirement)
M1/M2	Antilog of previous value
M2/Mo	Ratio of mass at end of supersonic cruise to take-off value (allows for fuel used previously, including transonic acceleration and supersonic climb)

End Supersonic phase:

Mss/Mo	Mass ratio at end of supersonic phase, including combat (allows for combat fuel)

End Subsonic Combat phase:

Msb/Mo	Mass ratio at end of subsonic combat (allows for subsonic combat fuel and disposal of payload)

Cl A3:

(Wf/Mg)o	Ratio of fuel used to take-off weight in second subsonic climb at Mn = 0.8 [Ch.7, Eq (7.24)]

Cr A3:

log (M1/M2)	Logarithm of mass ratio over return subsonic cruise at Mn = 0.8 and 11 km altitude [Ch.7, Eq (7.50b)]
M1/M2	Antilog of previous value
M2/Mo	Ratio of mass at end of return cruise to take-off value (allows for fuel used in climb to 11 km altitude, Cl A3)

Hover for 1 minute:

Reqd(T/Mg)	Thrust to weight ratio required to hover at end of return flight (based on previous M2/Mo value factored by 1.15 to allow for control requirements, Ch.7, para 7.3.5)
Av(T/Mg)	Installed dry static thrust to weight ratio, corrected for reduced mass at landing condition
Av/Req	Ratio of previous two values
(c)des	Design specific fuel consumption (dry) at hover condition [Ch.3, Eq (3.12a)]
(c)od	Off design specific fuel consumption [Ch.3, Eq (3.12b)]
(Wf/Mg)o	Ratio of fuel used for 1 min in hover to take-off weight

End Flight:

Ml/Mo	Final landing mass ratio
Fuel used	Ratio of weight of fuel used to take-off weight (difference of landing and take-off weights with allowance for disposal of payload)
Tot Fuel	Total fuel ratio required (1.053 times previous value to allow for reserve)
K*Mo	Sum of terms dependent on take-off mass (fuel, powerplant and systems)
x bar (0.25)/ root chrd	Location of 0.25 mean aerodynamic chord on centreline chord of wing [Ch.8, Eq (8.7b) using geometry to be defined]
Root chrd	Wing centreline chord (defined subsequently)
MAC	Wing mean aerodynamic chord (defined subsequently)

Mission B

Cruise B1- Mn = 0.6 at sea level (dry):

(Cd)zCrB1	Zero lift drag coefficient at Mn = 0.6 [Ch.6, Eq (6.12a)]
(Kv)CrB1	Induced drag factor at Mn = 0.6 [Ch.6, Eq (6.13a)]
CrB1Cl	Lift coefficient in cruise at Mn = 0.6 and sea level

Reqd (T/Mg)	Reciprocal of lift to drag ratio in case to give required thrust to weight ratio
Av(T/Mg)	Available thrust to weight ratio [uses Ch.3, Eq (3.7) and Table 3.2]
Av/Reqd	Ratio of previous two values
(c)des	Design specific fuel consumption (dry) [Ch.3, Eq (3.12a)]
(c)od	Off design specific fuel consumption [Ch.3, Eq (3.12b)]
Log (M1/M2)	Logarithm of mass ratio in cruise over specified range [Ch.7, Eq (7.50b)]
M1/M2	Antilog of previous value
M2/Mo	Ratio of mass at end of Mn = 0.6 cruise to take-off value (allows for fuel used in take-off)

Cruise B2 -	Mn = 0.8 at sea level (dry):
CrB2Cl	Lift coefficient at Mn = 0.8 and sea level
(T/Mg)Reqd	Required thrust to weight ratio, the reciprocal of case lift to drag ratio (drag terms previously evaluated)
(T/Mg)Av	Available thrust to weight ratio [Ch.3, Eq (3.7) and Table 3.2]
Av/Reqd	Ratio of previous terms
(c)des	Design specific fuel consumption (dry) [Ch.3, Eq (3.12a)]
(c)od	Off design specific fuel consumption [Ch.3, Eq (3.12b)]
log (M1/M2)	Logarithm of mass ratio in cruise over specified range [Ch.7, Eq (7.50b)]
M1/M2	Antilog of previous value
M2/Mo	Ratio of mass at end of Mn = 0.8 cruise to take-off value (allows for fuel used previously)

Combat at sea level and Mn = 0.9 (dry):

D bar	Drag factor in this case [Ch.7, Eq (7.13e)]
Reqd(T/Mg)	Required thrust to weight ratio in manoeuvre [Ch.7, Eq (7.40a) allows for fuel used up to combat phase]
Av(T/Mg)	Available thrust to weight ratio [Ch.3, Eq (3.7) and Table 3.2]
Av/Reqd	Reciprocal of previous two values
(c)des	Design specific fuel consumption (dry) [Ch.3, Eq (3.12a)]
(c)od	Off design specific fuel consumption [Ch.3, Eq (3.12b)]
Dist	Distance covered in four 360° sustained 8*g* turns at Mn = 0.9 at sea level
(Wf/Mg)o	Ratio of weight of fuel used in combat to take-off weight

Return Cruise B3 - Mn = 0.8 at sea level (dry):

M/Mo	Mass ratio at start of return cruise (allows for all fuel used previously and disposal of payload)
CrB3Cl	Lift coefficient at mass M in cruise at Mn = 0.8 and sea level

Reqd(T/Mg)	Reciprocal of lift to drag ratio in return cruise B3 (drag terms previously evaluated)
Av(T/Mg)	Available thrust to weight ratio [uses Ch.3, Eq (3.7) and Table 3.2]
Av/ Reqd	Ratio of previous two terms
(c)des	Design specific fuel consumption (dry) [Ch.3, Eq (3.12a)]
(c)od	Off design specific fuel consumption[Ch.3, Eq (12b)]
log (M1/M2)	Logarithm of mass ratio in Mn = 0.8 return cruise [Ch.7, Eq (7.50b)]
M1/M2	Antilog of previous value
M2/Mo	Ratio of mass at end of Mn = 0.8 return cruise to take-off value

Return Cruise B4- Mn = 0.6 and sea level (dry):

CrB4Cl	Lift coefficient at Mn = 0.6 and sea level (allows for all fuel used)
Reqd(T/Mg)	Required thrust to weight ratio as reciprocal of lift to drag ratio
Av(T/Mg)	Available thrust to weight ratio [Ch.3, Eq (3.7) and Table 3.2]
Av/Reqd	Ratio of two previous values
(c)des	Design specific fuel consumption (dry) [Ch.3, Eq (3.12a)]
(c)od	Off design specific fuel consumption [Ch.3, Eq (3.12b)]
log (M1/M2)	Logarithm of mass ratio in Mn = 0.6 return cruise at sea level [Ch.7, Eq (7.50b)]
M1/M2	Antilog of previous value
M2/Mo	Ratio of mass at end of return cruise to take-off value

Hover for 1 minute:

Reqd(T/Mg)	Required hover thrust to weight ratio at end of return flight (based on previous value, M2/Mo, factored by 1.15 for control requirements, Ch.7, para 7.3.5)
Av(T/Mg)	Available thrust to weight ratio (static thrust with weight corrected to M2 mass value)
Av/Reqd	Reciprocal of previous two values
(c)des	Design specific fuel consumption (dry) [Ch.3, Eq (3.12a)]
(c)od	Off design specific fuel consumption [Ch.3, Eq (3.12b)]
(Wf/Mg)o	Fuel used in 1 min hover as ratio of take-off weight

End Flight:

M1/Mo	Final landing mass ratio
Fuel used	Ratio of fuel used in mission to take-off weight (difference of landing and take-off values with allowance for load disposed)
Tot Fuel	Total fuel ratio required (1.05 times previous value)
K* Mo	Sum of terms dependent upon take-off mass (fuel, powerplant and systems)

M fixed	Mass fixed (sum of fuselage structure, payload and operational items) (note this is the initial payload and so it reduces when weapons are used)
x bar*(0.25)/ root chord	Location of 0.25 mean aerodynamic chord on wing root chord [Ch 8, Eq (8.7b) using geometry to be defined]
Root chord	Wing centreline (root) chord (defined subsequently)
MAC	Wing mean aerodynamic chord (defined subsequently)

AD5.5.5.8 Second stage analysis: optimisation

It is necessary to optimise the design for each of the two missions independently. Thus there are two sets of Analysis and Results Boxes. The first part of the second stage analysis uses the calculated values to determine the actual take-off mass for a given mission, which is optimised. The second part uses the resulting component mass values to make a first estimate of the wing position along the body which brings the centre of gravity to the 0.25 mean aerodynamic chord position. Use of the assumed volume coefficients then enables the sizes of the horizontal and vertical tails to be predicted. The layout of the procedure is identical for both missions:

(Mo)est1	A first estimate of the take-off mass derived by assuming that the lifting surface terms are 12% of the total mass
(Mo)est2	Initially the same value as (Mo)est1 entered directly as a number, but subsequently changed on optimisation
Kappa*Mo	Mass of terms directly proportional to take-off mass (product of K*Mo and (Mo)est2)
M lift sur	Mass of lifting surfaces [uses cl bar with (Mo)est2]
(Mo) calc	Initially this is the sum of (M fixed), (Kappa*Mo) and (M lift sur) to give the total mass. This is the target cell used for optimisation by invoking SOLVER©. The value is minimised, in this case by changing the aspect ratio, thickness to chord ratio, wing loading and (Mo)est2. The constraints imposed are: Structural Parameter, SP to be less than 18 (Mo)calc to be equal to (Mo)est2 Wing loading to be less than or equal to the landing and two instantaneous turn conditions Wing loading to be greater than gust sensitivity condition (when applicable)
error	The difference between (Mo)calc and (Mo)est2 as a check on the optimiser

l CG	Location of 0.25 aerodynamic mean chord point aft of the fuselage nose, derived from longitudinal balance of all the mass items, using fuselage layout where appropriate (assumed position of aircraft c.g.)
l WG APX	Location of leading edge of wing centreline chord aft of the nose [derived from l CG and x bar (0.25)/root chord]
l TL ARM	Distance between the centre of gravity and nominal centres of pressure of vertical and horizontal tails (derived from lCG and l TAIL)
S Hor Tail S Vert Tail	Areas of horizontal and vertical tail surfaces (derived from wing area, wing mean chord, l TL ARM and assumed volume coefficients)

The optimiser can handle only one set of conditions on a given spreadsheet thus it is necessary to apply it to each of the two missions in turn. However, application to a given mission can alter the values of the other and hence the following procedure is adopted:

i) Optimise *Mission A* and separately record (Mo)calc in the cell below the Analysis Box.

ii) Optimise *Mission B* and separately record the appropriate (Mo)calc as before.

iii) Select the mission which gives the critical case, that is the highest required take-off mass. In this case it is found to be *Mission B*. (*Mission A* optimum mass is 16,052 kg and *Mission B* is 20,352 kg).

iv) Adjust the assumed values of S^-0.1 and S/l^2 in the knowledge of the first set of optimised values for *Mission B* and re-optimise until the assumed and output values of these two parameters are coincident to three significant figures.

v) Using the final value of the operating empty mass for *Mission B*, recalculate the take-off mass for *Mission A* using the "Goal Seeker"© tool. The result of this is shown in the Revised Mass - *Mission A* Box (final *Mission A* mass is 19,092 kg).

AD5.5.5.9 Summary of final results.

The summary of the final results for the *Mission B* design is shown in the relevant box.

The data have been derived directly from the previous calculations as follows:

i) Wing mass has been isolated from the total lifting surface mass by use of the coefficient, c_5.

ii) The wing geometry defined from the assumed and derived values of aspect ratio, *A*, taper ratio, lambda and sweepback, (Del 0.25).

iii) Wing area follows from the take-off wing loading, (Mg/S)o.

iv) Of the total landing gear allowance of 4% of the aircraft mass some 15% has been allocated to the nose unit.

AD5.5.6 Finalised configuration

The lifting surface data and centre of gravity evaluation for *Mission B* enables the configuration of the aircraft to be completed. This is shown in Figure AD5.9, to which reference has already been made. It is inevitable that with such a complex design as this advanced combat aircraft further stages of analysis will result in changed values of many of the assumed terms. Nevertheless the initial synthesis process can be expected to provide a sound basis for a detailed design study.

AD5.6 Conclusions

The five case studies, that given in Chapter 8 and the four in this Addendum, cover the great majority of aircraft types, or can be readily adapted to do so. For example the single seat piston engine trainer can be simply changed to cover twin-engine types. The turboprop feeder aircraft may be altered to cover turboprop trainers and the high altitude subsonic example also covers subsonic military aircraft generally providing the engine is of low bypass ratio. Higher bypass ratio subsonic types may be studied by modification of the short haul transport example. Finally the supersonic V/STOL combat aircraft spreadsheet has been arranged to cover other supersonic types, including airliners.

Index

SPREADSHEETS

The spreadsheets referred to in Chapter 8 and Addendum 5 are contained on the diskette entitled "Aircraft conceptual design synthesis".

The format used is Microsoft EXCEL© version 7 (MS Office 95). The listing of the spreadsheets is as follows:

Chapter 8 Spreadsheet 1 - Subsonic flight - Fan engine 1 (First stage of analysis).
Spreadsheet 2 - Subsonic flight - Fan engine 2 (Second stage of analysis using the graphical approach).
Spreadsheet 3 - Subsonic flight - Fan engine Optimiser (Second stage of analysis using the overall optimisation approach).

Addendum 5 Spreadsheet AD5.1 - Piston engine - Two seat, aerobatic, trainer aircraft.
Spreadsheet AD5.2 - Turboprop - Feeder line / Commuter aircraft.
Spreadsheet AD5.3 - Low bypass ratio engine - Uninhabited, high altitude, long range, reconnaissance aircraft.
Spreadsheet AD5.4 - Supersonic - short take-off / vertical landing, combat aircraft.

The software within the spreadsheets is primarily intended for instructional purposes but may be suitable for more general application. It should be pointed out, however, that it has not been validated.

Please note that all the spreadsheets can now be found at

www.wiley.com/g0/howe_aircraft

Lightning Source UK Ltd.
Milton Keynes UK
23 May 2010

154525UK00001B/30/A